D1766816

Encyclopaedia of Mathematical Sciences
Volume 102

Mathematical Physics III

Subseries Editors:
J. Fröhlich   S. P. Novikov   D. Ruelle

Christian Bonatti
Lorenzo J. Díaz
Marcelo Viana

Dynamics Beyond
Uniform Hyperbolicity

A Global Geometric and Probabilistic Perspective

 Springer

Authors

Christian Bonatti
Institut de Mathématiques de Bourgogne
UMR 5584 du CNRS
Université de Bourgogne
B.P. 47870
21078 Dijon Cedex
France
e-mail: bonatti@u-bourgogne.fr

Lorenzo J. Díaz
Departamento de Matemática, PUC-Rio
Rua Marquês de São Vicente, 225
Edifício Cardeal Leme
Gávea – Rio de Janeiro
Brazil
22453-900
e-mail: lodiaz@mat.puc-rio.br

Marcelo Viana
IMPA
Estrada Dona Castorina, 110
Jardim Botânico – Rio de Janeiro
Brazil
22460-320
e-mail: viana@impa.br

Founding editor of the Encyclopaedia of Mathematical Sciences:
R. V. Gamkrelidze

Mathematics Subject Classification (2000): 37XX, 37Cxx, 37Dxx, 37Exx
ISSN 0938-0396
ISBN 3-540-22066-6 Springer Berlin Heidelberg New York

Springer is a part of Springer Science+Business Media
springeronline.com
© Springer-Verlag Berlin Heidelberg 2005
Printed in Germany

The use of general descriptive names, registered names, trademarks, etc. in this publication does not imply, even in the absence of a specific statement, that such names are exempt from the relevant protective laws and regulations and therefore free for general use.

Typesetting: by the authors using a Springer LATEX macro package
Production: LE-TEX Jelonek, Schmidt & Vöckler GbR, Leipzig
Cover Design: E. Kirchner, Heidelberg, Germany
Printed on acid-free paper   46/3142 YL   5 4 3 2 1 0

To IMPA, for bringing us together.

Preface

What is Dynamics about?

In broad terms, the goal of Dynamics is to describe the long term evolution of systems for which an "infinitesimal" evolution rule is known. Examples and applications arise from all branches of science and technology, like physics, chemistry, economics, ecology, communications, biology, computer science, or meteorology, to mention just a few.

These systems have in common the fact that each possible state may be described by a finite (or infinite) number of observable quantities, like position, velocity, temperature, concentration, population density, and the like. Thus, the space of states (*phase space*) is a subset M of an Euclidean space $\mathbb{R}^m$. Usually, there are some constraints between these quantities: for instance, for ideal gases pressure times volume must be proportional to temperature. Then the space M is often a manifold, an n-dimensional surface for some $n < m$.

For *continuous time* systems, the evolution rule may be a differential equation: to each state $x \in M$ one associates the speed and direction in which the system is going to evolve from that state. This corresponds to a vector field $X(x)$ in the phase space. Assuming the vector field is sufficiently regular, for instance continuously differentiable, there exists a unique curve tangent to X at every point and passing through x: we call it the *orbit of x*.

Even when the real phenomenon is supposed to evolve in continuous time, it may be convenient to consider a *discrete time* model, for instance, if observations of the system take place at fixed intervals of time only. In this case the evolution rule is a transformation $f : M \rightarrow M$, assigning to the present state $x \in M$ the one $f(x)$ the system will be in after one unit of time. Then the *orbit of x* is the sequence x_n obtained by iteration of the transformation: $x_{n+1} = f(x_n)$ with $x_0 = x$.

In both cases, one main problem is *to describe the behavior as time goes to infinity for the majority of orbits*, for instance, for a full probability set of initial states. Another problem, equally important, is *to understand whether that limit behavior is stable under small changes of the evolution law*, that is, whether it remains essentially the same if the vector field X or the transformation f are slightly modified. It is easy to see why this is such a crucial question, both conceptually and for the practical applications: mathematical models are always simplifications of the real system (a model of a chemical reaction, say, taking into account the whole universe would be obviously unpractical ...) and, in the absence of stability, conclusions drawn from the model might be specific to it and not have much to do with the actual phenomenon.

It is tempting to try to address these problems by "solving" the dynamical system, that is, by looking for analytic expressions for the trajectories, and indeed that was the prevailing point of view in differential equations until little more than a century ago. However, that turns out to be impossible in most cases, both theoretically and in practice. Moreover, even when such an analytic expressions can be found, it is usually difficult to deduce from them useful conclusions about the global dynamics.

Then, by the end of the 19th century, Poincaré proposed to bring in methods from other disciplines, such as topology or ergodic theory, *to find qualitative information on the dynamics without actually finding the solutions.* A beautiful example, among many others, is the Poincaré-Birkhoff theorem stating that an area preserving homeomorphism of the annulus which rotates the two boundary circles in opposite directions must have some fixed point. This proposal, which was already present in Poincaré's early works and attained full maturity in his revolutionary contribution to Celestial Mechanics, is usually considered to mark the birth of Dynamics as a mathematical discipline.

Hyperbolicity and stability.

This direction was then pursued by Birkhoff in the thirties. In particular, he was much interested in the phenomenon of transverse *homoclinic points,* that is, points where the stable manifold and the unstable manifold of the same fixed or periodic saddle point intersect transversely. This phenomenon had been discovered in the context of the N-body problem by Poincaré, who immediately recognized it as a major source of dynamical complexity. Birkhoff made this intuition much more precise by proving that any transverse homoclinic orbit is accumulated by periodic points. A definitive understanding of this phenomenon unfolded at the beginning of the sixties, when Smale introduced the *horseshoe,* a simple geometric model whose dynamics can be understood rather completely, and whose presence in the system is equivalent to the existence of transverse homoclinic points.

The horseshoe, and other robust models containing infinitely many periodic orbits, such as Thom's cat map (hyperbolic toral automorphism), were unified by Smale's notion of uniformly *hyperbolic set*: a subset of the phase space invariant under the dynamical system and such that the tangent space at each point splits into two complementary subspaces that are uniformly contracted under, respectively, forward and backward iterations. Then Smale also introduced the notion of *uniformly hyperbolic dynamical system* (Axiom A) which essentially means that the limit set, consisting of all forward or backward accumulation points of orbits, is a hyperbolic set. These ideas much influenced contemporary remarkable work of Anosov where it was shown that the geodesic flow on any manifold with negative curvature is ergodic.

Another major achievement of uniform hyperbolicity was to provide a characterization of structurally stable dynamical systems. The notion of *structural stability*, introduced in the thirties by Andronov, Pontrjagin, means that the whole orbit structure remains the same when the system is slightly modified: there exists a homeomorphism of the ambient manifold mapping orbits of the initial system into orbits of the modified one, and preserving the time arrow. Indeed, uniform hyperbolicity proved to be the key ingredient of structurally stable systems, together with a transversality condition, as conjectured by Palis, Smale.

In the process, a theory of uniformly hyperbolic systems was developed, mostly from the sixties to the mid eighties, whose importance extended much beyond the original objectives. It was part of a revolution in our vision of determinism, strongly driven by observations originating from experimental sciences, which shattered the classical opposition between deterministic evolutions and random evolutions. The uniformly hyperbolic theory provided a mathematical foundation for the fact that deterministic systems, even with a small number of degrees of freedom, often present chaotic behavior in a robust fashion. Thus, it led to the almost paradoxical conclusion that "chaos" may be stable.

On the other hand, structural stability and uniform hyperbolicity were soon realized to be less universal properties than was initially thought: there exist many classes of systems that are robustly unstable and non-hyperbolic and, in fact, that is often the case for specific models coming from concrete applications. The dream of a general paradigm in Dynamics had to be postponed.

Beyond uniform hyperbolicity.

The next years saw the theory being extended in several distinct directions:

- The study of specific classes of systems, such as quadratic maps, Lorenz flows, and Hénon attractors, which introduced a host of new methods and ideas.
- Bifurcation theory including, in particular, the study of the boundary of uniformly hyperbolic systems, and of the local and global mechanisms leading to chaotic behavior, especially homoclinic bifurcations.
- New developments in the ergodic theory of smooth systems and, especially, the theory of non-uniformly hyperbolic systems (Pesin theory).
- Weaker formulations of hyperbolicity, still with a uniform flavor but where one allows for invariant "neutral" directions (partial hyperbolicity, projective hyperbolicity or existence of a dominated splitting).

- The converse implication in the stability conjecture (hyperbolicity is necessary for stability), which led to the introduction of new perturbation lemmas (ergodic closing lemma, connecting lemma).

Building on remarkable progress obtained in these directions, especially in the eighties and early nineties, several ideas have been put forward and a new point of view has emerged recently, which again allow us to dream of a global understanding of "most" dynamical systems. Initiated as a survey paper requested to us by David Ruelle, the present work is an attempt to put such recent developments in a unified perspective, and to point open problems and likely directions of further progress.

Two semi-local mechanisms, very different in nature but certainly not mutually exclusive, have been identified as the main sources of persistently non-hyperbolic dynamics:

- What we call here "critical behavior", corresponding to critical points in one-dimensional dynamics and, more generally, to homoclinic tangencies, and which is at the heart of Hénon-like dynamics. This is now reasonably well understood, in terms of non-uniformly hyperbolic behavior. Moreover, recent results show that this type of behavior is always present in connection to non-hyperbolic dynamics in low dimensions.
- In higher dimensions, dynamical robustness (robust transitivity, stable ergodicity) extends well outside the uniformly hyperbolic domain, roughly speaking associated to coexistence of uniformly hyperbolic behavior with different unstable dimensions. It requires some uniform geometric structure (transverse invariant bundles: partial hyperbolicity, dominated decomposition) that we refer to as "non-critical behavior".

On the other hand, new perturbation lemmas permitted to organize the global dynamics of generic dynamical systems, by breaking it into elementary pieces separated by a filtration. A great challenge is to understand the dynamics on (the neighborhood of) these elementary dynamical pieces, which should involve a deeper analysis of the two mechanisms mentioned previously. Indeed, a good understanding has already been possible in several cases, especially at the statistical level.

What is this book, and what is it not?

The text is aimed at researchers, both young and senior, willing to get a quick yet broad view of this part of Dynamics. Main ideas, methods, and results are discussed, at variable degrees of depth, with references to the original works for details and complementary information.

We assume the reader is familiar with the fundamental objects of smooth Dynamics, like manifolds or C^r diffeomorphisms and vector fields, as well as with the basic facts in the local theory of dynamical systems close to a hyperbolic periodic point, such as the Hartman-Grobman linearization theorem and the stable manifold theorem. This material is covered by several books, like Bowen [86], Irwin [225], Palis, de Melo [342], Ruelle [394], Katok, Hasselblatt [232], or Robinson [382].

Familiarity with the classical theory of uniformly hyperbolic systems is also desirable, of course. This is also covered by a number of books, including Bowen [86], Shub [411], Mañé [281], Palis, Takens [345], and Katok, Hasselblat [232]. For the reader's convenience, in Chapter 1 we review the main conclusions of the theory that are relevant for our purposes. In that chapter we also give an introductory discussion of robust mechanisms of non-hyperbolicity, and other key issues outside the hyperbolic set-up. This is to be much expanded afterwards, so at that point our presentation is sketchier than elsewhere.

Apart from these pre-requisites, we have tried to keep the text self-contained, giving the precise definitions of all relevant non-elementary notions. Occasionally, this is done in an informal fashion at places where the notion is first needed in a non-crucial way, with the formal definition appearing at some later section where it really is at the heart of the subject. This is especially true about Chapter 1, as explained in the previous paragraph.

Although we have used parts of this book as a basis for graduate courses, it is certainly not designed as a text book that could be used for that purpose all alone. The properties of the main notions are often only stated, and most results are presented with just an outline of the proof.

The book is also not meant to be an exhaustive presentation of the recent results in Dynamics. We are only too conscious of the many fundamental topics we left outside, or touched only briefly. Deciding where to stop could be one of the most difficult and most important problems in this kind of project, and no answer is entirely satisfactory.

How should this book be used and what does it contain?

The 12 chapters are organized so as to convey a global perspective of dynamical systems. The 5 appendices include several other important results, older and new, which we feel should not be omitted, either because they are used in the text or because they provide complementary views of some aspects of the theory.

Although there is, naturally, a global coherence in the text, we have tried to keep the various chapters rather independent, so that the reader may choose to read one chapter without really needing to go through the previous ones. This means that we often recall main notions and statements introduced elsewhere, or else give precise references to where they can be found. On the other hand, the chapters often rely on ideas and results from the appendices.

The main text may be, loosely, split into the following blocks:

- Chapter 1 contains a brief review of uniformly hyperbolic theory and an introduction to main themes to be developed throughout the text.
- Chapters 2 to 4 are devoted to critical behavior in various aspects: one-dimensional dynamics, homoclinic tangencies, Hénon-like dynamics.

- Chapter 5 shows that, for low dimensional systems, far from critical behavior the dynamical behavior is hyperbolic.
- Chapters 6 to 9 treat non-critical behavior, especially the relation between robustness and existence of invariant splittings. While most of the text focusses on dissipative discrete time systems, Chapter 8 deals with conservative diffeomorphisms and Chapter 9 is devoted to flows.
- In Chapter 10 we try to give a global framework for the dynamics of generic maps, where critical and non-critical behavior could fit together.
- Chapter 11 presents some of the progress attained in describing the dynamics in ergodic terms, both in critical and in non-critical situations (either separate or coexisting). Lyapunov exponents are an important tool in this analysis, and Chapter 12 is devoted to their study and control.

Acknowledgements:

Input from several colleagues greatly helped shape this text and improve our presentation. Besides the referees, we are especially thankful to F. Abdenur, J. Alves, V. Araújo, A. Arbieto, M.-C. Arnaud, A. Avila, M. Benedicks, J. Bochi, S. Crovisier, V. Horita, C. Liverani, S. Luzzatto, L. Macarini, C. Matheus, W. de Melo, C. Morales, C. G. Moreira, K. Oliveira, M. J. Pacifico, J. Palis, E. Pujals, J. Rocha, R. Roussarie, M. Sambarino, M. Shub, A. Tahzibi, M. Tsujii, T. Vivier, A. Wilkinson, J.-C. Yoccoz, and D. Ruelle.

    Our collaboration was supported by the Brazil-France Agreement in Mathematics, CNPq-Brazil, CNRS-France, Faperj-Rio de Janeiro, in addition to our own institutions. To all of them we express our warm gratitude.

Dijon and Rio de Janeiro
May 31, 2004

Christian Bonatti
Lorenzo J. Díaz
Marcelo Viana

Contents

1

Hyperbolicity and Beyond

Uniformly hyperbolic systems are presently fairly well understood, both from the topological and the ergodic point of view. In Sections 1.1 through 1.3 we review some of their main properties (spectral decomposition, stability, physical invariant measures) that one would like to extend to great generality. Several very good references are available for this material, including the books of Shub [411], Palis, Takens [345, Chapter 0] and Katok, Hasselblat [232, Part 4], and Bowen [86] for the ergodic theory of this systems.

Outside the hyperbolic domain, two main phenomena occur: homoclinic tangencies and cycles involving saddles with different indices. These notions are introduced in Section 1.4 and 1.5 and serve as a guiding thread through the chapters that follow, where they will be revisited in much more detail. In Section 1.7, we present a conjecture of Palis pointing at a global description of most dynamical systems. Section 1.6 introduces a few fundamental notions involved in this conjecture and in most of our text.

1.1 Spectral decomposition

Let M be a compact manifold, and $f : M \to M$ be a diffeomorphism.

Definition 1.1. An invariant compact set $\Lambda \subset M$ is a *hyperbolic set* for $f : M \to M$ if the tangent bundle over Λ admits a continuous decomposition

$$T_\Lambda M = E^u \oplus E^s, \tag{1.1}$$

invariant under the derivative and such that $\|Df^{-1} \mid E^u\| \leq \lambda$ and $\|Df \mid E^s\| \leq \lambda$ for some constant $\lambda < 1$ and some choice of a Riemannian metric on the manifold.

A point z is *non-wandering for f* if for every neighborhood U of z there is $n \geq 1$ such that $f^n(U)$ intersects U. The set of non-wandering points is denoted $\Omega(f)$. It contains the set $\text{Per}(f)$ of periodic points, as well as the α-limit set and the ω-limit set of every orbit.

Definition 1.2. The diffeomorphism $f : M \to M$ is *uniformly hyperbolic*, or satisfies the *Axiom A*, if $\Omega(f)$ is a hyperbolic set for f and $\mathrm{Per}(f)$ is dense in $\Omega(f)$.

The definitions for smooth flows $f^t : M \to M$, $t \in \mathbb{R}$, are analogous, except that (unless Λ consists of equilibria) the decomposition (1.1) becomes

$$T_\Lambda M = E^u \oplus E^0 \oplus E^s, \tag{1.2}$$

where E^0 is 1-dimensional and collinear to the flow direction.

The spectral decomposition theorem of Smale [418] asserts that the limit set of a uniformly hyperbolic system splits into a finite number of pairwise disjoint *basic pieces* that are compact, invariant, and dynamically indecomposable. The precise statement follows.

We say that an f-invariant set is *indecomposable*, or *transitive*, if it contains some dense orbit $\{f^n(z) : n \geq 0\}$. An f-invariant set Λ is called *isolated*, or *locally maximal*, if there exists a neighborhood U of Λ such that

$$\Lambda = \bigcap_{n \in \mathbb{Z}} f^{-n}(U). \tag{1.3}$$

That is, Λ coincides with the set of points whose orbits remain in U for all times.

Theorem 1.3. *The non-wandering set $\Omega(f)$ of a uniformly hyperbolic diffeomorphism f decomposes as a finite pairwise disjoint union*

$$\Omega(f) = \Lambda_1 \cup \cdots \cup \Lambda_N$$

of f-invariant transitive sets Λ_i, that are compact and isolated. Moreover, the α-limit set and the ω-limit set of every orbit are contained in some Λ_i.

Here is a sketch of the proof. Consider the equivalence homoclinic relation defined in $\mathrm{Per}(f)$ by $p_1 \sim p_2 \Leftrightarrow$ the stable set of the orbit of each of the points has some transverse intersection with the unstable set of the orbit of the other. The stable manifold theorem implies that there are finitely many equivalence classes, and they are open in $\mathrm{Per}(f)$. The basic pieces Λ_i in the theorem are the closures of the equivalence classes. By construction, they are compact, invariant, and open in $\Omega(f) = \overline{\mathrm{Per}}(f)$. The latter implies that they are isolated sets, because $\overline{\mathrm{Per}}(f)$ is an isolated set if it is hyperbolic. Moreover, the stable and the unstable manifold of any periodic point in some Λ_i are dense in Λ_i. This implies that Λ_i is transitive. Finally, if the α- or ω-limit set of some orbit intersected more than one Λ_i, there would be non-wandering points outside the union of the basic pieces, a contradiction.

Transitive sets and isolated sets of flows $f^t : M \to M$, $t \in \mathbb{R}$, are defined in the same way. Theorem 1.3 remains true for uniformly hyperbolic flows $\{f^t : t \in \mathbb{R}\}$.

Remark 1.4. A question dating from the late sixties asked whether every hyperbolic set is contained in an isolated one. This was recently solved by Crovisier [143], who constructed a transitive diffeomorphism of $M = \mathbb{T}^4$ having a hyperbolic set which is not contained in any isolated hyperbolic set Λ: Crovisier shows that Λ would have to be the whole torus, which is not a hyperbolic set because the diffeomorphism has saddles with different indices. This has been improved by Fisher [183], using different methods: he obtains robust examples in any dimension ≥ 2, and for dimension 4 or higher his examples are also transitive.

1.2 Structural stability

A smooth dynamical system is called *structurally stable* [15] if it is equivalent to any other system in a C^1 neighborhood. In the discrete-time case, equivalence means conjugacy by a global homeomorphism. In the case of flows this notion is too restrictive: it forces all periods of closed orbits to be preserved under perturbation. Instead, one asks for the existence of a global homeomorphism sending orbits of one system to orbits of the other, and preserving the direction of time. More generally, replacing C^1 by C^r neighborhoods, any $r \geq 1$, one obtains the notion of C^r structural stability.

The stability conjecture of Palis-Smale [343] proposes a complete characterization of structurally stable systems. Namely, they should coincide with the hyperbolic systems having the property of strong transversality: every stable and unstable manifolds of points in the non-wandering set should be transversal. In fact their conjecture is for C^r structural stability, any $r \geq 1$.

Robbin [377], de Melo [146], and Robinson [379, 380] proved that these are sufficient conditions for structural stability. Strong transversality is also necessary [378]. These results hold in the C^r topology, any $r \geq 1$. The hardest part was to prove that stable systems must be hyperbolic. This was achieved by Mañé [283] in the mid-eighties, for C^1 diffeomorphisms, and extended about ten years later by Hayashi [208] for C^1 flows. Thus

Theorem 1.5. *A C^1 diffeomorphism (or flow) on a compact manifold is structurally stable if and only if it is uniformly hyperbolic and verifies the strong transversality condition.*

A weaker property, called Ω-*stability* is defined requiring conjugacy (respectively, equivalence) only restricted to the non-wandering set. The Ω-stability conjecture in [343] proposes a characterization of the Ω-stable systems: they should be the hyperbolic systems having no *cycles*, that is, no basic pieces in their spectral decompositions cyclically related by intersections of the corresponding stable and unstable sets.

The Ω-stability theorem of Smale [418] states that these properties are sufficient. The proof uses the following notion:

Definition 1.6. A *filtration* for a diffeomorphism $f : M \to M$ is a finite family $M_1, M_2, \ldots, M_k$ of submanifolds with boundary and with the same dimension as M, such that

- $M_1 = M$ and M_{i+1} is contained in the interior of M_i for every $1 \leq i < k$.
- $f(M_i)$ is contained in the interior of M_i for all $1 \leq i \leq k$.

The open sets $L_i = \mathrm{int}(M_i \setminus M_{i+1})$ are the *levels of the filtration* (set $M_{k+1} = \emptyset$).

The first step is top show that if f is hyperbolic and has the no-cycles property, then it admits a filtration such that each basic piece coincides with the set of orbits contained in some level: up to reordering,

$$\Lambda_i = \bigcap_{n \in \mathbb{Z}} f^n(L_i) \qquad \text{for all } i.$$

Let g be any diffeomorphism C^r-close to f. Then $M_1, , \ldots, M_k$ is also a filtration for g. Therefore, $\Omega(g)$ is contained in $\Lambda_1(g) \cup \cdots \cup \Lambda_k(g)$, where

$$\Lambda_i(g) = \bigcap_{n \in \mathbb{Z}} g^n(L_i).$$

Stability of hyperbolic sets gives that each $f \mid \Lambda_i$ is conjugate to $g \mid \Lambda_i(g)$. Then, every $\Lambda_i(g)$ is contained in $\Omega(g)$, and $f \mid \Omega(f)$ is conjugate to $g \mid \Omega(g)$. Thus f is Ω-stable, in the C^r sense.

Palis [339] proved that the no-cycles condition is necessary for Ω-stability, in any C^r topology. Necessity of hyperbolicity for Ω-stability was proved by Palis [340], based on Mañé [283], for C^1 diffeomorphisms, and extended to C^1 flows by Hayashi [208].

1.3 Sinai-Ruelle-Bowen theory

A basic piece Λ_i is a *hyperbolic attractor* if the stable set

$$W^s(\Lambda_i) = \{x \in M : \omega(x) \subset \Lambda_i\}$$

contains a neighborhood of Λ_i. In this case we call $W^s(\Lambda_i)$ the *basin* of the attractor Λ_i, and denote it $B(\Lambda_i)$. When the Axiom A system is of class C^2, a basic piece is an attractor if and only if its stable set has positive Lebesgue measure. Thus, the union of the basins of all attractors is a full Lebesgue measure subset of M. This remains true for a residual (dense G_δ) subset of C^1 uniformly hyperbolic diffeomorphisms and flows. See Bowen [86, 87].

The following fundamental result, due to Sinai [416], Ruelle, Bowen [86, 90, 390] says that, no matter how complicated it may be, the behavior of typical orbits in the basin of a hyperbolic attractor is completely well-defined at the statistical level:

Theorem 1.7. *Every attractor Λ of a C^2 uniformly hyperbolic diffeomorphism (or flow) supports a unique invariant probability measure μ such that*

$$\lim_{n \to \infty} \frac{1}{n} \sum_{j=0}^{n-1} \varphi(f^j(z)) = \int \varphi \, d\mu \tag{1.4}$$

for every continuous function φ and Lebesgue almost every point $x \in B(\Lambda)$.

One way to construct μ is starting with normalized Lebesgue measure m_D over a compact domain D inside any leaf of the unstable foliation of the attractor. A distortion argument, using the fact that f is C^2, gives that the sequence of iterates $\{f_*^n(m_D) : n \geq 1\}$ has a property of *uniform* absolute continuity along the unstable foliation. Contracting behavior in the transverse direction, together with minimality of the unstable foliation, are used to show that the limit $\mu = \lim_{n \to \infty} f_*^n(m_D)$ exists and is an ergodic measure whose support coincides with Λ. Very important, because of the previous observation μ disintegrates into conditional measures along unstable foliation (see Appendix C) that are equivalent to the Lebesgue measure of each leaf.

Ergodicity gives (1.4) for μ-almost every point in Λ. Using conditional measures we get it for a full Lebesgue measure subset L of some unstable leaf. To prove the much more interesting fact that (1.4) is true for Lebesgue almost every point in the basin of attraction, a whole open set, one uses the observation that time-averages of continuous functions are constant on stable manifolds. The stable manifolds of points in the attractor foliate the whole $B(\Lambda)$. Absolute continuity of this stable foliation (see Appendix C) ensures that the stable manifolds of the points in L cover a full Lebesgue measure of the basin. Theorem 1.7 follows.

Property (1.4) means that the Sinai-Ruelle-Bowen measure μ may be explicitly computed, meaning that the weights of subsets may be found with any degree of precision, as the sojourn-time of any orbit picked "at random" in the basin of attraction:

$$\mu(V) = \text{ fraction of time the orbit of } z \text{ spends in } V$$

for any typical subset V of M (the boundary of V should have zero μ-measure), and for Lebesgue almost any point $z \in B(\Lambda)$. For this reason μ is called a *physical measure*.

There is another sense in which this measure is "physical" and that is that μ is the zero-noise limit of the stationary measures associated to the stochastic processes obtained by adding small random noise to the system. This property is called *stochastic stability*; a formal definition will appear later. For uniformly hyperbolic systems it is due to Sinai [416], Kifer [240, 242], and Young [457]. The model of small stochastic perturbations to represent external influences, too small or too complex to express in deterministic terms, goes back to Kolmogorov and Sinai.

1.4 Heterodimensional cycles

Although uniform hyperbolicity was originally intended to encompass a residual, or at least dense subset of all dynamical systems, it was soon realized
that this is not true. There are two main mechanisms that yield robustly non-
hyperbolic behavior, that is, whole open sets of non-hyperbolic systems. Not
surprisingly, they are at the heart of recent developments that we are going
to review in the next sections.

Historically, the first one was the coexistence of periodic points with different Morse indices (dimensions of the unstable manifolds) inside the same
transitive set. See Figure 1.1. This is how the first examples of C^1-open subsets
of non-hyperbolic diffeomorphisms were obtained by Abraham, Smale [4, 414]
on manifolds of dimension $d \geq 3$. It was also the key in the constructions by
Shub [410] and Mañé [277] of non-hyperbolic yet robustly transitive diffeomorphisms, that is, such that every diffeomorphism in a C^1 neighborhood has
a dense orbit. The examples of Shub and Mañé are outlined in Section 7.1.

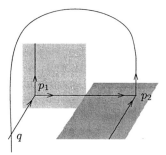

Fig. 1.1. A heterodimensional cycle

For flows this mechanism may assume a novel form, because of the in-
terplay between regular periodic orbits and singularities (equilibrium points).
That is, robust non-hyperbolicity may stem from the coexistence of regular
and singular orbits in the same transitive set. The first and very striking example was the geometric Lorenz attractor proposed by Afraimovich, Bykov,
Shil'nikov [5] and Guckenheimer, Williams [201, 452] to model the behavior
of the Lorenz equations [267]. This is a main theme in Chapter 9.

1.5 Homoclinic tangencies

Heterodimensional cycles may exist only in dimension 3 or higher. The first robust examples of non-hyperbolic diffeomorphisms on surfaces were constructed
by Newhouse [319], exploiting the second of the mechanisms we mentioned:

homoclinic tangencies, or non-transverse intersections between the stable and the unstable manifold of the same periodic point. See Figure 1.2.

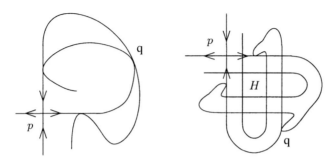

Fig. 1.2. Homoclinic tangencies

It is important to observe that individual homoclinic tangencies are easily destroyed by small perturbations of the invariant manifolds. To construct open examples of surface diffeomorphisms with *some* tangency, Newhouse started from systems where the tangency is associated to a periodic point inside an invariant hyperbolic set with rich geometric structure. See the right hand side of Figure 1.2. His argument requires a very delicate control of distortion, as well as of the dependence of the fractal dimension on the dynamics. Actually, for this reason, his construction is restricted to the C^r topology for $r \geq 2$. By comparison, robust examples of heterodimensional cycles in higher dimensions are obtained by much more elementary transversality and hyperbolicity arguments.

1.6 Attractors and physical measures

Several attempts were made, specially in the seventies, to weaken the definitions of hyperbolicity and structural stability, while keeping their topological flavor, so that they could encompass a residual set of dynamical systems. One such extension was the notion of Ω-stability that we mentioned before.

In parallel, a more probabilistic approach was being proposed, especially by Sinai, Ruelle, Eckmann, where one focus on the statistical behavior of typical orbits, and its stability under perturbations. See [176] for a detailed exposition.

Definition 1.8. A set $\Lambda \subset M$ is an *attractor* for a diffeomorphism (or a flow) on a manifold M if it is invariant and transitive, and the *basin of attraction*,

$$B(\Lambda) = \{x \in M : \omega(x) \subset \Lambda\}$$

has positive Lebesgue measure.

Definition 1.9. A *physical measure*, or *SRB measure*, of a diffeomorphism on a manifold M is an invariant probability measure μ on M, such that the time average of every continuous function $\varphi : M \to \mathbb{R}$ coincides with the corresponding space-average with respect to μ,

$$\lim_{n\to\infty} \frac{1}{n} \sum_{j=0}^{n-1} \varphi(f^j(z)) = \int \varphi \, d\mu \qquad (1.5)$$

for a set of initial points z with positive Lebesgue measure. We call this set the *basin* of μ, and denote it $B(\mu)$.

For flows (1.5) is replaced by

$$\lim_{T\to\infty} \frac{1}{T} \int_0^T \varphi(f^t(z)) \, dt = \int \varphi \, d\mu.$$

Problem 1.10. For most dynamical systems, does Lebesgue almost every point have a well-defined time average? Are there SRB measures whose basins cover almost all M?

This is the case for C^2 hyperbolic systems, by Theorem 1.7. But the answer can not always be affirmative. A simple counter-example, due to Bowen, is described in Figure 1.3: time averages diverge over any of the spiraling orbits in the region bounded by the saddle connections. However, no robust counter-example is known.

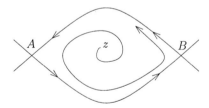

Fig. 1.3. A planar flow with divergent time averages

The following semi-global version of the previous problem also goes back to Sinai and Ruelle:

Problem 1.11. (basin problem) Let Λ be an attractor for a diffeomorphism, or a flow, supporting a unique SRB measure μ. Does

$$B(\mu) = B(\Lambda) \quad \text{up to a zero Lebesgue measure set?} \qquad (1.6)$$

More generally, does $B(\Lambda)$ coincide with the union of the basins of the SRB measures supported in Λ, up to zero Lebesgue measure?

Again, the answer can not always be positive: [41] shows how Bowen's example can be inserted into an attractor, so that the basin property (1.6) is violated. Note also that general (transitive) attractors may support more than one SRB measure. See Kan [229] and Section 11.1.1. But no robust examples of either of these phenomena have been found to date.

1.7 A conjecture on finitude of attractors

A program towards a global understanding of complex dynamical behavior has been proposed by Palis [341]. Here we quote some of the conjectures embodying his program, others will appear later.

Conjecture 1.12. (finitude of attractors [341]) There is a C^r, $r \geq 1$, dense set subset $\mathcal{D}$ of dynamical systems on any compact manifold that exhibit a finite number of attractors whose basins cover Lebesgue almost all of the manifold.

Conjecture 1.13. (physical measures and stochastic stability [341]) For any element of $\mathcal{D}$ all the attractors support SRB measures and have the basin property. Moreover, any element of $\mathcal{D}$ is stochastically stable on the basin of each of the attractors.

Stochastic stability means that adding small random noise to the system has little effect on its statistical behavior. For discrete-time systems $f : M \to M$, one considers *random orbits* $\{z_j : j \geq 0\}$ where each z_{j+1} is chosen at random in the ε-neighborhood of $f(z_j)$. Then, for ε small, the time averages of continuous functions over almost every random orbit should be close to the corresponding time averages over typical orbits of the original f. There is a corresponding notion for flows, where noise takes place at infinitesimal intervals of time: random orbits are solutions of a stochastic differential equation. Precise definitions are given in Appendix D.

Conjecture 1.14. (metric stability of basins of attraction [341]) Given any element of $\mathcal{D}$ and any of its attractors, then for almost all C^r small perturbations along generic k-parameter families there is a finite set of attractors whose basins cover most (a fraction close to 1 in volume) of the original basin, and these attractors also support physical measures.

A key novelty in the formulation of this conjecture, and in the whole scenario proposed by Palis, is to allow the existence of pathological phenomena, e.g. related to cycles, occupying a small volume in the ambient space. Indeed, cycles have been a main obstruction to the realization of previous global scenarios for Dynamics.

For one-dimensional systems these conjectures take a stronger form: finiteness of attractors, with the nice properties above and with their basins containing Lebesgue almost all points, should correspond to full Lebesgue measure in parameter space, for generic parametrized families. This has been fully verified for C^2 unimodal maps of the interval, as we shall see in Chapter 2. In this case the attractor is unique. There is also substantial partial progress in higher dimensions, some of which is reported throughout the text.

Beyond finiteness?

Palis' program represents an ambitious attempt to achieve a global description of dynamical systems, and has been inspiring much work in this area. As we have just mentioned, its conclusions have been fully confirmed in the context of unimodal maps of the interval, and it should certainly hold for general smooth maps in dimension 1. At the present stage this program also seems a realistic goal for surface diffeomorphisms. The main remaining difficulty for understanding these systems is given by Newhouse's phenomenon of coexistence of infinitely many attractors or repellers. The previous conjectures are compatible with this phenomenon, except if it occurs robustly, that is, for every diffeomorphism in a whole open set. This is not known at the time, but seems unlikely. Another important related open question is whether this phenomenon may correspond to positive Lebesgue probability in parameter space, on generic parametrized families of diffeomorphisms. In higher dimensions, the situation is presently much less understood, and the previous conjectures remain long term goals.

Even if parts of this program turn out no to be confirmed, investigation of these questions will certainly lead to important further progress. For instance, if coexistence of infinitely many SRB measures does occur for an open subset of systems, but their basins cover a full volume set, and the measures vary continuously with the dynamics, then one will still get a very satisfactory variation of the conclusion in Conjecture 1.14. Also, coexistence of infinitely many independent ergodic behaviors is known to be robust in the conservative setting: by KAM (Kolmogorov, Arnold, Moser) theory many conservative systems exhibit positive volume sets consisting of invariant tori (see for instance [454]). Notwithstanding the existence of a whole continuum of ergodic behaviors, these systems can be understood to some extent. This leaves hope for dissipative systems as well, even if the general finiteness paradigm turns out not to be dense.

Robust non-convergence of time averages for many initial states would, perhaps, be a more disturbing difficulty. We have seen a codimension-2 example in Section 1.6 and it is also known that non-convergence occurs with codimension-1 in the setting of interval maps (see for instance [217]). Different approaches may be envisaged to handling such a difficulty.

One of them, going back to Kolmogorov, is to consider zero-noise limits of stationary measures associated to random perturbations of the system. Indeed, stationary measure exist in great generality (see for instance [18, 19]) and, for small noise levels, they may be considered to provide a certain "physical" perception of the system's behavior. This approached was undertaken in [19] for Bowen's example in Figure 1.3, where the zero-noise limit is an average of the Dirac measures on the two saddle points.

A very different approach is to consider convenient resumations of the time averages, that might be convergent even when the time averages themselves do not. For instance, one may consider higher order averages

$$a_n^{(k)} = \frac{1}{n} \sum_{j=0}^{n-1} a_j^{(k-1)} \qquad a_j^{(0)} = \varphi(f^j(x)).$$

This would yield a stratification of many dynamical systems reflecting, in some sense, their statistical complexity. In this direction we propose the following example, as a test case for both aforementioned approaches:

Problem 1.15. Let X be a vector field on $[0,1]$ which vanishes exactly at the endpoints. Consider the map $f : S^1 \times [0,1] \rightarrow S^1 \times [0,1]$ defined by $f(x,y) = (2x \mod \mathbb{Z}, X^{\sin(2\pi x)}(y))$. Observe that the second coordinate of $f^n(x,y)$ is given by the time-T_n map of the vector field X, where $T_n = \sum_{j=0}^{n-1} \sin(2^{n+1}\pi x)$. The Birkhoff time averages $a_n^{(1)}$ of f should diverge almost everywhere. Do second order time averages converge? Are there zero-noise limits of stationary measures? Do they provide the same information?

Notice that it may happen that all these higher order averages $a_k^{(n)}$ diverge; indeed, that seems to be the case in Bowen's example.

2

One-Dimensional Dynamics

The study of one-dimensional transformations is a classical subject going back to Poincaré. Here these systems interest us primarily as models for dynamical behavior in higher dimensions. We focus on continuous smooth maps with finitely many critical (multimodal maps) exhibiting complicated behavior. This area of Dynamics has been experiencing remarkable progress since the late seventies. Recent surveys have been written by Graczyk, Swiatek [199] and Lyubich [275], and the book of de Melo, van Strien [147] remains the fundamental expository text on the subject.

Hyperbolicity is the simplest form of behavior for one-dimensional maps. We recall the definition and characterization in Section 2.1. A special feature of these maps is that the dynamics is always very much hyperbolic far from critical points, as we comment upon in Section 2.2. It had been conjectured for a long time, and was only recently established, that hyperbolic dynamics is C^r-dense among these maps. This remarkable fact, which has no counterpart in higher dimensions, is discussed in Section 2.3. On the other hand, the presence of critical points may also give rise to non-uniform hyperbolicity persistently in parameter space. This is the theme of Section 2.4.

Most of the recent progress in this area has been driven by the effort to understand the renormalization operator, and explain the universality phenomena discovered by Coullet, Tresser [138] and Feigenbaum [181] in cascades of bifurcations. See Section 2.5. Among the main recent achievements, let us mention the proof that *almost all maps inside typical families of unimodal maps are either hyperbolic (regular) or chaotic (stochastic) and, in either case, stochastically stable.* This is discussed in Section 2.6.

2.1 Hyperbolicity

Throughout we consider C^r maps $f : M \to M$, where M is either the circle S^1 or the compact interval, and $r \geq 1$. In the interval case we assume that $f(\partial M) \subset \partial M$ and any fixed point on the boundary is repelling.

The map f is called *hyperbolic* if the periodic attractors are all hyperbolic, and the complement $\Sigma(f)$ of the basins of attraction is a hyperbolic set for f. The latter means that there are $C > 0$ and $\sigma > 1$ such that $|Df^n(x)| \geq C\sigma^n$ for every $x \in \Sigma(f)$ and $n \geq 1$. The reason one prefers to give the definition in terms of $\Sigma(f)$, rather than the non-wandering set as in the previous chapter, is that $\Sigma(f)$ is always fully invariant, meaning that $f^{-1}(\Sigma(f)) = \Sigma(f)$, which is usually not be the case for $\Omega(f)$ if the map is not invertible.

For C^2 maps without critical points, a satisfactory characterization of hyperbolicity is provided by the following

Theorem 2.1 (Mañé [280]). *Let $f : M \to M$ be a C^2 map of the circle or the interval, without critical points and whose periodic points are all hyperbolic. Then either f is hyperbolic or f is (conjugate to) an irrational rotation of $M = S^1$.*

Hyperbolicity of all periodic points is, of course, a necessary condition for hyperbolicity of the map. Among circle diffeomorphisms, the hyperbolic ones form an open dense subset. In contrast, by a celebrated theorem of Herman [212], irrational rotations correspond to positive Lebesgue measure inside generic parametrized families. This kind of dichotomy is not uncommon in one-dimensional dynamics, as we shall see.

Example 2.2. The Arnold family $h_{a,b} : S^1 \to S^1$ of circle maps is defined by [32]

$$h_{a,b}(\theta) = \theta + a + b\sin 2\pi\theta \quad \mod \mathbb{Z}. \tag{2.1}$$

For $|b| < 1/2\pi$, the map $h_{a,b}$ is a diffeomorphism. There is an open dense subset of values of a for which the rotation number is rational. However, by Herman's theorem, for a positive Lebesgue measure subset of values of a, the diffeomorphism $h_{a,b}$ is differentiably conjugate to an irrational rotation.

For $|b| = 1/2\pi$ the map develops a critical point and for $|b| > 1/2\pi$ it is no longer a homeomorphism. These situations have also been intensively studied, and found applications in higher dimensions, e.g. in [159, 325]. See [147, Section I.7] for references and more information on the Arnold family.

If f is hyperbolic and C^2, the set $\Sigma(f)$ has zero Lebesgue measure, unless it coincides with the whole ambient space. The second case is possible only if $M = S^1$ and f has no critical points. Moreover, f is topologically conjugate to a linear covering map of the circle [409], and there is a unique SRB measure, which is equivalent to Lebesgue measure. In the first case, the typical dynamics is really very simple: almost every orbit just converges to some periodic attractor; the SRB measures are the Dirac measures supported on those attractors.

In general, hyperbolic maps are structurally stable, that is, they are topologically conjugate to all nearby maps. See [147, Chapter 3]. The converse is not known in general, but it may be deduced in some situations, most notably, whenever the subset of hyperbolic maps is known to be dense.

We call $f : M \to M$ a *multimodal map* if it has a finite, non-zero number of critical points, all non-flat (finite-order contact). *Unimodal* means that the critical point is unique. *Infinite-modal* maps are studied in [338].

Example 2.3. The main model of unimodal maps is the quadratic family

$$q_a : [0, 1] \to [0, 1], \quad q_a(x) = ax(1 - x),$$

where a is a parameter in $(0, 4]$. For all a, the critical point $c = 1/2$ is quadratic: $D^2 q_a(c) \neq 0$. Quadratic maps are sometimes written in different, but conjugate, forms e.g. $x \mapsto 1 - ax^2$ or $x \mapsto x^2 + a$.

The dynamics of multimodal maps is strongly influenced by the behavior of critical orbits. This is already apparent in the following remarkable result, that extends Theorem 2.1:

Theorem 2.4 (Mañé [280]). *Let $f : M \to M$ be a C^2 multimodal map of the circle or the interval. If all the periodic points are hyperbolic and every critical point is contained in the basin of a periodic attractor, then f is hyperbolic.*

In fact, Mañé proves a stronger statement: A compact invariant set of a C^2 multimodal map is hyperbolic if (and only if) it contains neither critical points nor non-repelling periodic points. See Theorem 2.5 below.

In the special case of maps with negative Schwarzian derivative,

$$Sf = \frac{f'''}{f'} - \frac{3}{2}\left(\frac{f''}{f'}\right)^2 < 0 \qquad (2.2)$$

including quadratic maps, every non-hyperbolic periodic point is either an attractor (the basin of attraction contains a neighborhood) or a semi-attractor (the basin contains an interval with the periodic point on its boundary) and, in either case, the basin contains some critical point. This statement is due to Singer [417] and we recall the proof in the next paragraph, to illustrate one use of the Schwarzian derivative in one-dimensional dynamics. Therefore, the map is hyperbolic if and only if every critical orbit is contained in the basin of some hyperbolic attractor. Moreover, if the map is unimodal, hyperbolicity is equivalent to existence of a hyperbolic attractor.

The first step in the proof of Singer's theorem is a minimum principle: for maps with negative Schwarzian derivative local minima of $|f'|$ can only occur at critical points. Indeed, at minima with $f' > 0$ we must have $f'' = 0$ and $f''' \geq 0$, which implies $Sf \geq 0$, and the case $f' < 0$ is analogous. We also need the observation, easily proved by induction, that $Sf < 0$ implies $Sf^k < 0$ for all $k \geq 1$. Now suppose p is a periodic attractor, hyperbolic or not, and let $k \geq 1$ be its period. Let J be the maximal neighborhood of p contained in the basin of attraction. The endpoints of J are also periodic points of period k, or else they are boundary points of M. In the first case we must have $|(f^k)'| \geq 1$ at the boundary points, whereas $|(f^k)'(p)| \leq 1$. This implies that

$|(f^k)'|$ has some local minimum in J. From the minimum principle applied
to f^k we get that this minimum corresponds to some critical point of f^k.
This proves that the forward orbit of J contains some critical point of f, as
stated. The second case, when some endpoint of J is a boundary point of M,
is easily reduced to the first one: recall that we assumed that $f(\partial M) \subset \partial M$
and any periodic point on the boundary is repelling. The case when p is a
semi-attractor is similar: take J a maximal interval inside the basin having
p as an endpoint and note that $|(f^k)'| \geq 1$ at both endpoints. Hence, once
more, J must contain some local minimum of $|(f^k)'|$, and this minimum must
correspond to a critical point. Finally, the same kind of arguments proves that
there can be no non-hyperbolic periodic repellers. Suppose there was such a
point p, with period $k \geq 1$, and let J be its unstable set J of p, that is, the
union of all forward iterates under f^k of a sufficiently small neighborhood of
p. Then J would contain some local minimum of $|(f^k)'|$ but no critical points,
contradicting the minimum principle.

2.2 Non-critical behavior

Theorems 2.1 and 2.4 may be seen as predecessors to results of Pujals, Sambarino [371] about surface diffeomorphisms, that we are going to review in
Chapter 5. We sketch their proofs, because ideas from them proved to be
fruitful in higher dimensions too. The following statement contains both theorems:

Theorem 2.5 (Mañé [280]). *Let f be a C^2 map of the circle or the interval,
and Λ be an f-invariant compact set containing no critical points of f. Then*

*(a) the periods of the non-hyperbolic periodic points of f in Λ are bounded
    above, and all such points are contained in finitely many periodic intervals,
    restricted to which f is a diffeomorphism (permuting the intervals);*
*(b) if Λ does not contain any non-hyperbolic periodic point then Λ is a hyper-
    bolic set, or else $\Lambda = S^1$ and f is conjugate to an irrational rotation.*

We start the sketch of the proof with an alternative characterization of
hyperbolicity of an invariant set Λ of a C^1 one-dimensional map $f : M \to M$,
in terms of the *natural extension*

$$\hat{f} : \hat{\Lambda} \to \hat{\Lambda}, \quad \hat{f}(\dots, x_n, \dots, x_0) = (\dots, x_n, \dots, x_0, f(x_0))$$

of f restricted to Λ. Here $\hat{\Lambda}$ is the set of all pre-orbits of points in Λ, that is,
all sequences $(x_n)_{n \geq 0}$ in Λ such that $f(x_{n+1}) = x_n$ for all $n \geq 0$.

Lemma 2.6. *Let $f : M \to M$ be a C^1 map and Λ be a compact f-invariant
set such that*

$$\lim_{n \to \infty} |(f^n)'(x_n)| = \infty$$

for all $\hat{x} \in \hat{\Lambda}$. Then Λ is a hyperbolic set for f.

A *pre-orbit* of an open interval $I \subset M$ is a sequence $(\phi_n)_n$ where each ϕ_n is a branch of $f^{-n} \mid I$ satisfying $f \circ \phi_{n+1} = \phi_n$ for all n. The pre-orbit is *associated* to the invariant set Λ if there is $x \in I \cap \Lambda$ such that $\phi_n(x) \in \Lambda$ for all $n \geq 0$. An interval $I \subset M$ is *adapted* to Λ if, for any $x \in I \cap \Lambda$ and any pre-orbit $(x_n)_n$ of $x = x_0$ there is a pre-orbit $(\phi_n)_n$ of I such that $\phi_n(x) = x_n$, the intervals $\phi_n(I)$ are uniformly far from the critical points of f, and each $\phi_n(I)$ is either contained or disjoint from I.

Lemma 2.7. *Consider a C^2 map $f : M \to M$ which is not topologically equivalent to an irrational rotation. Let Λ be a compact invariant subset that does not meet the boundary of M nor contains critical points of f.*

If $\cap_{n \geq 0} f^n(\Lambda)$ contains non-periodic points, then there is an adapted interval J and a constant $K > 0$ such that every pre-orbit $(\phi_n)_n$ of J associated to Λ satisfies

$$\sum_{i=1}^{\infty} |\phi_i'(x)| \leq K \quad and \quad |\phi_i'(x)| \leq K \, |\phi_i'(y)|$$

for any points x and y in J and every $i \geq 1$. Moreover, for every non-periodic point $x \in \cap_{n \geq 0} f^n(\Lambda)$ there is an interval J adapted to Λ and containing x.

The two lemmas above imply that every f-invariant compact set Λ that does not contain critical points, sinks or non-hyperbolic periodic points is hyperbolic or f is equivalent to an irrational rotation. The proof is by contradiction as follows.

Assume f is not a rotation and Λ is not hyperbolic. Consider the family $\mathcal{N}$ of non-hyperbolic compact f-invariant subsets of Λ, ordered by inclusion. The intersection of the sets of any totally ordered sequence in $\mathcal{N}$ is also non-hyperbolic, thus it belongs to $\mathcal{N}$. So, by Zorn's lemma, the family $\mathcal{N}$ has some minimal element Λ_0. The compact set Λ_0 can not be a union of periodic orbits: otherwise, it would be hyperbolic. We also claim that $\Lambda_0 = f(\Lambda_0)$: if $f(\Lambda_0)$ were properly contained in Λ_0, then it would be hyperbolic, which would imply the hyperbolicity of Λ_0. This shows that the set $\cap_n f^n(\Lambda_0) = \Lambda_0$ contains some non-periodic point. Let J be an interval adapted to Λ_0 as given by Lemma 2.7. From the existence of such an interval, we are going to deduce that Λ_0 is hyperbolic. This contradiction will show that the family $\mathcal{N}$ must be empty, and so Λ must be hyperbolic.

By Lemma 2.6, it suffices to prove that $\lim_{n \to \infty} |(f^n)'(x_n)| = \infty$ for every pre-orbit $(x_n)_n \in \hat{\Lambda}_0$ of a point $x_0 \in \Lambda_0$. Denote by Γ the closure of the sequence $(x_n)_n$. This is a compact invariant set. If Γ is properly contained in Λ_0 then it is hyperbolic, and so $\lim_n |(f^n)'(x_n)| = \infty$. Otherwise, $\Gamma = \Lambda_0$ and so there is $k \geq 0$ such that $x_k \in J$. Replacing x by x_k, we may suppose $k = 0$. Since J is adapted to Λ_0, it has an (associated) pre-orbit $(\phi_n)_n$ such that $\phi_n(x_0) = x_n$ for all n. The summability property in Lemma 2.7 implies that $\lim_n |\phi_n'(x_0)| \to 0$. Since ϕ_n is an inverse branch of f^n with $\phi_n(x_0) = x_n$, this means that $\lim_n |(f^n)'(x_n)| \to \infty$. This ends our sketch of the proof of Theorem 2.5.

2.3 Density of hyperbolicity

The set of hyperbolic maps is C^r open in the space of C^r multimodal maps. For $r = 1$, it was proved by Jakobson [227] that this set is also dense. Whether this remains true for $r \geq 2$ has been one of the main problems in one-dimensional dynamics:

Conjecture 2.8. The set of hyperbolic maps is C^r dense in the space of multimodal maps, for any $r \geq 2$.

A classical transversality argument, combined with Theorem 2.4, shows that any multimodal map whose critical points are all contained in basins of hyperbolic periodic attractors is C^r approximated by a hyperbolic one. So, Conjecture 2.8 is equivalent to showing that the set of multimodal maps whose critical points are all in basins of hyperbolic attractors is C^r dense.

A major step was the solution of this problem for quadratic maps:

Theorem 2.9 (Graczyk, Swiatek [198], Lyubich[273]). *The set of parameters a for which $q_a(x) = ax(1 - x)$ is hyperbolic is dense in $(0, 4]$.*

The core of the proofs was to show that any two quadratic maps whose periodic points are all repelling are quasi-symmetrically conjugate if they are topologically conjugate. A homeomorphism h is *quasi-symmetric* if there exists $K > 0$ such that any two adjacent intervals of equal length are mapped by h onto intervals whose lengths differ by a factor $\leq K$. The arguments also relied on substantial information specific to the quadratic family, especially the fact that the topological entropy or, equivalently, the kneading invariant [293] of q_a depends monotonically on the parameter a.

So, it came as a surprise when Kozlovski announced, only shortly afterwards, that he could extend the conclusion to the general context of unimodal maps:

Theorem 2.10 (Kozlovski [246, 247]). *The set of hyperbolic maps is C^r dense in the space of all unimodal maps, for any $r \geq 2$.*

Thus, Smale's original conjecture that hyperbolicity should be topologically prevalent was confirmed in this low-dimensional setting! Density of structural stability among unimodal maps is easier and was known before. See [147, section III.2]. Concerning general multimodal maps, Shen [407] recently proved C^2 density of hyperbolicity. Even more recently, Kozlovski, Shen, van Strien [248] announced a complete proof of Conjecture 2.8: hyperbolicity is C^r dense among all multimodal maps, for every $r \geq 2$.

2.4 Chaotic behavior

If a multimodal map f is hyperbolic and C^2, almost every orbit converges to some periodic attractor, and the whole dynamics is quite simple. A theorem of Jakobson states that maps with complicated (non-hyperbolic) behavior are also abundant, in the probabilistic sense, in one-dimensional dynamics:

Theorem 2.11 (Jakobson [227]). *For an open class of families of C^2 unimodal maps, including the quadratic family, the set of parameters for which the map has an absolutely continuous invariant probability μ has positive Lebesgue measure.*

This remarkable result opened the way to much progress in non-uniformly hyperbolic dynamics.

A general strategy to construct an absolutely continuous invariant measure is to consider the forward iterates $f_*^n m$ of Lebesgue measure m. Any Cesaro weak limit

$$\lim_k \frac{1}{n_k} \sum_{j=0}^{n_k-1} f_*^j m$$

of a subsequence $(n_k)_k$ is an invariant measure. To ensure that some of these limit measures is absolutely continuous, one must show that iterates of f do not distort Lebesgue measure too much. For multimodal maps, distortion occurs, primarily, close to the critical points.

Jakobson exploited the method of *inducing*. He showed how to find, for Lebesgue almost every point x, an integer $n(x)$ such that during the time interval $[0, n(x)]$ iterates $f^i(x)$ close to the critical point c are compensated for by iterates far from c, so that $Df^{n(x)}(x)$ is larger than 1 in norm. More precisely, he constructed a countable partition $\{I_j\}$ of a full Lebesgue measure subset of M into intervals, such that $n(\cdot)$ is constant on each I_j, and every $f^{n(I_j)} \mid I_j$

- maps I_j onto some fixed interval J, which coincides with the union of a subfamily of I_j's (Markov property);
- is uniformly expanding and has bounded distortion: the logarithm of the derivative is uniformly Lipschitz.

The construction is by induction, and requires parameter exclusions at each step. A crucial part of the proof is to show that the exclusions decrease sufficiently fast, so that a set of parameters with positive Lebesgue measure remains at the end.

The restriction g of $f^{n(\cdot)}$ to J is a *Markov map*. A well-known result states that every Markov map has an absolutely continuous invariant measure ν, which is the limit of forward iterates of Lebesgue measure. Then

$$\mu_0 = \sum_{k=1}^{\infty} \sum_{i=0}^{k-1} f_*^i(\nu|\{n(\cdot) = k\})$$

defines an f-invariant measure which is also absolutely continuous. To ensure that μ_0 is finite, one needs the set of points with large inducing time $n(\cdot)$ to have small measure:

$$\sum_{k=1}^{\infty} k\,\nu(\{n(\cdot) = k\}) < \infty. \tag{2.3}$$

Actually, it suffices to check the corresponding property for Lebesgue measure m, because ν has bounded density. Finally, $\mu = \mu_0/\mu_0(M)$ is the absolutely continuous invariant probability of f.

We call a unimodal map *chaotic* if it has an absolutely continuous invariant probability. For unimodal maps with negative Schwarzian derivative, such a probability is unique and ergodic, and so it is an SRB measure. See Blokh, Lyubich [62] and [147, Section V.1].

Jakobson's theorem was much extended, and alternative approaches have been proposed for proving chaotic dynamics in dimension 1. See [147, Chapter V] and references therein. The first of these, due to Collet, Eckmann [130], focuses on proving that the map is exponentially expanding along the critical orbit: there are $C > 0$ and $\sigma > 1$ such that

$$|Df^n(c_1)| \geq C\sigma^n \quad \text{for all } n \geq 1, \tag{2.4}$$

where $c_1 = f(c)$ is the critical value. Indeed, in [131] they showed that, for unimodal maps with negative Schwarzian derivative, condition (2.4) implies the existence of absolutely continuous invariant probability. As a matter of fact, Collet, Eckmann needed a second condition, involving the backward orbit of the critical point, but this was shown by Nowicki [328] to be a consequence of the first one. This was further improved by Nowicki, van Strien [330]: if ℓ denotes the order of the critical point, the summability condition

$$\sum_n |Df^n(c_1)|^{-1/\ell} < \infty, \tag{2.5}$$

suffices for existence of an absolutely continuous invariant probability.

2.5 The renormalization theorem

Much of the impressive development of the theory of unimodal maps over the last years had to do with understanding the partition of the space into classes of topological conjugacy, and how this relates with finer equivalence classes, like real-analytic, differentiable, or quasi-symmetric conjugacy (rigidity phenomena). Two main tools were the kneading theory of Milnor, Thurston [293], which reduces essentially all the topological dynamics to symbolic terms, and renormalization theory, which provides an extremely efficient way to deal with strongly non-linear behavior in small scales.

The idea of renormalization comes from Physics, and was proposed by Coullet, Tresser [138] and by Feigenbaum [181] to explain the so-called universality phenomena they observed in bifurcation cascades of unimodal maps.

Renormalization operators proved to be powerful tools to reveal the fine geometric structure of phase space, not only for these systems but also in other situations of Dynamics, like critical circle maps, the Herman-Yoccoz theory of circle diffeomorphisms, or rational transformations of the Riemann sphere. In the unimodal setting, they led to the proof of universal geometries in phase and in parameter spaces. More recently, generalized renormalization techniques have been very fruitful in the analysis of statistical properties of these maps.

A unimodal map f is *renormalizable* if there is $p > 1$ and a family $I_0, \ldots,$ I_{p-1}, $I_p = I_0$ of compact disjoint intervals such that I_0 contains the critical point in its interior, and $f(I_{j-1}) \subset I_j$ for every $1 \leq j \leq p$. The *renormalization operator* $f \mapsto \mathcal{R}(f)$ is defined, in the subset of renormalizable maps, by

$$\mathcal{R}(f) = \phi \circ f^p \circ \phi^{-1}, \tag{2.6}$$

where one chooses p and I_0 smallest as above, and $\phi : I_0 \to M$ is the affine rescaling of I_0 to the original interval M. Then $\mathcal{R}(f)$ is also a unimodal map. We say that f is *infinitely renormalizable* if $\mathcal{R}^n(f) = (\mathcal{R} \circ \cdots \circ \mathcal{R})(f)$ is defined for all $n \geq 1$.

The renormalization operator is much better behaved in the real-analytic case: in particular, it is differentiable and injective. In addition, successive renormalizations of any C^2 map come closer and closer to real-analytic maps. See [147, Theorem VI.1.1]. For these reasons, one often considers the restriction of $\mathcal{R}$ to real-analytic unimodal maps. In what follows we also restrict ourselves to the case of quadratic critical points. The following statement is a global reformulation of the universality conjectures of Coullet, Tresser and Feigenbaum:

Conjecture 2.12 (Renormalization Conjecture). The renormalization operator $\mathcal{R}$ has an invariant uniformly hyperbolic set $\mathcal{H}$ consisting of infinitely renormalizable maps, with one-dimensional expanding direction and topological conjugacy classes as (codimension-one) stable manifolds, and such that the restriction of $\mathcal{R}$ to the set $\mathcal{H}$ is a homeomorphism conjugate to the full two-sided shift with countably many symbols.

The proof of this conjecture was a major achievement in the whole theory of one-dimensional dynamics. Sullivan [423] characterized the stable manifolds as conjugacy classes and proved sub-exponential contraction, in the case of bounded combinatorics (bounded p). McMullen [290] proved that the contraction is indeed exponential. Lyubich [272, 274] proved exponential expansion along the unstable direction, and treated also the unbounded case, thus completing the proof.

2.6 Statistical properties of unimodal maps

Hyperbolic sets of sufficiently smooth systems in finite dimension have zero volume, unless they coincide with the whole ambient. Thus, although $\mathcal{R}$ acts in infinite dimension, hyperbolicity of the renormalization horseshoe $\mathcal{H}$ may motivate the following (non-trivial) consequence proved by Lyubich [272]: the set of infinitely renormalizable quadratic maps has zero Lebesgue measure in parameter space.

On the other hand, Lyubich [276] and Martens, Nowicki [287] prove that Lebesgue almost all finitely renormalizable quadratic maps without periodic attractors satisfy the summability condition (2.5), and so they have an absolutely continuous invariant probability. Combining these two results one gets the following remarkable conclusion:

Theorem 2.13 (Lyubich [272]). *For Lebesgue almost every $a \in (0, 4]$, the quadratic map q_a is either hyperbolic or chaotic.*

Recently, Avila, Lyubich, de Melo [36] extended the conclusion of Theorem 2.13 to most real-analytic families of unimodal maps. The key to this extension is the following structure theorem for the space of non-hyperbolic maps:

Theorem 2.14 (Avila, Lyubich, de Melo [36]). *Non-hyperbolic topological conjugacy classes are codimension-one submanifolds of the space of real-analytic unimodal maps, with the possible exception of countably many classes. The lamination defined by these submanifolds is transversely quasi-symmetric: its holonomy maps extend to quasi-symmetric homeomorphisms.*

The possible exceptional cases in the statement correspond to the existence of indifferent periodic points. The quadratic family is everywhere transverse to the laminae. Thus, using the holonomies, one can transfer information from the quadratic family to any other that is fairly transverse to the lamination. In this way, Avila, Lyubich, de Melo [34, 36] prove that for every *non-trivial* family $(f_a)_a$, that is, any family such that no parameter interval is contained in the same non-hyperbolic conjugacy class, almost every map f_a is either hyperbolic or chaotic.

Let us point out that this extension of Theorem 2.13 is not a direct consequence of the statement of Theorem 2.14, one must appeal to the proof itself. One reason is that *the lamination is not transversely absolutely continuous:* the holonomies do not preserve the class of zero Lebesgue measure sets. We shall return to this point in a while.

In a series of very recent works, Avila, Moreira enhance this strategy, to provide a very complete picture of the statistical properties of unimodal maps. To begin with, they prove

Theorem 2.15 (Avila, Moreira [34, 39, 38]). *Let $(f_a)_a$ be a non-trivial real-analytic family of unimodal maps. Then, for Lebesgue almost every parameter a, either f_a is hyperbolic or it satisfies the Collet-Eckmann condition (2.4) together with slow recurrence of the critical orbit: given any $\delta > 0$,*

$$|f_a^n(c) - c| > n^{-(1+\delta)} \quad \text{for every large } n. \tag{2.7}$$

They deduce that almost every f_a has a finite number of SRB measures (exactly one if the Schwarzian derivative is negative), after proving that f_a has some renormalization conjugate to a quadratic map. For hyperbolic parameters, almost every point is in the basin of some periodic attractor. In the non-hyperbolic case, there exist $N \geq 1$ and an interval J such that $F_a = f_a^N$ defines a non-renormalizable unimodal map from J to J, and there is an absolutely continuous invariant probability μ_a supported inside the orbit of J. Almost every point in J is in the basin of μ_a, and almost every orbit that never enters J is in the basin of some periodic attractor.

Combining Theorem 2.15 with Keller, Nowicki [239] and Young [458], Avila, Moreira conclude that (F_a, μ_a) has exponential decay of correlations in the space of functions with bounded variation, for almost all non-hyperbolic parameter a. Actually, [239, 458] assume negative Schwarzian derivative, but this can be by-passed using the Collet-Eckmann condition. See Appendix E for definitions and background material on decay of correlations.

It had been shown by Baladi, Viana [43] that the Collet-Eckmann condition and a recurrence condition weaker than (2.7) imply stochastic stability, in a strong sense: L^1 convergence of densities. They considered additive absolutely continuous noise. Weak stochastic stability was proved in [55, 233], under somewhat different hypotheses. In view of the easy fact that hyperbolic maps are stochastically stable, Theorem 2.15 and [43] imply that almost all maps f_a are strongly stochastically stable.

In fact, these results extend to generic C^r families of unimodal maps, any $r \geq 2$. The description of the generic set is less explicit in this case, but Avila, Moreira still get that, for such families and for Lebesgue almost every parameter, either f_a is hyperbolic or it satisfies the Collet-Eckmann condition and has sub-exponential recurrence of the critical orbit. See [34]. Then, f_a has finitely many SRB measures, with stochastic stability (strong if $r \geq 3$) and exponential decay of correlations. In particular, the conjectures of Palis from Section 1.7 are fully verified in the context of unimodal maps!

The next theorem answers another old question in this field:

Theorem 2.16 (Avila, Moreira [37]). *Let $(f_a)_a$ be a non-trivial real-analytic family of unimodal maps. For Lebesgue almost every non-hyperbolic parameter a, the critical point is in the basin of the absolutely continuous invariant measure μ_a, and*

$$\lim_n \frac{1}{n} \log |Df^n(c_1)| = \int \log |Df| \, d\mu_a,$$

that is, the Lyapunov exponent of the critical orbit coincides with that of μ_a.

This leads to the following surprising conclusion [37]: for almost every map in any non-trivial real-analytic family, *the eigenvalue of every repelling periodic orbit in the support of μ_a is determined, through an explicit formula, by the itineraries (kneading sequences) of the periodic orbit and the critical orbit.*

Here is a quick explanation of this connection between eigenvalues and itineraries. For a periodic point p in the support of μ_a one considers neighborhoods V_k of p such that each V_k is mapped to V_{k-1} by the per(p):th iterate. On the one hand, each $\mu_a(V_k)$ may be computed as the frequency of visits of $f^n(c)$ to V_k, because the critical point c is in the basin of μ_a, and that is readily translated as the frequency inside the itinerary of the critical orbit of periodic blocks duplicating the itinerary of p. On the other hand, using absolute continuity of the SRB measure, one shows that $\mu_a(V_k)^{-1/k}$ converges to the eigenvalue of p when $k \to \infty$. This would follow from elementary arguments if $\mu_a(J)$ were proportional to length $|J|$ for every interval J in the support. However, the density is not bounded from above and, in general, we only have $c_1|J| \leq \mu_a(J) \leq c_2|J|^{1/2}$, which is not enough for the convergence statement. In short terms, one uses certain intervals related to gaps of hyperbolic Cantor sets, satisfying $\mu_a(J) \leq C_\delta|J|^{1-\delta}$ with $\delta > 0$ arbitrarily small.

Topologically conjugate unimodal maps are real-analytically conjugate if they have the same eigenvalues on corresponding periodic orbits. Therefore, the previous result implies that, for a set $\mathcal{E}$ of unimodal maps intersecting every non-trivial family on a full measure subset, *any two maps in $\mathcal{E}$ are topologically conjugate if and only if they are real-analytically conjugate on the support of their SRB measures.* This striking rigidity statement means that the "full measure" subset $\mathcal{E}$ intersects each topological conjugacy class in a subset with infinite codimension (contained in a single real-analytic conjugacy class!). This shows that the lamination into topological conjugacy classes is highly non-absolutely continuous.

3

Homoclinic Tangencies

A point is called *homoclinic* if it is in the intersection of the stable manifold and the unstable manifold of the same hyperbolic periodic point p. One speaks of *homoclinic tangency* if the intersection is not transverse. See Figure 1.2 for two examples; in the second one p is part of a non-trivial hyperbolic set H. It is well known since Poincaré that homoclinic points are a characteristic feature of systems with complicated dynamical behavior. Moreover, homoclinic tangencies have long been recognized as a main mechanism for instability and non-hyperbolicity. This is especially true about low dimensional systems such as surface diffeomorphisms, where instability is always associated to homoclinic tangencies as we shall see in Chapter 5. So, it is fundamental to understand as much as possible the great variety of dynamical phenomena associated to the creation and destruction of homoclinic points through tangencies. In particular,

Problem 3.1. (frequency of hyperbolicity) How common are hyperbolic diffeomorphisms close to one exhibiting a homoclinic tangency?

A geometric approach developed by Newhouse, Palis, Takens over the seventies and the eighties, and sharpened more recently by Moreira, Yoccoz, led to a surprisingly precise answer to this problem for surface diffeomorphisms, in terms of fractal invariants such as the Hausdorff dimension.

The first major result was the proof, by Newhouse, that close to a homoclinic tangency there may be whole open sets of non-hyperbolic diffeomorphisms. In fact, this is always the case for generic tangencies on surfaces and, even more striking, maps with infinitely many sinks or sources are dense in those open sets. In Section 3.1 we introduce the basic point of view to prove this and other results in this area, and in Section 3.2 we comment more closely on the proof as well as on related results. Concerning more directly Problem 3.1, we shall see in Section 3.3 that *the frequency of hyperbolicity close to a homoclinic tangency is governed by the fractal dimension of the associated horseshoe:* Hausdorff dimension less than 1 yields large relative measure (full

Lebesgue density) of hyperbolicity, and a converse is also true. See also the book of Palis, Takens [345] for a detailed broad presentation of these topics.

The two previous aspects of the theory have been unified during the last decade, as we shall see in Section 3.4: *in a generic parametrized family of surface diffeomorphisms going through a homoclinic tangency, the union of hyperbolic dynamics with persistent tangencies has full Lebesgue density at the tangency parameter.* Extension of this result to higher dimensions is currently on the way. Some recent progress is discussed in Section 3.5.

The theme of tangencies will return in Chapter 4, where we analyze how homoclinic phenomena participate in the dynamics of certain strange attractors.

3.1 Homoclinic tangencies and Cantor sets

The basic strategy, initiated by Newhouse [319], is to use invariant foliations to translate the dynamical problem, usually with some degree of simplification, into questions about certain dynamically defined Cantor sets.

Let $f : M \to M$ be a diffeomorphism with a homoclinic tangency q associated to a hyperbolic periodic point p, and H be a hyperbolic basic set of f containing p. That is, H is transitive, isolated, and periodic points are dense in it. In the simplest case H is just the orbit of p. We take all maps to be at least C^2. Moreover, for the time being *we suppose that the ambient manifold M is a surface.* This is crucial to ensure that the families

$$\mathcal{W}^s = \{W^s_{loc}(x) : x \in H\} \quad \text{and} \quad \mathcal{W}^u = \{W^u_{loc}(x) : x \in H\}$$

of local stable manifolds and local unstable manifolds extend to C^1 foliations $\mathcal{F}^s$ and $\mathcal{F}^u$ defined on a neighborhood of H. In fact, the extensions may be chosen so that the tangent bundles of $\mathcal{F}^s$ and $\mathcal{F}^u$ are generated by C^1 vector fields. See [345, Appendix 1]. Consequently, the *holonomies*, i.e. the projections along leaves between any pair of cross-sections of $\mathcal{F}^s$ or $\mathcal{F}^u$ are C^1 maps.

We may extend these foliations – by negative iteration of $\mathcal{W}^s$, $\mathcal{F}^s$ and positive iteration of $\mathcal{W}^u$, $\mathcal{F}^u$ – so that their domains contain the homoclinic point q. We are going to assume that the tangency is quadratic, meaning that the curvatures of the stable manifold and the unstable manifold at q are different. Then the leaves of $\mathcal{F}^s$ and $\mathcal{F}^u$ intersect transversely near q, except for a well-defined curve ℓ of mutual tangencies containing q. See Figure 3.1. This follows from an implicit function argument applied to the tangent bundles of $\mathcal{F}^s$ and $\mathcal{F}^u$, which also shows that the line of tangencies ℓ is of class C^1.

Let K^s and K^u be the sets of intersections of ℓ with the leaves of $\mathcal{W}^s$ and $\mathcal{W}^u$, respectively. K^s and K^u are Cantor sets, as long as the hyperbolic set H is non-trivial. Indeed, K^s is the image of $W^u_{loc}(p) \cap H$ under $\mathcal{W}^s$-holonomy,

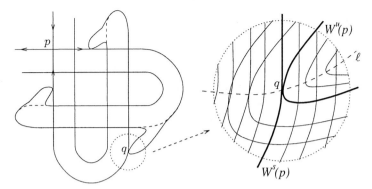

Fig. 3.1. Invariant foliations close to a quadratic homoclinic tangency

and similarly for K^u. Recall that these holonomies are C^1 diffeomorphisms [1]. Clearly, q is in the intersection of K^s and K^u. In the example of Figure 3.1 this is the only point of intersection (K^s is to the left of $W^s(p)$ and K^u is to the right of $W^u(p)$), but this needs not be so in general.

Now we analyze the evolution of this picture as the parameter varies. Let g be a diffeomorphism close to f. By robustness of hyperbolic objects, g has a hyperbolic basic set H_g close to $H = H_f$ and containing the continuation p_g of the saddle point $p = p_f$. Then we may perform for H_g the same construction as above for H. Denote by $\ell_g, \mathcal{W}_g^*, \mathcal{F}_g^*, K_g^*, * = s, u$, the corresponding objects.

Observe that every $z \in K_g^s \cap K_g^u$ is a point of tangency between some stable manifold $W^s(g, x)$ and some unstable manifold $W^u(g, y)$ of points $x, y \in H_g$. It is not difficult to see that any point in the intersection of $W^s(H_g)$ with $W^u(H_g)$ is non-wandering, e.g. using the fact that such a point is accumulated by the stable manifold and the unstable manifold of any periodic point in H_g.

Thus, *when such a tangency does exist the diffeomorphism g can not be hyperbolic.* A much deeper converse was observed in [324, 344]: *if the two Cantor sets are sufficiently far away from each other then the dynamics of g related to the tangency is hyperbolic.* This will be made more precise later. These two observations show that *the hyperbolicity problem is directly related to determining how frequently do K_g^s and K_g^u have points of (quasi) intersection.*

3.2 Persistent tangencies, coexistence of attractors

It was a remarkable discovery of Newhouse [319] that the Cantor sets K^s and K^u may intersect each other in a robust fashion, that is for a whole open set of diffeomorphisms close to the original f. Here openness is relative to

---

[1] This fact, together with the local product property, implies that the hyperbolic set H is locally C^1-*diffeomorphic* to a product $\left(W_{loc}^u(x) \cap H\right) \times \left(W_{loc}^s(x) \cap H\right)$ close to any point $x \in H$.

the C^2 topology, for reasons that will become apparent from the explanation that follows. Perhaps even more striking, assuming the initial map is area-dissipative at the periodic point p, *the generic map in such an open set of persistent tangencies has an infinite number of coexisting periodic attractors.*

For the construction, Newhouse begins by introducing the following notion of thickness of a Cantor set [2]. Let $K \subset \mathbb{R}$ be a Cantor set and $\hat{K}$ be its convex hull, i.e. the smallest compact interval containing K. Let $\{G_n : n \geq 1\}$ be the gaps of K, i.e. the connected components of $\hat{K} \setminus K$, ordered by nonincreasing length. The *thickness* of K is

$$\tau(K) = \inf \left\{ \frac{\text{length}\,(B(u))}{\text{length}\,(G_n)} : u \in \partial G_n \text{ and } n \geq 1 \right\}$$

where $B(u)$ denotes the connected component of $\hat{K} \setminus (G_1 \cup \cdots \cup G_n)$ containing the point u. To interpret this definition one should think of the Cantor set as obtained from $\hat{K}$ by successive removal of gaps. Then K has large thickness if every gap G_n is much smaller than both adjacent connected components of the set $\hat{K} \setminus (G_1 \cup \cdots \cup G_n)$ that remains after G_n is removed. What makes this concept so useful is the following

Lemma 3.2 (gap lemma [319]). *Let $K^1, K^2 \subset \mathbb{R}$ be two Cantor sets such that $\hat{K}^1 \cap \hat{K}^2 \neq \emptyset$ and neither of the two sets is contained in a gap of the other. If $\tau(K^1)\tau(K^2) > 1$ then $K^1 \cap K^2$ is non-empty.*

Proof. Let $\{G_n^i : n \geq 1\}$ be the gaps of K^i, $i = 1, 2$, in nonincreasing order of length. We say that two open intervals are linked if each one contains exactly one endpoint of the other. The proof of the lemma is by contradiction. Assuming the intersection is empty, we construct sequences $G_{m_i}^1$, $G_{n_i}^2$ of gaps such that every $G_{m_i}^1$ is linked to $G_{n_i}^2$ and the indices m_i and n_i converge to ∞. The latter implies that $\text{diam}\,(G_{m_i}^1 \cup G_{n_i}^2)$ converges to zero. Then any accumulation point of $G_{m_i}^1 \cup G_{n_i}^2$ belongs to K_1 and to K_2, and so the intersection is not empty after all.

To begin the construction, let G_m^1 be any gap of K^1 that intersects K^2. Each endpoint of G_m^1 is either outside $\hat{K}^2$ or in some gap of K^2, because we suppose that $K_1 \cap K_2$ is empty. They can not be both outside the convex hull, because K^2 is not contained in G_m^1, and they can not be in the same gap of K^2, because G_m^1 intersects K^2. So, we can always find a gap G_n^2 linked to G_m^1 and such that the component $B(v_n^2)$ is contained in G_m^1 for some endpoint v_n^2 of G_n^2.

Let u_m^1 be the endpoint of G_m^1 contained in G_n^2. The component $B(u_m^1)$ can not be contained in G_n^2: otherwise,

$$\frac{\text{length}\,(B(u_m^1))}{\text{length}\,(G_m^1)} \frac{\text{length}\,(B(v_n^2))}{\text{length}\,(G_n^2)} < 1,$$

---

[2] A rediscovery really: the same notion had been used by Hall [204] in a rather different context of number theory.

which would imply $\tau(K^1)\tau(K^2) < 1$. Therefore, $B(u_m^1)$ must contain the other endpoint u_n^2 of G_n^2. It follows that u_n^2 is contained in some gap G_s^1 of K^1 with $s > n$. Then the intervals G_n^2 and G_s^1 are linked, and the component $B(v_s^1)$ is contained in G_n^2 for some endpoint v_s^1 of G_s^1. This means that we recovered the situation in the previous paragraph, with G_n^2, G_s^1, v_s^1 in the place of G_m^1, G_n^2, v_n^2, respectively. Iterating this procedure we construct sequences $G_{m_i}^1$ and $G_{n_i}^2$ as announced.   □

Going back to the context of homoclinic tangencies, we may now define thicknesses of K^s and K^u, by fixing a differentiable parametrization of the line of tangencies ℓ so that the two Cantor sets may be considered as subsets of $\mathbb{R}$. In the same way we may define thicknesses of $W_{loc}^u(p) \cap H$ and $W_{loc}^s(p) \cap H$, by means of parametrizations of the stable manifold and the unstable manifold. These definitions depend on the choice of the parametrizations, because the thickness is preserved by a diffeomorphism up to a factor only, but that is not important for what follows [3].

Homoclinic tangencies as in Figure 3.1 may be found with $\tau(W_{loc}^s(p) \cap H)$ and $\tau(W_{loc}^u(p) \cap H)$ arbitrarily large, and the same for $\tau(K^s)$ and $\tau(K^u)$. In particular, we may have such examples satisfying the hypothesis of

Theorem 3.3 (Newhouse [319, 320]). *Suppose that $\tau(K^s)\tau(K^u) > 1$. Then*

1. *f belongs to the closure of an open subset $\mathcal{U}$ of $\mathrm{Diff}^2(M^2)$ such that every $g \in \mathcal{U}$ may be approximated by a diffeomorphism exhibiting a homoclinic tangency associated to the continuation of p; in particular no element of $\mathcal{U}$ can be uniformly hyperbolic;*
2. *if f is area-dissipative at p, meaning that $|\det Df^{\mathrm{per}(p)}(p)| < 1$, then there exists a residual subset $\mathcal{R} \subset \mathcal{U}$ such that every $g \in \mathcal{R}$ has an infinite number of periodic attractors.*

The first step is to perturb the initial diffeomorphism f with the homoclinic tangency, in order to obtain an open set $\mathcal{U}$ of diffeomorphisms g, containing f in its closure, for which the interiors $\hat{K}_g^s$ and $\hat{K}_g^u$ are linked. Then neither K_g^s nor K_g^u are contained in a gap of the other. Another crucial ingredient in the proof, besides the gap lemma, is the continuity of thickness as a function of the dynamical system:

Lemma 3.4. *The functions $g \mapsto \tau(K_g^s)$ and $g \mapsto \tau(K_g^u)$ are continuous on a neighborhood of f, relative to the C^2 topology.*

So, reducing $\mathcal{U}$ if necessary, $\tau(K_g^s)\tau(K_g^u) > 1$ and so $K_g^s \cap K_g^u \neq \emptyset$ for every $g \in \mathcal{U}$. Thence, there exists a tangency between some stable manifold and some unstable manifold of points in H_g. Using that the stable manifold and

---

[3] One way to avoid this dependence is to consider instead *local thickness*, which is invariant under diffeomorphisms. See [321] or [345, Chapter IV].

the unstable manifold of p_g are dense in $W^s(H_g)$ and $W^u(H_g)$, respectively, one gets the first part of Theorem 3.3.

The second part is a consequence, using that the unfolding of a homoclinic tangency associated to a dissipative saddle always yields periodic attractors:

Lemma 3.5. *Suppose $f : M \to M$ has a homoclinic tangency q associated to an area-dissipative saddle point p. For every $\varepsilon > 0$ there exist diffeomorphisms g arbitrarily close to f exhibiting a periodic attractor contained in the ε-neighborhood of the f-orbit of q.*

The subset R_n of diffeomorphisms $g \in \mathcal{U}$ with at least n periodic attractors is open, since periodic attractors are robust. Using Lemma 3.5 one gets that every R_{n+1} is dense in R_n. Hence, $\mathcal{R} = \cap_n R_n$ is residual, as claimed.

If f is area-expansive at p, meaning that $|\det Df^{\text{per}(p)}| > 1$, the same argument applied to f^{-1} produces a residual subset of diffeomorphisms in $\mathcal{U}$ with an infinite number of periodic repellers.

Another remarkable result of Newhouse shows that the phenomena in parts (1) and (2) in Theorem 3.3 are actually much more general: they occur close to *every* surface diffeomorphism with a homoclinic tangency, it is not necessary to assume that p is part of a thick hyperbolic set.

Theorem 3.6 (Newhouse [321]). *Let $f : M \to M$ be any surface diffeomorphism with a homoclinic tangency associated to a saddle-point p. Then*

1. *there exists an open subset $\mathcal{U}$ of $\text{Diff}^2(M^2)$ containing f in its closure, such that every $g \in \mathcal{U}$ may be approximated by a diffeomorphism with a homoclinic tangency associated to the continuation of p;*
2. *if f is area-dissipative, respectively area-expansive, at p then there exists a residual subset $\mathcal{R} \subset \mathcal{U}$ such that every $g \in \mathcal{R}$ has an infinite number of periodic attractors, respectively periodic repellers.*

For the proof one successively perturbs the initial map in order to reach a situation similar to Theorem 3.3. A first step allows us to suppose that p belongs to a non-trivial hyperbolic basic set H_1: otherwise we just perturb f so as to create transverse homoclinic orbits together with a new homoclinic tangency associated to p. The thickness $\tau^u(H_1) := \tau(W^s_{loc}(p) \cap H_1)$ is always positive, and further perturbations will be small enough so that H_1 and $\tau^u(H_1)$ remain essentially unchanged. Up to an additional perturbation we may suppose that the tangency is quadratic, and the Jacobian of f on the orbit of p is different from 1. We consider the area-dissipative case; for the area-expanding case it suffices to replace f by its inverse.

The crucial step in the proof of Theorem 3.6 consists in showing that *the generic unfolding of a homoclinic tangency always generates (small) thick hyperbolic sets which are involved in new homoclinic tangencies.* One way such thick hyperbolic sets can be constructed is through the renormalization scheme that we outline in the next couple of paragraphs. More details can be found in Section 4.6.1 and [345, Section III.4].

Let $(f_\mu)_\mu$ be a curve of diffeomorphisms with $f_0 = f$ and such that the stable manifold and the unstable manifold of the continuation p_μ of p move with non-zero relative velocity close to the tangency as the parameter μ varies. One finds parameter values $\mu_n \to 0$ and small domains $Q_n \subset M$ converging to the point of homoclinic tangency, such that Q_n returns close to itself under $f_{\mu_n}^n$ and, in convenient n-*dependent* coordinates, the return maps $f_{\mu_n}^n | Q_n$ converge in the C^2 topology to

$$\phi(x, y) = (1 - 2x^2, x)$$

as $n \to \infty$. The fact that the limit ϕ has zero Jacobian is no surprise: most iterates of the return map take place close to the saddle p, where our diffeomorphisms are area-dissipative.

The most important feature of ϕ in this context is that it admits invariant hyperbolic Cantor sets with arbitrarily large thickness in the x-direction. This can be seen, for instance, from the fact that ϕ is smoothly conjugate to the piecewise affine map $\psi(x, y) = (1 - 2|x|, x)$, for which such thick invariant sets may be exhibited explicitly. By hyperbolic continuation and continuity of the thickness, it follows that for every large n the map $f_{\mu_n}^n | Q_n$ admits invariant hyperbolic sets H_2 with large thickness transversely to the stable direction: we may take H_2 and a periodic point $p_2 \in H_2$ so that $\tau^s(H_2) := \tau(W_{loc}^u(p_2) \cap H_2)$ satisfies

$$\tau^u(H_1)\tau^s(H_2) > 1. \tag{3.1}$$

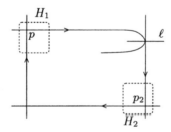

Fig. 3.2. Heteroclinic tangencies

The hyperbolic basic sets H_1 and H_2 are *heteroclinically related*, i.e. there are transverse intersections between their stable manifolds and unstable manifolds, and a tangency between $W^u(p)$ and $W^s(p_2)$ may be created by another small perturbation of the diffeomorphism. See Figure 3.2. This is possible because the map ϕ itself has homoclinic tangencies: the unstable set [4] of the fixed point $P = (-1, 0)$ has non-transverse intersections with the stable set of P, for instance at $q = (1, 0)$. The continuation of P for the diffeomorphism

---

[4] The unstable set is the union of all $\phi^n(V)$, $n \geq 1$, for some small neighborhood V of P, and the stable set is the union of all pre-images $\phi^{-n}(P)$, $n \geq 0$.

$f_{\mu_n}^n$ is heteroclinically related to both H_1 and H_2 and it is close to having a homoclinic tangency. Then the stable and unstable manifolds of p and p_2 accumulate on the corresponding invariant manifolds of that continuation, and a heteroclinic tangency associated to p and p_2 may be created by a small perturbation, as claimed.

At this point we may use the gap lemma and (3.1) in much the same way as for Theorem 3.3. We conclude that there exists an open set $\mathcal{U} \subset$ Diff$^2(M)$ with persistent heteroclinic tangencies involving H_1 and H_2, and so also with persistent homoclinic tangencies associated to p. The same Baire type argument as in Theorem 3.3 gives that maps in a residual subset of $\mathcal{U}$ have infinitely many coexisting attractors.

Remark 3.7. This renormalization scheme also provides a quick proof of Lemma 3.5: there exist $\phi_a(x,y) = (1 - ax^2, x)$ with a arbitrarily close to 2 admitting hyperbolic periodic attractors; by robustness, this remains true for f_μ^n with μ close to μ_n. So f_μ has hyperbolic periodic attractors for μ arbitrarily close to zero.

3.2.1 Open sets with persistent tangencies

Some three decades after their existence was proved, the phenomena of persistent tangencies and coexistence of infinitely many sinks or sources remain a mystery, and we shall be referring to them often as we attempt to the develop a global picture of dynamics beyond hyperbolicity. Here we review some recent results and extensions that also give a measure of their inherent complexity.

We begin by noticing that the two phenomena are intimately related. We have already seen how persistence of tangencies leads to coexistence of sinks or sources. A converse, in the C^1 topology, follows from results of Pujals, Sambarino [371] that we shall review in Chapter 5: if an open set of Diff$^1(M^2)$ contains a dense subset corresponding to infinitely many coexisting sinks or sources, then it also contains a dense subset corresponding to homoclinic tangencies.

Let us also mention that Newhouse's argument can not work in the C^1 topology. Indeed, Ures [434] shows that hyperbolic sets of a C^1 residual subset of diffeomorphisms have thickness equal to zero, in both stable direction and unstable directions. He also deduces that for generic C^1 families unfolding a homoclinic tangency, the set of parameters corresponding to hyperbolicity has full Lebesgue density at the tangency parameter. Indeed, the following basic problem remains wide open:

Problem 3.8. Are uniformly hyperbolic systems dense among all C^1 diffeomorphisms on a surface?

On the other hand, Bonatti, Díaz [72] constructed C^1 open sets with coexistence of infinitely many sinks, for a residual subset of C^1 diffeomorphisms on any manifold of dimension $d \geq 3$. See Section 10.6.2 below. An extension of this construction, in [68], also gives coexistence of infinitely many non-trivial compact invariant sets where the dynamics is uniquely ergodic (an adding machine). See Section 10.6.3. It was recently remarked by Bonatti, Moreira that adding machines are also present, as quasi-attractors [5], in the unfolding of homoclinic tangencies in dimension 2. This is because the unfolding of a homoclinic tangency creates small disks strictly invariant by some iterate and containing new homoclinic tangencies. By iterating this procedure one obtains a nested sequence of disks, the intersection of which contains the adding machine.

In addition, it is interesting to point out that the phenomenon of coexistence of periodic attractors has been extended by Colli [133] for strange attractors. Indeed, he constructs open sets of diffeomorphisms containing dense subsets for which there are infinitely many coexisting Hénon-like attractors. Observe that the argument must be much more delicate than in the periodic case, because Hénon-like attractors are persistent in a measure-theoretical sense only (see Theorem 4.4).

Conjecture 3.9 ([426, 341]). In the unfolding of a homoclinic tangency associated to an area-dissipative (sectionally dissipative in higher dimensions, see Section 3.5), the set of parameter values corresponding to coexistence of infinitely many periodic or Hénon-like attractors has Lebesgue measure zero. Similarly for quasi-attractors.

Also very interesting, Theorem 3.6 has been extended to conservative maps by Duarte [171, 172]: any area preserving map with a homoclinic tangency is C^2 approximated by an open domain in the space of area preserving maps exhibiting persistent tangencies; generic diffeomorphisms on such a domain have infinitely many "independent" elliptic islands (which take the role of attractors or repellers in the non-conservative case).

Another striking property of Newhouse's domains with persistent tangencies has been discovered recently by Kaloshin: super-exponential growth of the number of periodic orbits. Indeed, he proves

Theorem 3.10 (Kaloshin [228]). *Let $\mathcal{U} \subset \mathrm{Diff}^2(M^2)$ be an open set with persistent tangencies. Given any sequence $(\eta_n) \to \infty$ there exists a residual subset of $\mathcal{U}$ for which the number of periodic points of period n is larger than η_n for all large n.*

In a few words, Kaloshin exploits the fact that diffeomorphisms with highly degenerate saddle-nodes (high contact with the identity along the central direction) are dense in the Newhouse domain $\mathcal{U}$ (Gonchenko, Shil'nikov, Turaev [193]), to produce abnormal rates of growth of the number of periodic

---

[5]A *quasi-attractor* is a decreasing intersection $\Lambda = \cap_n \Lambda_n$ where each Λ_n is a topological attractor, that is, the maximal invariant set in an open neighborhood U_n such that the closure of $f(U_n)$ is contained in U_n.

points. Quite in contrast, Hunt, Kaloshin [221] announce that the number of periodic points of period n is bounded by const $\exp(n^{1+\delta})$, for a "full probability" subset of diffeomorphisms.

3.3 Hyperbolicity and fractal dimensions

We are going to see that a refinement of the previous analysis leads to a deep connection between frequency of hyperbolicity in the unfolding of a homoclinic tangency and fractal dimensions of invariant sets.

Let $(f_\mu)_\mu$ be a smooth parametrized family of diffeomorphisms such that $f = f_0$ has a quadratic homoclinic tangency q associated to a periodic point p, and this tangency is *generically unfolded* as the parameter μ varies: the stable manifold and the unstable manifold of the continuation p_μ of p move with respect to each other with non-zero velocity near the point q, so that the tangency gives rise to a pair of transverse intersections when $\mu > 0$. As before, we suppose that p is part of a hyperbolic basic set H, which may or may not be reduced to the orbit of p.

We are going to suppose that $\mu = 0$ is a *first bifurcation*, meaning that f_μ is uniformly hyperbolic for every $\mu < 0$ close to zero. This is for simplicity only: the conclusions hold in general, restricted to the dynamics of f_μ related to the homoclinic tangency of f. More precisely, in the general case one must refer to the maximal invariant set

$$\Sigma_\mu = \bigcap_{n \in \mathbb{Z}} f_\mu^n(U \cup V_\mu) \tag{3.2}$$

inside a neighborhood U of H union a $C\mu$-neighborhood V_μ of the orbit of tangency. See [345, Chapter V]. In the first bifurcation case the non-wandering set $\Omega(f_\mu)$ is contained in the union of Σ_μ with a finite number of hyperbolic basic sets, so that the diffeomorphism is uniformly hyperbolic if Σ_μ is.

As we have seen, hyperbolicity is directly related to existence of intersections, or almost intersections, between the Cantor sets $K_\mu^* := K_{f_\mu}^*$, $* = s, u$. Let us again look at the way these Cantor sets vary with the parameter μ. Refer to Figure 3.1. The stable foliation $\mathcal{F}_\mu^s$ and the unstable foliation $\mathcal{F}_\mu^u$ may be chosen such that their tangent bundles vary in a C^1 fashion with μ. Then so does the curve of tangencies ℓ_μ. We fix parametrizations of these curves, jointly C^1 in phase space and parameter space. Of course, we may choose the point of $W^u(p_\mu) \cap \ell_\mu$ close to the tangency to lie at the origin of the coordinates induced in ℓ_μ by this parametrization.

The main effect on the variation of the two Cantor sets comes from the generic unfolding of the tangency: K_μ^s is translated with respect to K_μ^u when the parameter varies, so that their convex hulls become linked when $\mu > 0$. With our choices of coordinates, and rescaling the parameter μ is necessary,

$$K_\mu^u \approx K^u \quad \text{and} \quad K_\mu^s \approx K^s + \mu$$

for all μ close to zero. Thus, K_μ^u and K_μ^s are close to intersecting each other if and only if K^u and $K^s + \mu$ are close to intersecting each other. In this way, one is led to studying the *arithmetic difference*

$$K^u - K^s = \{\mu \in \mathbb{R} : \mu = \kappa_u - \kappa_s \text{ for some } \kappa_u \in K^u \text{ and } \kappa_s \in K^s\}$$
$$= \{\mu \in \mathbb{R} : K^u \cap (K^s + \mu) \text{ is non-empty}\}.$$

And Problem 3.1 directly points to:

Problem 3.11. Let K^1 and K^2 be two Cantor sets in the real line. How big is their arithmetic difference $K^1 - K^2$?

The question is of special interest in the case of dynamically defined Cantor sets, that is, arising in connection with hyperbolic sets of diffeomorphisms in the way described in Section 3.1 (a formal definition will appear later).

A very precise answer to this problem may be given in terms of metric invariants of K^1 and K^2. To begin with, it follows from the gap lemma 3.2 that if $\tau(K_1)\tau(K_2) > 1$ then the arithmetic difference has non-empty interior. More refined statements are in terms of fractal dimensions, like the Hausdorff dimension and the limit capacity (box dimension).

Let X be a compact metric space and $N(X, \varepsilon)$ denote the minimum number of ball of radius $\varepsilon > 0$ required to cover X. The *limit capacity* of X is

$$c(X) = \inf\{d > 0 : N(X, \varepsilon) \leq \varepsilon^{-d} \text{ for every small } \varepsilon > 0\}.$$

Now we define the Hausdorff dimension of X. Given any $d > 0$ the *Hausdorff d-measure* of X is

$$m_d(X) = \lim_{\varepsilon \to 0} \left(\inf \sum_{U \in \mathcal{U}} \operatorname{diam}(U)^d \right)$$

where the infimum is taken over all coverings $\mathcal{U}$ of X by sets U with diameter less than $\varepsilon > 0$. It is easy to see that there is a unique number $0 \leq \operatorname{HD}(X) \leq \infty$, the *Hausdorff dimension* of X, such that

$$m_d(X) = \infty \text{ if } d < \operatorname{HD}(X) \quad \text{and} \quad m_d(X) = 0 \text{ if } d > \operatorname{HD}(X).$$

Detailed information about these notions can be found in [178, 353] and [345, Chapter IV].

It is easy to show that $\operatorname{HD}(X) \leq c(X)$, and $0 \leq \operatorname{HD}(X) \leq c(X) \leq d$ for every compact subset of a d-dimensional manifold. Moreover,

$$\operatorname{HD}(Y) \leq \nu^{-1}\operatorname{HD}(X) \quad \text{and} \quad c(Y) \leq \nu^{-1}c(X)$$

if Y is the image of X by a ν-Hölder continuous transformation. In particular, both dimensions are preserved by bi-Lipschitz homeomorphisms. Since stable and unstable holonomies are C^1, in this 2-dimensional setting, it follows that

$$c(K_\mu^s) = c\big(W_{loc}^u(p_\mu) \cap H_\mu\big) \quad \text{and} \quad \mathrm{HD}(K_\mu^s) = \mathrm{HD}\big(W_{loc}^u(p_\mu) \cap H_\mu\big) \quad (3.3)$$

and analogously for K_μ^u and $W_{loc}^s(p_\mu) \cap H_\mu$.

In fact, our dynamical sets are much more regular than arbitrary Cantor subsets of $\mathbb{R}$. In particular, $c(K_\mu^*) = \mathrm{HD}(K_\mu^*) \in (0,1)$ for $* = s, u$ and

$$c(H_\mu) = \mathrm{HD}(H_\mu) = \mathrm{HD}(K_\mu^s) + \mathrm{HD}(K_\mu^u) \in (0,2).$$

The second equality follows from the fact that H_μ is locally bi-Lipschitz homeomorphic to the product $K_\mu^s \times K_\mu^u$, and

$$c(X \times Y) \le c(X) + c(Y) \quad \text{and} \quad \mathrm{HD}(X \times Y) \ge \mathrm{HD}(X) + \mathrm{HD}(Y)$$

(so that we have equality whenever $c = \mathrm{HD}$). See [178, 345]. Moreover, the fractal dimension $c(K_\mu^*) = \mathrm{HD}(K_\mu^*)$ varies continuously with the diffeomorphism f_μ, in the C^1 topology. See [345, Chapter IV] and [285, 346].

The following simple lemma shows that the arithmetic difference of sets with small limit capacity is a zero Lebesgue measure set. The proof is an exercise.

Lemma 3.12. *Let $K^1, K^2 \subset \mathbb{R}$ be compact sets. Then*

$$c(K^1 - K^2) \le c(K^1 \times K^2) \le c(K^1) + c(K^2).$$

In particular, if $c(K^1) + c(K^2) < 1$ then $m(K^1 - K^2) = 0$.

Starting from this observation, Palis, Takens show that *if H has fractal dimension less than 1 then most values of μ close to $\mu = 0$ correspond to hyperbolicity*. The case when the set H reduces to the orbit of p had been treated in [324].

Theorem 3.13 (Palis, Takens [344]). *If $c(H) < 1$ then $\mu = 0$ is a Lebesgue density point for the set $\mathcal{H}$ of parameter values for which f_μ is uniformly hyperbolic:*

$$\lim_{\varepsilon \to 0} \frac{m(\mathcal{H} \cap [0, \varepsilon])}{\varepsilon} = 1.$$

Conversely, Palis, Yoccoz prove that *if the fractal dimension is bigger than 1 then non-hyperbolic parameters correspond to a sizable portion of parameter space near $\mu = 0$*. The statement is slightly more involved than in the case of Theorem 3.13. They consider smooth two-parameter families of surface diffeomorphisms $(f_{\mu,\nu})_{\mu,\nu}$ such that $(f_{\mu,0})_\mu$ unfolds generically a homoclinic tangency associated to a hyperbolic set H of $f_{0,0}$. For every ν small the family $(f_{\mu,\nu})_\mu$ unfolds generically a homoclinic tangency associated to the continuation of H; we may suppose that the tangency always takes place at $\mu = 0$. Let $\Sigma_{\mu,\nu}$ be the maximal invariant set of $f_{\mu,\nu}$ defined as in (3.2). Then

Theorem 3.14 (Palis, Yoccoz [348]). *If* $HD(H) > 1$ *then*

$$\liminf_{\varepsilon \to 0} \frac{m(\mathcal{H}_\nu \cap [0, \varepsilon])}{\varepsilon} < 1, \text{ for Lebesgue-almost every } \nu \text{ close to zero,}$$

where $\mathcal{H}_\nu$ *is the set of values of* μ *for which* $f_{\mu,\nu}$ *is uniformly hyperbolic.*

We shall later see that the lim sup is also smaller than 1. At the heart of the proof of Theorem 3.14 is the following deep result which provides a converse to Lemma 3.12. The need for the extra parameter ν in Theorem 3.14 is related to the presence of the parameter λ in Marstrand's theorem:

Theorem 3.15 (Marstrand [286]). *Let* K^1 *and* K^2 *be compact subsets of* $\mathbb{R}$ *with* $HD(K^1) + HD(K^2) > 1$. *Then for Lebesgue almost every* $\lambda \in \mathbb{R}$ *there exists a positive Lebesgue measure set* T_λ *such that for every* $t \in T_\lambda$

$$HD(K^1 \cap (\lambda K^2 + t)) \geq d > 0, \qquad d = HD(K^1) + HD(K^2) - 1.$$

In particular, $T_\lambda \subset K^1 - \lambda K^2$ *and so* $m(K^1 - \lambda K^2) > 0$ *for almost every* $\lambda \in \mathbb{R}$.

This admits the following geometric interpretation. Let $\pi_\lambda : \mathbb{R}^2 \to \mathbb{R}$ be the projection along co-slope λ: $\pi_\lambda(x_1, x_2) = x_1 - \lambda x_2$. For almost every λ the image $\pi_\lambda(K^1 \times K^2) = K^1 - \lambda K^2$ has positive Lebesgue measure. Actually, Marstrand proves that *for almost every* λ *the projection* $(\pi_\lambda)_*(m_{HD})$ *of the Hausdorff measure* m_{HD} *on* $K^1 \times K^2$ *is absolutely continuous, with* L^2 *density.* Moreover, there even exists a positive Lebesgue measure subset T_λ such that for $t \in T_\lambda$ the fiber

$$\pi_\lambda^{-1}(t) \cap (K^1 \times K^2) \approx K^1 \cap (\lambda K^2 + t)$$

has Hausdorff dimension at least d.

Remark 3.16. The hypothesis of Theorem 3.14 is more general than that of Theorem 3.3: thick Cantor sets always have Hausdorff dimension close to 1, in particular,

$$\tau(K^s)\tau(K^u) > 1 \quad \Rightarrow \quad HD(H) = HD(K^s) + HD(K^u) > 1.$$

See Palis, Takens [345, Chapter 4] for a discussion of relations between thickness and fractal dimensions.

Let us also mention that Díaz, Ures [161] show that saddle-node horseshoes with small Hausdorff dimension may lead to persistent homoclinic tangencies immediately after the bifurcation. Moreover, Rios [376] has extended a good part of the previous theory, including persistence of tangencies and control of hyperbolicity through fractal dimensions, to the more subtle situation of homoclinic tangencies inside the limit set (accumulated by periodic orbits).

3.4 Stable intersections of regular Cantor sets

In the context of these results, Palis has proposed the following ambitious answer to Problem 3.11:

Conjecture 3.17. The arithmetic difference $K^1 - K^2$ of generic dynamically defined Cantor sets either has zero Lebesgue measure or contains an interval.

This conjecture was proved by Moreira, Yoccoz [316, 315], in a very strong form. From it they deduced a statement about homoclinic tangencies that improves Theorem 3.14 in more than one way. These outstanding results are the subject of the present section. Notice that the statement above can not be true for *all* pairs of dynamically defined Cantor sets, by [403].

A Cantor set $K \subset \mathbb{R}$ is called *dynamically defined*, or *regular*, if it is the limit set

$$K = \bigcap_{n=0}^{\infty} \bigcup f_{j-n} \circ \cdots \circ f_{j-1}(D_{j-1}) \qquad (3.4)$$

of a Markov family of contractions, that is, a family $f = \{f_j : j = 1, \ldots, N\}$ of C^r, $r \geq 1$, diffeomorphisms $f_j : D_j \to R_j$ between compact subintervals D_j and R_j of $I = [0,1]$, with derivative bounded by some constant $\lambda < 1$, and such that every domain D_j is the convex hull of some subset of ranges R_i. The union in (3.4) is over all *admissible sequences* $j_{-n}, \ldots, j_{-1}$, that is, such that the composition is defined.

The Cantor sets K^s, K^u, $W_{loc}^u(p)$, $W_{loc}^s(p)$ introduced above are regular with $r > 1$, that is, the contractions f_j may be chosen so that their derivatives are at least Hölder continuous. See [345, Chapter IV]. In what follows we fix $r > 1$. The space of regular Cantor sets inherits a natural topology from the C^r topology in the space of such families (with fixed number of elements N). This induces a natural C^r topology in the space $\mathcal{K}$ of pairs of regular Cantor sets (K^1, K^2).

Definition 3.18 (Moreira [313]). K^1 and K^2 have *stable intersection* if for every pair $(\tilde{K}_1, \tilde{K}_2) \in \mathcal{K}$ close to (K_1, K_2) we have $\tilde{K}_1 \cap \tilde{K}_2 \neq \emptyset$.

Translates of a regular Cantor set K are also regular, and they are C^r close to K if the translation is small. So, if K^1 and K^2 have stable intersection or, more generally, if K^1 and $K^2 + t$ have stable intersection for some $t \in \mathbb{R}$, then $K^1 - K^2$ contains an interval. Moreira, Yoccoz prove that *an open dense subset of pairs of regular Cantor sets whose sum of the Hausdorff dimensions is larger than 1 have stable intersection after convenient translation*. In view of Lemma 3.12, and the fact that Hausdorff dimension and limit capacity coincide in this setting, this result implies the Palis Conjecture 3.17.

Theorem 3.19 (Moreira, Yoccoz [316]). *Let Ω be the set of $(K^1, K^2) \in \mathcal{K}$ such that* $\mathrm{HD}(K^1) + \mathrm{HD}(K^2) > 1$. *Then there exists an open and dense subset $\mathcal{W}$ of Ω such that for every $(K^1, K^2) \in \mathcal{W}$ the set*

$$S(K^1, K^2) = \{t \in \mathbb{R} : K^1 \text{ and } K^2 + t \text{ have stable intersection}\}$$

is open and dense in $K^1 - K^2$ *. In addition, the complement has zero Lebesgue measure: more than that,* $\mathrm{HD}\big((K^1 - K^2) \setminus S(K^1, K^2)\big)$ *is less than 1.*

Let $(f_\mu)_\mu$ be a smooth family of surface diffeomorphisms generically unfolding a homoclinic tangency associated to a hyperbolic set H of $f = f_0$. Adapting the proof of Theorem 3.19 and using results from [313], Moreira, Yoccoz show that, if the Hausdorff dimension of H is larger than 1, then persistent tangencies correspond to a definite fraction of parameters close to zero and the union of such parameters with those corresponding to hyperbolicity has full Lebesgue density at the bifurcation. More formally,

Theorem 3.20 (Moreira, Yoccoz [315]). *Assume* $\mathrm{HD}(H) > 1$*. Then there exists an open set* $\mathcal{N}$ *in parameter space corresponding to persistent tangencies such that*

$$\liminf_{\varepsilon \to 0} \frac{m(\mathcal{N} \cap [0, \varepsilon])}{\varepsilon} > 0 \quad and \quad \liminf_{\varepsilon \to 0} \frac{m((\mathcal{N} \cup \mathcal{H}) \cap [0, \varepsilon])}{\varepsilon} = 1,$$

where $\mathcal{H}$ *is the set of parameters for which* f_μ *is uniformly hyperbolic.*

The key to proving Theorem 3.19 is the fact, remarkable in itself, that for most pairs of regular Cantor sets the same geometric pattern is repeated over and over again at arbitrarily small scales. More precisely, Moreira, Yoccoz introduce a family of renormalization (or scale refinement) operators $\mathcal{R}_{\alpha,\tilde{\alpha}}$ and prove that, up to small perturbation of the Cantor sets, this family admits a compact invariant set. In the sequel we are going to describe the main steps in the proof in more detail. We begin by explaining the meaning and giving the precise definition of the renormalization operators.

3.4.1 Renormalization and pattern recurrence

For simplicity we consider Cantor sets dynamically defined by contractions $f_j : I \to I$ defined on the whole $I = [0, 1]$. Then, clearly, every sequence $\theta = (\ldots, \theta_{-n}, \ldots, \theta_{-1})$ with values in $\{1, \ldots, N\}$ is admissible. Let Σ be the space of such sequences. For every $n \geq 1$ denote $f_\theta^n = f_{\theta_{-n}} \circ \cdots \circ f_{\theta_{-1}}$. This is defined also when θ is a finite sequence with length at least n. Endow Σ with the metric

$$d(\theta, \theta') = \text{length}\left(f_\beta^k(I)\right), \tag{3.5}$$

where $\beta = (\theta_{-k}, \ldots, \theta_{-1})$ if $\theta_{-i} = \theta'_{-i}$ for $1 \leq i \leq k$ and k is maximum.

Given regular Cantor sets (K, f) and $(\tilde{K}, \tilde{f})$, sequences θ and $\tilde{\theta}$, with values in $\{1, \ldots, N\}$, and given integers m and $\tilde{m}$, the interval $f_\theta^m(I)$ is a connected component of the m:th stage of the construction of K, and analogously for $\tilde{K}$, $\tilde{f}$, $\tilde{\theta}$, $\tilde{m}$. Let α and $\tilde{\alpha}$ be two finite sequences of length k and $\tilde{k}$, respectively, with values in $\{1, \ldots, N\}$. Roughly, the renormalization operator is meant to assign

$$\mathcal{R}_{\alpha,\tilde{\alpha}} : \left(f_\theta^m(I), \tilde{f}_{\tilde{\theta}}^{\tilde{m}}(I)\right) \mapsto \left(f_{\theta\alpha}^{m+k}(I), \tilde{f}_{\tilde{\theta}\tilde{\alpha}}^{\tilde{m}+\tilde{k}}(I)\right) \quad \text{after rescaling.}$$

Here $\theta\alpha$ and $\tilde{\theta}\tilde{\alpha}$ represent the concatenations of the corresponding pairs of sequences. Observe that the intervals on the right hand side are connected components of deeper stages of the construction of the two Cantor sets.

For the formal definition one takes a kind of limit when m, $\tilde{m}$ go to ∞. For each $\theta \in \Sigma$ and $n \geq 1$ let h_θ^n be the orientation preserving affine map from $f_\theta^n(I)$ to the interval I. The limit diffeomorphism

$$\kappa_\theta = \lim(h_\theta^n \circ f_\theta^n)$$

exists for every $\theta \in \Sigma$ and depends Hölder continuously on θ, because our maps are differentiable and the derivatives are Hölder continuous. The images $\kappa_\theta(K)$ of K under the maps κ_θ are the *limit geometries* of the Cantor set. A direct important consequence of the definition is that for any sequence α with length $k \geq 1$, the set $\kappa_\theta(f_\alpha^k(K)) = \kappa_\theta(K) \cap \kappa_\theta(f_\alpha^k(I))$ is the affine image of another limit geometry, namely $\kappa_{\theta\alpha}(K)$.

For our purposes it is important to keep track also of relative positions, which are forgotten by the quotient by the affine group involved in the definition of limit geometries. The *relative position* of two oriented intervals J_1, J_2 is the pair $(\lambda, t) \in \mathbb{R}^* \times \mathbb{R}$ defined by

$$h(J_2) = \lambda h(J_1) + t,$$

where h is the orientation preserving affine map sending J_1 onto $[0, 1]$, and $\lambda < 0$ if the orientations of J_1 and J_2 are opposite. Given any finite sequences $\alpha, \tilde{\alpha}$, the associated *renormalization operator* is defined by

$$\mathcal{R}_{\alpha,\tilde{\alpha}} : \Sigma \times \tilde{\Sigma} \times \mathbb{R}^* \times \mathbb{R} \to \Sigma \times \tilde{\Sigma} \times \mathbb{R}^* \times \mathbb{R}, \quad (\theta, \tilde{\theta}, \lambda, t) \mapsto (\theta\alpha, \tilde{\theta}\tilde{\alpha}, \lambda_1, t_1)$$

where (λ_1, t_1) is the relative position of $\kappa_\theta(f_\alpha^k(I))$ and $\lambda\tilde{\kappa}_{\tilde{\theta}}(\tilde{f}_{\tilde{\alpha}}^{\tilde{k}}(I)) + t$. See Figure 3.3.

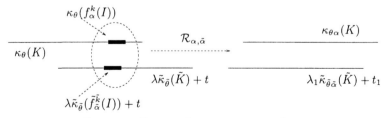

Fig. 3.3. Renormalization operator $\mathcal{R}_{\alpha,\tilde{\alpha}}$

Lemma 3.21. *Suppose there exists a compact subset $\mathcal{L}$ of $\Sigma \times \tilde{\Sigma} \times \mathbb{R}^* \times \mathbb{R}$ such that for every $(\theta, \tilde{\theta}, \lambda, t) \in \mathcal{L}$ there exist α, $\tilde{\alpha}$ such that*

$$\mathcal{R}_{\alpha, \tilde{\alpha}}(\theta, \tilde{\theta}, \lambda, t) \in \text{int}(\mathcal{L}).$$

Then $\kappa_\theta(K)$ and $\lambda \tilde{\kappa}_{\tilde{\theta}}(\tilde{K}) + t$ have C^r stable intersection [6] for all $(\theta, \tilde{\theta}, \lambda, t) \in \mathcal{L}$.

The idea of the proof is the following: by successive renormalizations one constructs a sequence $(\theta^{(n)}, \tilde{\theta}^{(n)}, \lambda^{(n)}, t^{(n)})$ inside the compact set $\mathcal{L}$. In particular, the sequence $t^{(n)}$ is bounded: the intervals involved are at bounded distance, *in terms of relative positions*. Since the diameters are going to zero as n increases, this really means that these intervals get closer and closer. In this way we get a common point of accumulation, and this point is in the intersection of the two Cantor sets.

The problem is now how to find such a *recurrent set* $\mathcal{L}$ as in the hypothesis of Lemma 3.21, for every pair of Cantor sets $(K, \tilde{K})$ in some open and dense subset $\mathcal{W}$ of Ω. Density is the main issue, of course. Fix $\varepsilon > 0$. We are going to explain that a recurrent set does exist, after some 10ε-perturbation of the original pair. The main difficulty comes from the fact that the behavior of the renormalization operators along the variable λ is mostly neutral; in contrast, the operators are strongly expanding along the t variable, corresponding to the rescalings involved in the definition. To see this, observe that the renormalization operator involves a rescaling, where the size ε of the selected components (see the overlapping intervals in Figure 3.3) is normalized to approximately 1. Under this rescaling, the effect of translations is great amplified: $dt_1/dt \approx \varepsilon^{-1}$. On the other hand, λ acts by homothetic transformation, which commutes with the rescaling, and so this variable is essentially unaffected by the operator: $\lambda_1 \approx \lambda$. For this reason, the construction is carried out in two steps.

3.4.2 The scale recurrence lemma

The first step is to find a compact recurrent set $\mathcal{L}_*$ for the renormalization operators acting on $\Sigma \times \tilde{\Sigma} \times \mathbb{R}^*$, that is, taking only the scale variable λ in consideration. This is handled by the following crucial *scale recurrence lemma*.

Choose a set $\Sigma(\varepsilon)$ of finite sequences α, with variable lengths k_α, such that length $(f_\alpha^{k_\alpha}(I)) \approx \varepsilon$ and the intervals $f_\alpha^{k_\alpha}(I)$ cover K. Choose $\tilde{\Sigma}(\varepsilon)$ analogously for $\tilde{K}$. Taking ε small,

$$\#\big(\Sigma(\varepsilon) \times \tilde{\Sigma}(\varepsilon)\big) \approx \varepsilon^{-c(K) - c(\tilde{K})} = \varepsilon^{-\text{HD}(K) - \text{HD}(\tilde{K})}.$$

Assume at least one of the Cantor sets, K say, is essentially non-affine: the corresponding expanding transformation ψ is C^2 and there exist θ_1, $\theta_2 \in \Sigma$ and $x \in K$ such that $(\kappa_{\theta_1}^{-1} \circ \kappa_{\theta_2})''(x) \neq 0$. The scale recurrence lemma says

---

[6] For any $r > 1$ such that the regular Cantor sets are C^r.

that *given a large $A > 0$ and a family $\{L_{\theta,\tilde{\theta}} : (\theta,\tilde{\theta}) \in \Sigma \times \tilde{\Sigma}\}$ of subsets of* $[-A, A]$ *with*

$$\text{Leb}([-A, A] \setminus L_{\theta,\tilde{\theta}}) \ll 1 \quad \text{for all } (\theta, \tilde{\theta}) \in \Sigma \times \tilde{\Sigma},$$

there exists a family $\{\mathcal{L}_{\theta,\tilde{\theta}} : (\theta,\tilde{\theta}) \in \Sigma \times \tilde{\Sigma}\}$ of compact sets, where each $\mathcal{L}_{\theta,\tilde{\theta}}$ is near $L_{\theta,\tilde{\theta}}$, such that

$$\text{Leb}([-A, A] \setminus \mathcal{L}_{\theta,\tilde{\theta}}) \ll 1 \quad \text{for all } (\theta, \tilde{\theta}) \in \Sigma \times \tilde{\Sigma},$$

and for any $\lambda \in \mathcal{L}_{\theta,\tilde{\theta}}$ there is a positive fraction [7] of pairs $(\alpha, \tilde{\alpha}) \in \Sigma(\varepsilon) \times \tilde{\Sigma}(\varepsilon)$ for which

$$\mathcal{R}_{\alpha,\tilde{\alpha}}(\theta, \tilde{\theta}, \lambda) \in \text{int}(\mathcal{L}_{\theta\alpha,\tilde{\theta}\tilde{\alpha}}). \tag{3.6}$$

To use this lemma, one considers sets $L_{\theta,\tilde{\theta}}$ as follows. Notice that (the κ_α are diffeomorphisms, and so preserve dimensions)

$$\text{HD}(\kappa_\theta(K)) + \text{HD}(\tilde{\kappa}_{\tilde{\theta}}(\tilde{K})) = \text{HD}(K) + \text{HD}(\tilde{K}) > 1,$$

and so Marstrand's theorem applies to $(\kappa_\theta(K), \tilde{\kappa}_{\tilde{\theta}}(K))$. Then, for almost every λ, the π_λ-projection down to $\mathbb{R}$ of the Hausdorff measure on the product of these Cantor sets has L^2 density. Consider $B \gg A \gg 0$ and let $L_{\theta,\tilde{\theta}}$ be the set of $\lambda \in [-A, A]$ for which the L^2 norm is less than B. Using the Cauchy-Schwarz inequality, we get that

$$B^2 \text{Leb}(\pi_\lambda(Y)) \geq m_{\text{HD}}(Y)^2 \quad \text{for every measurable set } Y \tag{3.7}$$

and any $\lambda \in L_{\theta,\tilde{\theta}}$. This relation will be important in the second step.

Then $\mathcal{L}_* = \{(\theta, \tilde{\theta}, \lambda) : \lambda \in \mathcal{L}_{\theta,\tilde{\theta}} \text{ and } (\theta, \tilde{\theta}) \in \Sigma \times \tilde{\Sigma}\}$ is a compact recurrent set for the renormalization operator in $\Sigma \times \tilde{\Sigma} \times \mathbb{R}^*$. The conclusion (3.6) contains

$$\lambda \in \mathcal{L}_{\theta,\tilde{\theta}} \implies \lambda_1 \in \mathcal{L}_{\theta\alpha,\tilde{\theta}\tilde{\alpha}}.$$

Observe that $\lambda_1 \approx \lambda$ because length $(\kappa_\theta(f_\alpha^{k_\alpha}(I))) \approx \varepsilon \approx$ length $(\tilde{\kappa}_{\tilde{\theta}}(\tilde{f}_{\tilde{\alpha}}^{k_{\tilde{\alpha}}}(I)))$. This corresponds to the fact we already mentioned that the renormalization operator acts in a neutral fashion along the λ variable.

Remark 3.22. The scale recurrence lemma implies the following beautiful formula:

$$\text{HD}(K^1 + K^2) = \min\{1, \text{HD}(K^1) + \text{HD}(K^2)\}$$

whenever K^1 and K^2 are regular Cantor sets, with K^1 being C^2 and essentially non-affine. See Moreira [312].

---

[7]More precisely, at least $c\varepsilon^{-\text{HD}(K) - \text{HD}(\tilde{K})}$ of them.

3.4.3 The probabilistic argument

The hypothesis of non-affinity above is the first condition in the definition of the open dense set $\mathcal{W}$ in Theorem 3.19. The remaining (much less explicit) conditions correspond to additional perturbations to ensure, from $\mathcal{L}_*$, the existence of an invariant set $\mathcal{L}$ for the full renormalization operator acting on $\Sigma \times \tilde{\Sigma} \times \mathbb{R}^* \times \mathbb{R}$.

Let $(\theta, \tilde{\theta}, \lambda) \in \mathcal{L}_*$. By Marstrand's theorem 3.15, there exists a positive measure set $T_{\theta,\tilde{\theta},\lambda}$ of values of t for which the Hausdorff dimension of

$$\kappa_\theta(K) \cap \left(\lambda \tilde{\kappa}_{\tilde{\theta}}(\tilde{K}) + t\right)$$

is at least $d = \mathrm{HD}(K) + \mathrm{HD}(\tilde{K}) - 1$. Since length $(\kappa_\theta(f_\alpha^{k_\alpha}(I))) \approx \varepsilon$, this implies (fix any $\tilde{d} < d$) that there are at least $c\varepsilon^{-\tilde{d}}$ pairs $(\alpha, \tilde{\alpha}) \in \Sigma(\varepsilon) \times \tilde{\Sigma}(\varepsilon)$ for which the corresponding intervals intersect. We shall refer to these as distinguished pairs and distinguished intervals. To be more precise, we satisfy also the conclusion (3.6) of the scale recurrence lemma. The reason we can do this is, essentially, the following. The pairs given by the recurrence lemma cover a definite fraction of the product Cantor set. Using (3.7) we see that this covered subset projects to a positive Lebesgue measure subset and, in fact, Marstrand's argument may be carried out within this subset. So we may pick $c\varepsilon^{-\tilde{d}}$ distinguished pairs from among those $(\alpha, \tilde{\alpha})$ given by the scale recurrence lemma.

Very roughly, we now take $\mathcal{L}_\varepsilon$ to be the set of $(\theta, \tilde{\theta}, \lambda, t) \in \Sigma \times \tilde{\Sigma} \times \mathbb{R}^* \times \mathbb{R}$ such that t is in the ε-neighborhood of $T_{\theta,\tilde{\theta},\lambda}$. We want to prove that this is a recurrent set: given any $(\theta, \tilde{\theta}, \lambda, t) \in \mathcal{L}_\varepsilon$ there is a distinguished pair $(\alpha, \tilde{\alpha})$ such that

$$\mathcal{R}_{\alpha,\tilde{\alpha}}(\theta, \tilde{\theta}, \lambda, t) \in \mathcal{L}_{\varepsilon/2} \tag{3.8}$$

at least after a 10ε-perturbation of the Cantor sets. To this end, we embed the expanding transformation ψ of K into a parametrized family with $p \approx \varepsilon^{-d/2}$ parameters with values in $[\ 10c, 10c]$, such that varying one parameter causes a corresponding interval $\kappa_\theta(f_\alpha^{k_\alpha}(I))$ to move with respect to the associated $\tilde{\kappa}_{\tilde{\theta}}(\tilde{f}_{\tilde{\alpha}}^{k_{\tilde{\alpha}}}(I))$, while keeping the other $p-1$ distinguished intervals unchanged [8].

This displacement causes the relative position of the pair of intervals to move across a whole interval, of uniform size, along the t-direction. Noting that the two intervals intersect for the initial Cantor sets $(K, \tilde{K})$, we conclude that the probability (that is, the Lebesgue measure in parameter space) that $\mathcal{R}_{\alpha,\tilde{\alpha}} \in \mathcal{L}_{\varepsilon/2}$ is bounded below by some uniform $\eta > 0$. Since we are dealing with $p \approx \varepsilon^{-d/2}$ distinguished pairs, and the behavior of each one is ruled by a different parameter in an independent fashion, the probability that (3.8) fail for all these pairs is bounded above by

---

[8] Such a family exists with C^1 bounded norm. The argument extends to the C^r topology, any $r \geq 1$, considering only $p \approx \varepsilon^{-1/(r+1)}$ distinguished intervals and parameters.

$$(1 - \eta)^{\varepsilon^{-d/2}} \ll 1.$$

Summarizing these steps, given any $(\theta, \tilde{\theta}, \lambda, t) \in \Sigma \times \tilde{\Sigma} \times \mathbb{R}^* \times \mathbb{R}$ the majority of values of the parameter gives rise to Cantor sets $(K_\xi, \tilde{K})$ such that $\mathcal{R}_{\alpha, \tilde{\alpha}}$ is in the interior of $\mathcal{L}_\varepsilon$ for some distinguished pair $(\alpha, \tilde{\alpha})$. We still need to find some parameter value such that this holds for all $(\theta, \tilde{\theta}, \lambda, t)$ simultaneously. This uses the following pair of observations.

On the one hand, the previous estimates have some room for accommodating variation of the point. Using continuity of the renormalization operators $\mathcal{R}_{\alpha, \tilde{\alpha}}$, one checks that (3.8) implies

$$\mathcal{R}_{\alpha, \tilde{\alpha}}(\theta', \tilde{\theta}', \lambda', t') \in \text{int}(\mathcal{L}_\varepsilon)$$

for every $(\theta', \tilde{\theta}', \lambda', t')$ in a ball of radius ε^3. On the other hand, we have $\text{HD}(\Sigma) = \text{HD}(K)$ and analogously for $\tilde{K}$. It follows that the Hausdorff dimension of $\Sigma \times \tilde{\Sigma} \times \mathbb{R}^* \times \mathbb{R}$ is less than 4, and so compact subsets are covered by not more than ε^{-12} balls of radius ε^3. Therefore, the overall probability of the set of bad parameters is bounded above by

$$\varepsilon^{-12}(1 - \eta)^{\varepsilon^{-d/2}} \ll 1.$$

This completes the outline of the proof of Theorem 3.19.

3.5 Homoclinic tangencies in higher dimensions

Extension of this theory to higher dimensions involves some fundamental difficulties. A first obstruction is the lack of regularity of invariant foliations with codimension bigger than 1: in general the holonomy maps are not more than Hölder continuous, regardless of the smoothness of the dynamics. Related to this, the following basic problem remains wide open, outside the codimension 1 case:

Problem 3.23. Let H be a hyperbolic transitive set. Are the local Hausdorff dimension $\text{HD}(H \cap W_{loc}^*(x))$ and the local limit capacity $c(H \cap W_{loc}^*(x))$, with $* \in \{u, s\}$, independent of the point $x \in H$?

Another main difficulty is that, instead of Cantor sets on the real line, one now has to deal with fractal sets living in higher dimensional spaces. In a sense, such sets are still dynamically defined, but the dynamics is usually not conformal, besides being only Hölder continuous, which makes the geometry much harder to analyze. The Hausdorff dimension and limit capacity no longer coincide, in general (see [50]), and the notion of thickness does not apply to such higher dimensional fractal sets. However, Buzzard [108] has extended Theorem 3.3 to holomorphic maps of $\mathbb{C}^2$, using an astute variation of the thickness strategy.

In addition, contrary to the surface case, the Hausdorff dimension and the limit capacity of hyperbolic sets of higher dimensional systems may vary discontinuously with the dynamics. The following simple example is taken from [77]:

Example 3.24. Let $f : \mathbb{R}^2 \to \mathbb{R}^2$ be a diffeomorphism exhibiting a horseshoe $\Sigma = \cap_{n \in \mathbb{Z}} f^n([-1,1]^2)$. For simplicity, suppose that $f^{-1}([-1,1]^2) \cap [-1,1]^2$ has two connected components, A_1 and A_2, and the restriction of f to each component is affine:

$$Df \mid A_i = \begin{pmatrix} \lambda_i & 0 \\ 0 & \sigma_i \end{pmatrix}, \qquad \|\lambda_i\|, \ \|\sigma_i^{-1}\| < 1/2.$$

Then $A_i = [-1,1] \times B_i$ for some interval B_i. Define $f_t : \mathbb{R}^3 \to \mathbb{R}^3$, $t \in [-1,1]$ by

$$f_t(x,y,z) = \begin{cases} \big(f(x,y), \sigma z\big) & \text{if } y \in B_1 \\ \big(f(x,y), \sigma z - t\big) & \text{if } y \in B_2. \end{cases}$$

Let Λ_t be the continuation for f_t of the hyperbolic set $\Lambda_0 = \Sigma \times \{0\}$ of f_0. In other words, Λ_t is the homoclinic class of the fixed point at the origin. Assuming $\sigma \in (1,2)$, one checks that the projection of Λ_t onto the z coordinate contains the interval $[0,t]$. Consequently, $c(\Lambda_t) \geq \mathrm{HD}(\Lambda_t) \geq 1$ for every $t \neq 0$. Choosing f so that $c(\Sigma) = \mathrm{HD}(\Sigma) < 1$ (this corresponds to taking λ_i and σ_i^{-1} small), the Hausdorff dimension and limit capacity of Λ_t are *not* upper semi-continuous at $t = 0$.

Problem 3.25. Are the Hausdorff dimension and the limit capacity of hyperbolic sets always lower semi-continuous functions of the dynamical system?

Some partial progress is reported in Section 3.5.2.

3.5.1 Intrinsic differentiability of foliations

Some of these difficulties have been dealt with by Palis, Viana [347], in their proof that Theorem 3.6 does extend to arbitrary manifolds. To find attractors, one assumes that the periodic saddle involved in the tangency is *sectionally dissipative*: the product of any two eigenvalues of the derivative at this point has norm less than 1; in particular, the unstable manifold is 1-dimensional.

Theorem 3.26 (Palis, Viana [347]). *Let $f : M \to M$ be a diffeomorphism on any manifold, having a homoclinic tangency associated to a sectionally dissipative saddle point p. Then there exists an open subset $\mathcal{U}$ of $\mathrm{Diff}^2(M)$ containing f in its closure, such that every $g \in \mathcal{U}$ may be approximated by a diffeomorphism with a homoclinic tangency associated to the continuation of p. Moreover, there exists a residual subset $\mathcal{R} \subset \mathcal{U}$ such that every $g \in \mathcal{R}$ has an infinite number of periodic attractors.*

For persistence of tangencies one does not need the dissipativeness condition. This was proved, independently, by Romero [387], who reduces the general case to the sectionally-dissipative one, and by Gonchenko, Shil'nikov, Turaev [193].

A key novelty in the proof of Theorem 3.26 is the use of a notion of *intrinsic differentiability* à la Whitney. A function $\varphi : X \mapsto \mathbb{R}^n$ on a subset $X \subset \mathbb{R}^m$ is *intrinsically* C^1 if there exists a continuous map $X \times X \ni (x, z) \mapsto \Delta\varphi(x, z) \in \mathcal{L}(\mathbb{R}^m, \mathbb{R}^n)$ such that

$$\varphi(x) - \varphi(z) = \Delta\varphi(x, z) \cdot (x - z) \quad \text{for every } (x, z) \in X \times X.$$

In simple terms, it is shown that if the hyperbolic set admits a strong-stable foliation, of codimension 1 inside the corresponding stable foliation, such that each leaf contains a unique point of the set, then the unstable holonomy maps are intrinsically C^1. A more precise statement is given below.

Let us explain in more detail how the proof of Theorem 3.26 goes. Keep in mind the 2-dimensional arguments outlined in the context of Theorem 3.6. Just as in the surface case, we may assume that p is part of a nontrivial hyperbolic set H_1. Another useful observation is that an arbitrarily small unfolding of the homoclinic tangency yields the formation of other homoclinic tangencies, associated to periodic saddle points having a unique weakest contracting eigenvalue. Thus, we may assume right from the start that the saddle p has a unique weakest contracting eigenvalue which, as a consequence, is a real number.

Then we may choose the hyperbolic set H_1 to be transverse to the strongest contracting directions of $Df^{\mathrm{per}(p)}(p)$, in the following sense: at every $x \in H_1$, the *intrinsic tangent space*

$$IT_x(H_1) = \mathrm{span}\left\{ v : \text{ there is } (x_n)_n \in H_1^{\mathbb{N}} \text{ so that } \frac{x_n - x}{\|x_n - x\|} \to v \right\}$$

(consider $M = \mathbb{R}^m$, for simplicity) is 2-dimensional and transverse to the codimension 2 subspace generated by the strongest contracting directions. For such an H_1 one proves that the projections $\pi \colon \Sigma_0 \cap W^u(H_1) \to \Sigma_1 \cap W^u(H_1)$ along unstable leaves of H_1 are intrinsically C^1 maps, where Σ_0 and Σ_1 are any transverse sections to the unstable foliation of H_1.

Now we may define the *local unstable thickness* of H_1 at a point $x \in H_1$ as

$$\tau^u(H_1, x) = \tau(\tilde{\pi}(W^u(H_1) \cap \Sigma_0), \tilde{\pi}(x)),$$

where Σ_0 is a transverse section to the unstable foliation of H_1 at the point x, and $\tilde{\pi} \colon W^u(H_1) \cap \Sigma_0 \to \mathbb{R}$ is any intrinsically differentiable map admitting an intrinsic derivative $\Delta\tilde{\pi}$ such that $\Delta\tilde{\pi}(x, x) \mid IT_x(W^u(H_1) \cap \Sigma_0)$ is a bijection. Intrinsic differentiability of the unstable foliation permits to check that this definition does not depend on the choices of Σ_0 and $\tilde{\pi}$, and to prove that $\tau^u(H_1, x)$ is strictly positive and independent of $x \in H_1$. Furthermore, it varies continuously with the diffeomorphism $f \in \mathrm{Diff}^2(M)$.

Next, by arbitrarily small perturbations leaving H_1 and $\tau^u(H_1)$ essentially unchanged, one obtains another hyperbolic basic set, H_2, with codimension 1 stable foliation and large stable thickness: $\tau^u(H_1)\tau^s(H_2) > 1$. This is done by a comparatively straightforward extension of the renormalization scheme used in the surface case. As before, H_1 and H_2 may be taken to be heteroclinically related, and to exhibit a tangency q between $W^u(p_1)$ and $W^s(p_2)$, for some periodic points $p_1 \in H_1$ and $p_2 \in H_2$. The stable foliation of H_2, being of codimension 1, admits an extension to a C^1 foliation $\mathcal{F}^s(H_2)$ of a neighborhood of H_2, which we may assume to contain q.

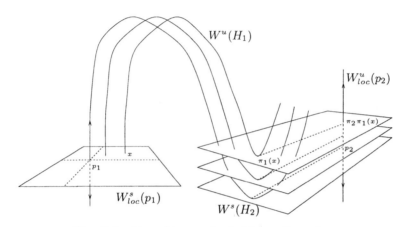

Fig. 3.4. Line of tangencies in higher dimensions

An implicit function argument shows that there exists an intrinsically differentiable map $\pi_1 \colon W^u(H_1) \cap W^s_{loc}(p_1) \to M$ such that $\pi_1(p_1) = q$ and each $\pi_1(x)$, $x \in W^u(H_1) \cap W^s_{loc}(p_1)$, is a point of tangency between leaves of $W^u(H_1)$ and $\mathcal{F}^s(H_2)$. See Figure 3.4. Let π_2 be the projection along the leaves of $\mathcal{F}^s(H_2)$ onto $W^u_{loc}(p_2)$, which we identify with an interval in $\mathbb{R}$. Then the theorem follows from an application of the gap lemma to the Cantor sets $\pi_2 \circ \pi_1(W^u(H_1) \cap W^s_{loc}(p_1))$ and $W^s(H_2) \cap W^u_{loc}(p_2)$. This finishes our sketch of the proof of Theorem 3.26.

3.5.2 Frequency of hyperbolicity

More recently, Moreira, Palis, Viana announced in [314] that the principle
    hyperbolicity prevails $\Leftrightarrow$ Hausdorff dimension is smaller than 1
remains valid for homoclinic bifurcations in any dimension.

They consider C^2 families of diffeomorphisms $f_\mu : M \to M$, $\mu \in (-1, 1)$, unfolding a generic homoclinic tangency associated to a periodic point p contained in a horseshoe Λ of $f = f_0$. It is assumed that $\mu = 0$ is the first bifurcation: f_μ is uniformly hyperbolic for all $\mu < 0$; this implies that the weakest contracting and weakest expanding eigenvalues of p are both real. The main theorem is stated in terms of the following notions of upper stable and unstable Hausdorff dimensions.

Let Λ be a horseshoe for a diffeomorphism $f : M \to M$. Let $\mathcal{V}_n$ be the set of vertical n-cylinders, that is, the subsets of Λ defined by prescribing the first n symbols in the backward itinerary with respect to some Markov partition. There is a dual notion of horizontal n-cylinder, where one considers forward itinerary instead. For each $V \in \mathcal{V}_n$ let

$$D_s(V) = \sup\{\mathrm{diam}(W^s_{loc}(x) \cap V) : x \in \Lambda \cap V\}.$$

Define λ_n by the relation

$$\sum_{V \in \mathcal{V}_n} D_s(V)^{\lambda_n} = 1.$$

The *upper stable dimension* of Λ is defined by $\bar{d}_s(\Lambda) = \lim_{n \to \infty} \lambda_n$. There is a dual notion of *upper unstable dimension* $\bar{d}_u(\Lambda)$, dealing with W^u_{loc} instead of W^s_{loc}.

It is not difficult to show that the limit always exists, and $\bar{d}_s$ is an upper semi-continuous function of f. Moreover,

$$\mathrm{HD}(W^s_{loc}(x) \cap \Lambda) \leq \bar{d}_s(\Lambda) \quad \text{for each } x \in \Lambda.$$

When $\bar{d}_s(\Lambda) < 1$ we even have equality, for a residual subset $\mathcal{R}$ of diffeomorphisms g in a neighborhood of f and every point x in the continuation Λ_g of Λ. In addition, $\mathcal{R}$ may be taken so that its elements are points of continuity of $\bar{d}_s$.

Theorem 3.27 (Moreira, Palis, Viana [314]). *There are open sets $\mathcal{U}$ and $\mathcal{V}$ of families of diffeomorphisms unfolding a first homoclinic tangency, such that $\mathcal{U} \cup \mathcal{V}$ is dense in the space of such families and*

1. *For families in $\mathcal{U}$ we have $\bar{d}_s(\Lambda) + \bar{d}_u(\Lambda) < 1$ and $\mathcal{H} = \{\mu \mid f_\mu \text{ is hyperbolic}\}$ has full Lebesgue density at $\mu = 0$:*

$$\lim_{\delta \to 0} \frac{m(\mathcal{H} \cap [0, \delta])}{\delta} = 1.$$

2. *For families in $\mathcal{V}$ we have $\bar{d}_s(\Lambda) + \bar{d}_v(\Lambda) > 1$ and $\mathcal{N} = \{\mu \mid f_\mu \text{ has persistent homoclinic tangencies associated to the continuation } \Lambda_\mu \text{ of } \Lambda_0\}$ has positive Lebesgue density and its union with $\mathcal{H}$ has full Lebesgue density at $\mu = 0$:*

$$\liminf_{\delta \to 0} \frac{m(\mathcal{N} \cap [0, \delta])}{\delta} > 0 \quad and \quad \lim_{\delta \to 0} \frac{m((\mathcal{H} \cup \mathcal{N}) \cap [0, \delta])}{\delta} = 1.$$

Let us say a few words about the proof of this result. The main step toward proving part (1), carried out in the next proposition, is to reduce to the case when the horseshoe Λ_0 does not intersect the strong-stable manifold, nor the strong-unstable manifold of the point p_0. Recall that these manifolds have codimension 1 inside the stable manifold and the unstable manifold (the weak contracting and weak expanding eigenvalues have multiplicity 1).

Proposition 3.28. *Assume $\bar{d}_s(\Lambda) < 1$. For an open and dense subset of maps in some neighborhood of f we have $\Lambda \cap W^{ss}(p) = \{p\}$. Analogously for $\bar{d}_u$ and W^{uu}.*

This proposition implies that $\Lambda \cap W^s(p)$ is contained in a cuspidal region around the weak-stable direction of p. Let K^s be the projection of $\Lambda \cap W^s(p)$ to this weak-stable direction along the strong-stable foliation of p. Then, for any $\delta > 0$ there exists $\varepsilon > 0$ such that each $x \in \Lambda \cap W^s(p)$ in the ε-neighborhood of p is within distance $\varepsilon\delta$ from some $y \in K^s$. We say that $\Lambda \cap W^s(p)$ is well-represented by K^s near p.

Then a similar fact is true replacing $W^s(p)$ by any other cross-section Σ to the unstable direction at any $q \in W^u(p)$: the intersection $W^u(p) \cap \Sigma$ is well-represented by K^s near q. Using also smoothness of the unstable leaves, together with Hölder continuity of their tangent bundle, we conclude that the intersection of $W^u(\Lambda)$ with a neighborhood of q is well-represented by the product of K^s by $T_q W^u(p)$ near q.

Consider q to be the point of homoclinic tangency. The previous paragraph, together with a dual statement for $W^s(\Lambda)$, mean that the unstable and the stable foliations of Λ are well-represented by affine models

$$K^s \times T_q W^u(p) \quad \text{and} \quad K^u \times T_q W^s(p)$$

near q. By continuity of the invariant foliations this remains true when the tangency is unfolded along a one-parameter family f_μ: the sets $W^u(\Lambda_\mu)$ and $W^s(\Lambda_\mu)$ are well-represented by translates

$$[K^s \times T_q W^u(p)] + \mu v_1 \quad \text{and} \quad [K^u \times T_q W^s(p)] + \mu v_2$$

for some vectors v_1, v_2. So, to ensure that the two foliations are transverse, with angles bounded from zero, it suffices to prove a corresponding fact for these affine models. Thus, the original situation is once more reduced to a problem of arithmetic difference of interval Cantor sets.

The main ingredient for proving part (2) of the theorem is the construction, perhaps after perturbation of f, of strong-stable and strong-unstable foliations (of codimension-1 inside the corresponding stable and unstable foliations) for hyperbolic subsets of Λ with almost the same upper Hausdorff dimensions. These foliations are used to (essentially) reduce the study of the geometries of the stable and unstable foliations near the initial homoclinic tangency to the 2-dimensional case. More precisely, in case (2) of Theorem 3.27,

Proposition 3.29. *Generically, there are disjoint subsets* Λ_1, Λ_2 *of* Λ, *invariant by some iterate* f^k, *with the following properties:*

1. Λ_1 *has a globally defined strong-stable foliation, and* Λ_2 *has a globally defined strong-unstable foliation, both invariant under* f_0.
2. $\bar{d}_s(\Lambda_1) < 1$, $\bar{d}_u(\Lambda_2) < 1$, *and* $\bar{d}_s(\Lambda_1) + \bar{d}_u(\Lambda_2) > 1$.
3. *For each* $p_1 \in \Lambda_1$ *the projection* $\pi_{ss} \mid (\Lambda_1 \cap W^s(p_1))$ *to the weak stable direction along strong-stable leaves is injective and its inverse is intrinsically* C^1 *with Hölder continuous intrinsic derivative. Analogously for* $p_2 \in \Lambda_2$ *and the projection* $\pi_{uu} \mid (\Lambda_2 \cap W^u(p_2))$ *along strong-unstable leaves.*
4. *Either* Λ_1 *contracts area in the weak direction or* Λ_2 *expands area in that weak direction.*

By weak direction we mean the 2-dimensional space generated by the eigenspaces associated to the weakest contracting and weakest expanding eigenvalues. Conditions (1) to (3) imply that the projection of $\Lambda_1 \cap W^s(p_1)$ onto a weak-stable (1-dimensional) manifold, along the strong-stable foliation, is a regular $C^{1+\varepsilon}$ Cantor set K^s, whose conjugacy class does not depend on $p_1 \in \Lambda_1$ nor on the choice of the weak-stable leaf. The same holds for $\pi_{uu}(\Lambda_2 \cap W^u(p_2)) = K^u$.

The next lemma concludes the reduction to a problem of stable intersection of Cantor sets. It is assumed that the eigenvalues of f at the point p satisfy a generic non-resonance condition.

Lemma 3.30. *If there are* $\lambda > 0$ *and* $t \in \mathbb{R}$ *such that* K^s *intersects* $\lambda K^u + t$ *stably, then* $(f_\mu)_\mu$ *persistently exhibits positive density of stable tangencies at* $\mu = 0$.

In order to get the stable intersection between K^s and $\lambda K^u + t$, we first prove that condition (4) implies that either K^s or K^u is a C^2-regular Cantor set on the central leaf. This is important to show that, generically, K^s and K^u satisfy the hypotheses of the scale recurrence lemma in Section 3.4.2. Afterwards, we consider again a family with a large number of parameters of local perturbations of the diffeomorphism f and show that for most parameters there are λ and t such that K^s stably intersects $\lambda K^u + t$.

3.6 On the boundary of hyperbolic systems

The approach to homoclinic tangencies described in this chapter started within a more general strategy for understanding non-hyperbolic dynamics, going back to the seventies. In a few words, the idea is to consider parametrized families of systems starting in the uniformly hyperbolic domain (Axiom A with no cycles) and to describe the forms of dynamics which are persistent, or even prevalent, in parameter space immediately after the boundary of hyperbolicity is crossed. See [324, 325, 421]. In this direction it is crucial to analyze how the crossing takes place, that is, which dynamical phenomena are responsible for the breakdown of hyperbolicity. Two main ones are known, and have been extensively studied. They are both robust, in the sense that they occur for an open set of 1-parameter families of diffeomorphisms or flows, in fact, locally they correspond to (codimension-1) hypersurfaces in the space of all dynamics:

Loss of hyperbolicity on periodic orbits: the formation of a non-hyperbolic periodic orbit. Generically, it is either a saddle-node (a unique non-hyperbolic eigenvalue, equal to 1), a period doubling (a unique non-hyperbolic eigenvalue, equal to −1), or a Hopf bifurcation (a unique pair of non-hyperbolic eigenvalues, non-real) [9].

Although the creation of a non-hyperbolic periodic orbit is a local bifurcation, it may lead to drastic global changes of the dynamics if that orbit is involved in non-trivial recurrence. In fact, there are many examples showing that such bifurcations may lead directly from hyperbolic to robustly non-hyperbolic systems, see for instance [277].

Loss of transversality on periodic orbits: the creation of some transverse intersection between the stable manifold and the unstable manifold of a pair of hyperbolic periodic orbits. This includes homoclinic tangencies (Chapter 3), heterodimensional cycles (Chapter 6), and singular cycles (Chapter 9).

Results in [319], [152, 151], [6, 311] show that each of these three situations may lead directly from hyperbolic systems to robustly non-hyperbolic ones. In the two last cases one may even start with Morse-Smale systems. It is not known whether the same is true for the first one:

Problem 3.31. Is there a local hypersurface in the space of diffeomorphisms or flows corresponding to a homoclinic tangency and separating Morse-Smale from robustly non-hyperbolic systems?

As we mentioned before, both mechanisms above correspond, locally, to hypersurfaces in the space of dynamics. They account for a large part of the boundary of the set of hyperbolic systems (of course, they may also take place in the interior of non-hyperbolic systems). However, the boundary is far from being smooth. Indeed, it has a very fractal structure, these hypersurfaces being combined in rather intricate manners. A fundamental question is whether there are other sizable parts of the boundary, that is, whether there are other mechanisms responsible for the breakdown of hyperbolicity in "many" situations. The following negative partial answer has been proposed by Newhouse, Palis [324]:

---

[9]The definitions also ask the lower order significant jet to be non-degenerate.

Conjecture 3.32. For generic 1-parameter families starting from Morse-Smale diffeomorphisms, the first bifurcation corresponds either to a (unique) non-hyperbolic periodic orbit or to a (unique) non-transverse intersection between invariant manifolds of a pair of periodic points.

In other words, generically, the system should remain hyperbolic (Morse-Smale) for as long as it remains Kupka-Smale [10]. In fact, [325] proved that this is true, restricted to the case when the limit set is still finite at the bifurcation. Apart from this, there has been essentially no progress in the direction of this conjecture. However, we believe some relevant cases may now be within reach, see Problem 3.35 below.

On the other hand, motivated by the advances in the theory of Hénon-like dynamics, Bonatti, Viana recently suggested that the problem should also be approached from a probabilistic point of view, and then the conclusion should be opposite: the previous two mechanisms should *not* account for *almost all* transitions to non-hyperbolicity:

Conjecture 3.33. There exists a "positive probability" subset of the boundary of hyperbolic systems formed by Kupka-Smale systems.

Recent results of Horita, Muniz, Sabini [219, 400] support this conjecture. We are going to explain their statement in the simplest possible setting: circle maps. This will also allow us to explain what is meant by positive probability in the conjecture (the boundary, being infinite-dimensional, has no distinguished measure supported in it).

Consider a 2-parameter family $(f_{a,\theta})_{a,\theta}$, $(a,\theta) \in [-1,1]^2$ of circle maps of class C^r, $r \geq 3$, such that

1. $f_{a,-1}$ is uniformly expanding;
2. $f'_{a,\theta} > 0$ at all points for $\theta < 0$;
3. $f_{a,0}$ has a unique critical point, of cubic type;
4. $f_{a,\theta}$ has two quadratic critical points for $\theta > 0$.

Recall that f is uniformly expanding if there exist $n \geq 1$ and $\sigma > 1$ such that $|(f^n)'| \geq \sigma$. An example (up to convenient reparametrization $(a, \theta) \to (c, \sigma)$) is the following modification of the Arnold family [32]:

$$f_{c,\sigma}(x) = kx + c + \sigma \sin(2\pi x) \mod \mathbb{Z} \qquad (3.9)$$

where k is an integer number with $|k| \geq 2$.

Observe that existence of a cubic critical point is a robust phenomenon in this context: for any family $g_{a,\theta}$ in a C^r neighborhood of $f_{a,\theta}$ and every $a \in [-1, 1]$ there exists of $\bar{\theta}(a) \approx 0$ such that $f_{a,\bar{\theta}(a)}$ has a cubic critical point. Up to reparametrization we may consider $\bar{\theta}(a) = 0$, and we do so in the next statement:

---

[10]A dynamical system is Kupka-Smale if all its periodic orbits (and singularities) are hyperbolic, and all their invariant manifolds are transverse.

Theorem 3.34. *There exists an open set of families $(f_{a,\theta})_{a,\theta}$ as before, such that there exists a positive Lebesgue measure set of values of a for which*

1. *$f_{a,\theta}$ is uniformly expanding for all $\theta < 0$.*
2. *$f_{a,0}$ has a critical point but is Kupka-Smale: all periodic points are hyperbolic repelling, and the critical point is not pre-periodic [11].*
3. *$f_{a,0}$ admits an ergodic invariant measure equivalent to Lebesgue measure on the circle.*

In fact, we may take the positive Lebesgue measure set in the statement so that the critical orbit is dense in the circle. Moreover, hyperbolicity of the periodic points is uniform: there exists $\sigma > 1$ such that the derivative $|(f_{a,0}^n)'(p)|$ is larger than σ^n at every periodic point p of period n.

Let us comment on the proof of this result, as well as on some related open questions. As θ increases from -1, expansion becomes weaker and the map must cease to be uniformly expanding at some stage in the parameter interval $(-1, 0]$, since the presence of a critical point is a definite obstruction to uniform expansion. The first goal is to localize the bifurcation in this parameter interval. It is clear that expansion may break down *before* the critical point c is created. For instance, if the critical point is close to being periodic then $f_{a,0}$ must have some periodic attractor nearby. Since periodic attractors are a robust phenomenon, that implies that $f_{a,\theta}$ was already not expanding for slightly negative values of θ.

The role of the parameter a is to allow us to control the critical trajectory and, up to parameter exclusions, ensure that the situation we have described does not occur. Namely, the positive Lebesgue measure set in the theorem is chosen in such a way that the critical orbit be slowly recurrent: in particular,

$$d(f_{a,0}^n(c), c) \geq e^{-\varepsilon n} \quad \text{for all large } n \quad (\exists \varepsilon > 0).$$

For such values of a one proves that the map $f_{a,\theta}$ does remain uniformly expanding all the way up to the formation of the critical point.

Problem 3.35. Does the set $\mathcal{K}$ corresponding to a unique non-hyperbolic periodic point or else a unique critical point that is pre-periodic contain an open and dense subset of the boundary of uniformly expanding circle maps? For generic parametrized families, does the union of $\mathcal{K}$ with the set of maps satisfying the conclusion of Theorem 3.34 have full probability on the boundary?

Problem 3.36. Consider the family (3.9) and, for each c let $\tilde{\sigma}(c)$ be the smallest value of σ for which $f_{c,\sigma}$ is *not* uniformly expanding. What is the Hausdorff dimension of the graph of $\tilde{\sigma}$? A similar question may be asked for any family $f_{a,\theta}$ as in Theorem 3.34.

---

[11] A point is *pre-periodic* if some forward iterate is periodic. Pre-periodic critical points play the role of homoclinic tangencies for 1-dimensional transformations.

Finally, let us note that [78, 120, 177] have constructed codimension-3 submanifolds of $\mathrm{Diff}^r(\mathbb{T}^2)$ contained in the boundary of Anosov diffeomorphisms and corresponding to existence of a cubic tangency between the stable and unstable manifolds of a pair of periodic points. It is currently not known whether the codimension can be reduced.

On the other hand, [400] contains a version of Theorem 3.34 for surface diffeomorphisms with a non-trivial hyperbolic attractor: he finds a positive probability subset inside the boundary [12] for which the map has cubic tangencies associated to some pair of *non-periodic* stable and unstable leaves and, yet, is Kupka-Smale. It is not known whether his result extends to the Anosov case:

Problem 3.37. Is there a positive probability subset of the boundary of Anosov diffeomorphisms formed by Kupka-Smale diffeomorphisms?

---

[12] A kind of "codimension-$(1+)$" subset of the space of all diffeomorphisms.

4

Hénon-like Dynamics

The two-parameter Hénon family of transformations of the plane

$$H_{a,b} : (x, y) \mapsto (1 - ax^2 + y, bx) \tag{4.1}$$

was initially proposed in [210, 211] as the simplest model of an invertible dynamical system with a strange attractor. It has become the prototype for a whole class of dynamics, which we call Hénon-like. Their most distinctive features are *non-hyperbolic behavior of homoclinic (folding) type*, in localized regions of phase space, *combined with a fair amount of hyperbolicity* away from those so-called critical regions. The former is associated to the presence of homoclinic tangencies, which very much influence the way the dynamics unfolds as the parameters vary.

Numerical simulations of (4.1) suggested the presence of a persistent non-hyperbolic (transitive) attractor. However, those observations could also be explained if the attractor were just a periodic orbit with very high period, and it turned out to be remarkably difficult to prove rigorously that this is not the case (as a matter of fact, periodic attractors do occur for many parameter values). Indeed, it was only by the end of the eighties that Benedicks, Carleson [52] were able to exhibit a positive Lebesgue measure subset of parameters (a, b) for which the map $H_{a,b}$ does have a non-periodic non-hyperbolic attractor.

More precisely, they focused on values of $b > 0$ very close to zero. Then $H_{a,b}$ is strongly dissipative, and may be thought of as a "thickened" version of a quadratic map of the interval. Simple arguments (see Section 4.1.1) show that, for $a \in (1, 2)$ say, there is a forward invariant region such that the corresponding topological attractor coincides with the closure of the unstable manifold of some fixed point. Benedicks, Carleson proved that, as long as b is small enough, there exists a positive Lebesgue measure subset of values of a for which this topological attractor is transitive and, in fact, contains a dense orbit on which the derivative grows exponentially fast in norm.

This type of dynamics turns out to appear in many different contexts. A renormalization scheme (see Section 4.6.1) permits to show that the generic unfolding of a homoclinic tangency associated to a dissipative saddle involves families of strongly dissipative diffeomorphisms close to maps of the interval. The arguments of Benedicks, Carleson have been extended to such *Hénon-like families* (see Theorem 4.4), thus proving that strange attractors always occur for a positive Lebesgue measure set of parameters in the unfolding of such a tangency (see Theorem 4.32).

The notion of Hénon-like families of maps is made precise in Section 4.1. Essentially, one asks the map to be "almost one-dimensional" and, consequently, strongly dissipative. A number of results obtained since the early nineties proved that Hénon-like maps are amenable to a very complete dynamical description, at least for a positive Lebesgue measure set of parameters, where one recovers most important results from the theory of one-dimensional maps. Namely, these systems

- admit non-hyperbolic attractors (Section 4.2)
- supporting SRB measures (Section 4.3.1)
- whose basins have the no-holes property (Section 4.3.2)
- are exponentially mixing (Section 4.4)
- and stochastically stable under random noise (Section 4.5).

In Section 4.6 we also discuss a number of related results and open questions. All these developments have, no doubt, played a key part in shaping a new view of dynamical systems beyond the uniformly hyperbolic realm.

On the other hand, the dynamics of Hénon-like families over larger parameter regions remains poorly understood and, in fact, several basic questions are still open (see, especially, Sections 4.2.3 and 4.6.3). In fact, this class of Hénon-like systems is a good test case for some fundamental problems, such as the Palis conjecture on finitude of attractors outlined in Section 1.7.

Although here we consider only real systems, let us mention that the complex counterpart to (4.1) has also been intensively studied over the last couple of decades, especially by Bedford and Smillie, Fornaess and Sibony, and Hubbard and his coauthors, as a major step towards extending the theory of iterations of rational maps on the sphere to higher dimensional holomorphic systems. See for instance [419] and references therein.

4.1 Hénon-like families

Most results on the dynamics of Hénon maps are based on a perturbative approach developed by Benedicks, Carleson [52], and have been proven when the Jacobian $-b$ is very close to zero. Notice that (4.1) is affinely conjugate to

$$(x, y) \mapsto (1 - ax^2, 0) + \sqrt{|b|}\,(y, \mathrm{sign}(b)\,x)$$

so that for $|b|$ small we may think of $f_a = H_{a,b}$ as a singular perturbation of the family of quadratic interval maps $\phi_a : x \mapsto 1 - ax^2$. This led Mora, Viana [297, 438] to consider more general families

$$(x, y) \mapsto f_a(x, y) = (1 - ax^2, 0) + R(a, x, y) \tag{4.2}$$

where R has small norm, in the C^3 topology say. One important motivation was that such families model a good part of the dynamics associated to dissipative homoclinic tangencies, as we shall comment upon later.

Other applications, like the unfolding of saddle-node cycles [159] or kick-forced invariant curves [448] required further generalization. It was observed by Díaz, Rocha, Viana [159] and by Wang, Young [447] that the properties of the quadratic family relevant in this context are shared by much more general C^3 one-parameter families $\phi_a : M \to M$ of multimodal maps of the circle or the interval. Indeed, the results we are going to discuss extend to the families $(f_a)_a$ of d-dimensional embeddings obtained as singular perturbations of such families $(\phi_a)_a$ satisfying conditions (α) and (β) that follow. Just what we mean by singular perturbation is explained in condition (γ).

The first condition in the definition of these *Hénon-like families* $(f_a)_a$ corresponds to the existence of a Misiurewicz parameter:

(α) There exists $a = \bar{a}$ such that $\phi_{\bar{a}}$ has negative Schwarzian derivative, all its critical points are quadratic, and the critical set is non-recurrent.

Then, all periodic points are hyperbolic and repelling: use Singer [417] and also recall that our definition of multimodal map includes $\phi_a(\partial M) \subset \partial M$.

The second condition requires the kneading sequences of the critical orbits of $\phi_{\bar{a}}$ to unfold generically with the parameter a close to $\bar{a}$. From the fact that $\bar{a}$ is a Misiurewicz parameter, we know that for each i and all nearby values of a there exists a unique point $x_i(a)$ whose kneading sequence (the itinerary of its ϕ_a-orbit relative to the critical points), coincides with the $\phi_{\bar{a}}$-itinerary of the critical value $\phi_{\bar{a}}(c_i(\bar{a}))$. Moreover, $x_i(a)$ depends differentiably on the parameter. The second condition is

(β) Let $c_i(a)$ be the continuation, for nearby parameters, of each critical point $c_i(\bar{a})$ of $\phi_{\bar{a}}$. Then,

$$\frac{d}{da} \phi_a(c_i(a)) \neq \frac{d}{da} x_i(a) \quad \text{at } a = \bar{a}.$$

For $d \geq 2$, let B be the unit ball of dimension $d - 1$ around zero. Then we consider one-parameter families of embeddings

$$f_a : M \times B \to M \times B, \qquad f_a(x, y) = (\phi_a(x), 0) + R(a, x, y), \tag{4.3}$$

such that the C^3 norm of R is small and the Jacobian of f_a is homogeneous, in the following sense:

(γ) For all a, x, y, we have $\|R(a, x, y)\|_{C^3} \leq C\sqrt{b}$,

$$C^{-1} b \leq |\det Df_a(x, y)| \leq Cb \quad \text{and} \quad \|D(\log |\det Df_a(x, y)|)\| \leq C$$

for some sufficiently small b (depending also on C, which is arbitrary).

For parameters a close to $\bar{a}$, the critical regions of the diffeomorphism f_a correspond to neighborhoods of the critical points of $\phi_{\bar{a}}$, whereas hyperbolic behavior away from the critical regions results from a variation of Theorem 2.5, as we shall see in a while.

In our presentation we focus on the original 2-dimensional setting (4.2), which contains most main ingredients, referring the reader to [159, 447] for those more general formulations. Hénon-like families in higher dimensional manifolds were studied in [438] and, more recently, [159, 317, 446].

4.1.1 Identifying the attractor

Let us consider $\phi_a(x) = 1 - ax^2$ with $a \in (1, 2)$. The map ϕ_a has exactly two fixed points q_a and p_a, with

$$q_a < -1 < 0 < p_a < 1 < -q_a, \quad \lim_{a \to 2} q_a = -1, \quad \lim_{a \to 2} p_a = 1/2,$$

and they are both repelling. Denoting $M_a = [q_a, -q_a]$, we have $\phi_a(M_a) \subset M_a$ and $\phi_a(\partial M_a) \subset \partial M_a$ (the fact that M_a depends on the parameter is easily disposed of by rescaling). Moreover, we may find an open interval $I \subset M_a$, depending only on a lower bound for $2 - a$, such that $\phi_a(\bar{I}) \subset I$: just take $I = (-r, r)$ with $1 < r < -q_a$. Given $\varepsilon > 0$, we have $f_a(\bar{I} \times [-1, 1]) \subset I \times (-\varepsilon, \varepsilon)$ for any map f_a of the form (4.3) such that the C^0 norm of R is sufficiently small. This means there is some compact attracting set

$$\Lambda_a = \bigcap_{n=0}^{\infty} f_a^n(\bar{I} \times [-1, 1])$$

containing all the asymptotic dynamics of f_a. Let us give a more explicit description of this set.

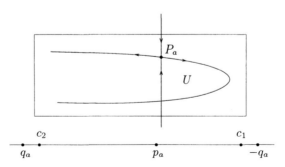

Fig. 4.1. Hénon-like attractor

Consider the fixed point p_a of ϕ_a and let $c_j = \phi_a^j(c)$ be the iterates of the critical point $c = 0$. The unstable set $W^u(p_a)$ of p_a, defined as the union of all forward iterates of some small neighborhood of p_a, coincides with the interval $[c_2, c_1]$. By hyperbolic continuation, p_a gives rise to a fixed hyperbolic saddle point $P_a \approx (p_a, 0)$ for the embedding f_a. The unstable manifold $W^u(P_a)$ is contained in $I \times (-1, 1)$, and so its closure is contained in Λ_a. It is easy to show that *the closure of $W^u(P_a)$ attracts whole open sets of orbits*, as we are going to see.

The first step is to note that, by continuous dependence of unstable sets on the dynamics, $W^u(P_a)$ is close to $W^u(p_a) \times \{0\}$ on compact parts. In particular, $W^u(P_a)$ must contain a parabola-shaped curve whose legs project to intervals close to $[c_2, c_1]$ on the horizontal axis. See Figure 4.1. Similarly, by continuous dependence of stable sets on the dynamics, the stable manifold $W^s(P_a)$ must contain a long nearly vertical segment through the fixed point. It follows that P_a has transverse homoclinic points, and so there are open regions U bounded by segments of the stable and the unstable manifolds of P_a. Using that f_a dissipates area (Jacobian close to zero) and contracts the stable manifold, we immediately get that the iterates $f_a^n(U)$ approach the unstable manifold $W^u(P_a)$ uniformly as $n \to \infty$, so that the ω-limit set of every point of U is contained in the closure of $W^u(P_a)$.

More elaborate versions of such arguments prove (see [54, 116, 439]) that in most situations the basin of attraction of the closure of $W^u(P_a)$ contains the whole $\bar{I} \times [-1, 1]$, and so $\Lambda_a = \overline{W^u(P_a)}$.

The main goal is to study the dynamics on this attracting invariant set. It turns out that this dynamics varies drastically with the value of a. A main reason us that Λ_a often contains homoclinic tangencies, which are unfolded as the parameter varies. Consequently, there are open sets of parameter values for which it contains periodic attractors. The main result (Theorem 4.3) is that, *for a positive Lebesgue measure set of parameters, Λ_a is a transitive attractor with some non-uniform hyperbolicity properties, in particular it contains no periodic attractors.* It should be noted that, in the surface case, non-periodic hyperbolic attractors are ruled out in this dissipative setting, by the classification theorem of Plykin [360].

4.1.2 Hyperbolicity outside the critical regions

The starting point for trying to understand the dynamics on the attractor is the observation that the orbits of Hénon-like maps behave in a uniformly hyperbolic fashion as long as they remain outside the critical region. At the origin is the following quantitative variation of Theorem 2.5:

Lemma 4.1. *Let $(\phi_a)_a$ be any C^3 family of multimodal maps satisfying condition (α) above. Then there exist $\sigma > 1$ and $c > 0$ such that for every small $\delta > 0$ there exists $c(\delta) > 0$ and a neighborhood $\Omega = \Omega(\delta)$ of the parameter $\bar{a}$ such that for every $a \in \Omega$,*

1. *$|(\phi_a^n)'(x)| \geq c(\delta)\sigma^n$ for any x such that $\phi_a^i(x)$ is outside the δ-neighborhood of the critical set for all $0 \leq i < n$.*

2. $|(\phi_a^n)'(x)| \geq c\sigma^n$ if, in addition, $\phi_a^n(x)$ is inside the δ-neighborhood of the critical set.

For the proof, begin by considering the Misiurewicz parameter $\bar{a}$. Let $\Delta(\delta)$ denote the δ-neighborhood of the critical set. A key ingredient is that for every small $\delta > 0$ there exist $m(\delta) \geq 1$ and $\sigma(\delta) > 1$ such that

$$|(\phi_{\bar{a}}^{m(\delta)})'(x)| > \sigma(\delta)^{m(\delta)}$$

for any x whose first $m(\delta)$ iterates remain outside $\Delta(\delta)$. This is similar to Theorem 2.5 and was first proved by Misiurewicz [294] in this context. The assumption that the critical points are non-recurrent also implies that the images of the monotonicity intervals of $\phi_{\bar{a}}^n$ have endpoints far from the critical set. Now a distortion argument, relying on the assumption that the Schwarzian derivative is negative, shows that there exists a uniform constant $c_0 > 0$ such that

$$|(\phi_{\bar{a}}^r)'(x)| > c_0 \quad \text{for any } r \geq 1 \text{ and } x \text{ such that } \phi_{\bar{a}}^r(x) \text{ is in } \Delta(\delta),$$

as long as δ is small enough. The claims in the lemma would follow directly, if it were not for the fact that we require the expansion rate to be independent of δ. To get this we proceed as follows. Fix $\delta_0 > 0$ and $1 < \sigma_0 < \sigma(\delta_0)$ and then choose a smaller $\delta_1 > 0$ such that for every $x \in \Delta(\delta_1)$ there exists some large $p = p(x) \geq 1$ such that $\phi_{\bar{a}}^i(x)$ is outside $\Delta(\delta_0)$ for all $1 \leq i < p$ and

$$|(\phi_{\bar{a}}^p)'(x)| > \sigma_0^p.$$

It is no restriction to suppose that $\sigma = \sigma(\delta_1)$ is smaller than σ_0. Now, given any x and $n \geq 1$ such that both x and $\phi_{\bar{a}}^n(x)$ are in $\Delta(\delta_1)$ and no intermediate iterate is in $\Delta(\delta_1)$, we may write $n = p + qm + r$ with $0 \leq r < m$, $m = m(\delta_1)$ and deduce that

$$|(\phi_{\bar{a}}^n)'(x)| > \sigma_0^p \sigma^{qm} c_0 > \sigma^n \qquad (4.4)$$

(reducing δ_1 increases p without affecting σ_0 or c_0). The conclusion of the lemma now follows, for all $\delta \leq \delta_1$, by breaking the time interval $\{0, 1, \ldots, n\}$ according to the passages of the orbit through $\Delta(\delta_1)$, and defining

$$c = (c_0 \sigma^{-m})^2 \quad \text{and} \quad c(\delta) = c_0 \sigma^{-m} \inf_{r,x} |(\phi_{\bar{a}}^r)'(x)| \sigma^{-r}$$

where the infimum is over all $1 \leq r < m$ and x whose first iterates are outside $\Delta(\delta)$. Finally, notice that m is fixed and p is bounded by some constant that depends only on δ, since the orbit avoids $\Delta(\delta)$. Therefore, all these estimates extend by continuity to any parameter a in a neighborhood $\Omega(\delta)$ of $\bar{a}$. This completes the argument.

Now consider a Hénon-like map $f_a(x, y) = (\phi_a(x), 0) + R(a, x, y)$ with b much smaller than δ. Given any x outside $\Delta(\delta)$ and any tangent vector $(1, v)$ with $|v|$ small, we have

$$Df_a(x, y) \cdot (1, v) \approx \phi'_a(x) (1, w)$$

where $|w|$ is again small. Thus, from Lemma 4.1 one easily gets that, over orbits avoiding the critical region $\Delta(\delta) \times B$, nearly horizontal vectors are uniformly expanded and mapped to nearly horizontal vectors:

Proposition 4.2. *Let $(\phi_a)_a$ be any C^3 family of multimodal maps satisfying condition (α). Then for every small $\delta > 0$ there exists $b(\delta) > 0$ such that if $(f_a)_a$ is a Hénon-like family satisfying (γ) with $b \leq b(\delta)$ then for any $a \in \Omega$ we have*

1. $\|Df_a^n(z)\| \geq c(\delta)\sigma^n$ *for any z such that $f_a^i(z) \notin \Delta(\delta) \times B$ for all $0 \leq i < n$.*
2. $\|Df_a^n(z)\| \geq c\sigma^n$ *if, in addition, $f_a^n(z)$ is in $\Delta(\delta) \times B$.*

Since these maps are area dissipative, this also implies uniform contraction in a complementary direction: tangent vectors that are mapped close to the vertical direction by the derivative are contracted by a factor $\approx (Cb/\sigma)^n$.

4.2 Abundance of strange attractors

The real difficulty with the dynamics of Hénon-like maps arises from the behavior in the critical region $\Delta(\delta) \times B$, where expanding (roughly horizontal) directions are mapped to contracting (roughly vertical) ones. As the unstable manifold of the fixed point P_a crosses the critical region it develops folds, that are then propagated under iteration. Since the local stable manifold of P_a is roughly vertical, one may expect that varying the parameter a will lead to the formation and unfolding of homoclinic tangencies, and such homoclinic bifurcations are indeed quite apparent in numerical experiments.

As we have seen, the unfolding of homoclinic tangencies of area dissipative maps leads to the formation of periodic attractors. The possible presence of periodic attractors turns out to be the major obstacle to proving Hénon's assertion that a global strange attractor occurs in (4.1): by [360], hyperbolic non-periodic attractors can not occur in this setting. In fact, this assertion remains unproven for parameter values close to $(a, b) = (1.4, 0.3)$ as he was considering originally (but "small" periodic strange attractors are known to exist for such parameters [184]).

4.2.1 The theorem of Benedicks-Carleson

However, by the end of the eighties, Benedicks, Carleson were able to prove that some non-uniformly hyperbolic attractor does exist in the Hénon family (4.1), for a positive Lebesgue measure set of parameters (a, b) with b close to zero:

Theorem 4.3 (Benedicks, Carleson [52]). *For any small $b > 0$, there exists a positive Lebesgue measure set of values of a for which the Hénon map $f_a = H_{a,b}$ has a strange (non-hyperbolic) attractor $\Lambda_a = \overline{W^u(P_a)}$: there exists a point $z_1 \in \Lambda_a$ whose forward orbit is dense in Λ_a and a constant $\sigma > 1$ such that*

$$\|Df_a^n(z_1)\| \geq \sigma^n \quad \text{for every } n \geq 1. \tag{4.5}$$

In particular, for these parameter values the closure of $W^u(P_a)$ contains no periodic attractors.

This historical result paved the way for much progress in understanding non-hyperbolic dynamical behavior. Shortly afterwards, Mora, Viana [297] extended this approach to the general setting of (4.2), thus freeing the arguments from any dependence on the explicit expression of the Hénon map:

Theorem 4.4 (Mora, Viana [297, 438]). *The conclusions of Theorem 4.3 extend to any Hénon-like family $(f_a)_a$ as in (4.2).*

They also established a connection between Hénon-like maps and the unfolding of homoclinic tangencies, that we shall review in Section 4.6.1. Moreover, Viana [438] extended the conclusions of [297] to arbitrary dimension. The ergodic theory of Hénon-like attractors was then developed, as we shall discuss in the next sections. At about the same time Díaz, Rocha, Viana [159] extended [297] to the strange attractors arising from saddle-node cycles and, in doing so, observed that the original approach applies to perturbations of very general families of unimodal or multimodal maps in one dimension, besides the quadratic family (see Section 4.6.2). More recently, Wang, Young [447] showed that the theory extends to the full generality of (α), (β), (γ). Moreover, Palis, Yoccoz [349] have been applying a related analysis to non-uniformly hyperbolic sets of saddle type arising from homoclinic tangencies. Similar ideas have been used in several other related situations: saddle-focus bifurcations of flows, by Pumariño, Rodriguez [373, 374]; saddle-node horseshoes, by Costa [136]; invariant curves subject to kick forcing, by Wang, Young [448]; Lorenz-like attractors without invariant foliations, by Luzzatto, Viana [269].

4.2.2 Critical points of dissipative diffeomorphisms

An extended survey of the arguments in [52, 297] to prove Theorems 4.3 and 4.4 has appeared recently, by Luzzatto, Viana [271]. Here we restrict ourselves to a few general comments.

Recurrence control in dimension 1:

The strategy introduced by Benedicks, Carleson for proving Theorem 4.3 is inspired from alternative approaches to the proof of Jakobson's Theorem 2.11, especially by Collet, Eckmann [130] and Benedicks, Carleson [51] themselves, based on controlling the recurrence of the critical orbits through parameter exclusions. Let $(\phi_a)_a$ be a family of one-dimensional maps satisfying condition (α) above, and $\mathcal{C} = \{c_i\}$ represent the set of critical points of each of the maps. The aim is to find a constant $\sigma > 1$ and a positive Lebesgue measure set of parameters close to $\bar{a}$ such that

$$|(\phi_a^n)'(\phi_a(c_i))| \geq \sigma^n \quad \text{for all } n \geq 1 \text{ and } c_i \in \mathcal{C}. \tag{4.6}$$

To achieve this one combines two ideas: on the one hand, one collects expansion during the iterates outside $\Delta(\delta)$ using Lemma 4.1; on the other hand, one forbids the critical set to return close to itself too fast: (α is a small positive constant)

$$d(\phi_a^n(c_i), c_j) \geq e^{-\alpha n} \quad \text{for all } n \geq 1 \text{ and } c_i, c_j \in \mathcal{C}, \tag{4.7}$$

to ensure that expansion is not completely lost when the critical orbit hits $\Delta(\delta)$. An important ingredient is a recovery period (bound period) following any return of a critical point c_i to $\Delta(\delta)$, at the end of which the expansion lost at the return time n has been completely compensated for. The idea is that the orbit of $\phi_a^n(c_i)$ remains close to the orbit of some critical point c_j (possibly the same) for some period of time, during which information about the latter may be transmitted to the former, by continuity and in an inductive fashion. The recurrence conditions (a large deviations type condition is needed in addition to (4.7)) are taken as restrictions on the parameter, and shown to hold for a positive Lebesgue measure set of values of a. For such parameters one gets (4.6). By Singer [417], this implies that the map has no periodic attractors. In fact, it even admits absolutely continuous invariant probability measures whose basins cover a full measure subset of phase space; see [131, 147, 328].

Critical points in higher dimensions:

The cornerstone of Benedicks, Carleson's approach to the Hénon family is a notion of critical point for surface diffeomorphisms which they use to organize the dynamics of Hénon maps in much the same way as before for interval transformations. In very simplified terms, a *critical point* corresponds to a tangency between Pesin stable and unstable manifolds (Appendix C): the main feature is that *the same tangent direction is contracted by both forward and backward iterates of the derivative.* For proving Theorem 4.3 it suffices to consider critical points inside the unstable manifold of the fixed point and, in fact, one needs only a subset $\mathcal{C}$. The precise choice of $\mathcal{C}$ and the very definition of these points turn out to be very delicate, though. Let us spend some time explaining this.

Critical points play a dual role in the one-dimensional arguments sketched above. Firstly, the distance to the nearest critical point is a measure of the norm of the derivative, which is the reason why a recurrence condition like (4.7) can be used to bound the loss of expansion at each return. Secondly, as explained before, information is passed on from the orbit of c_j to that of $\phi_a^n(c_i)$ during the recovery period in which the two orbits remain close to each other (*bound period*). In order to duplicate this in the surface case, one must be able to associate to each return $f_a^n(c_i)$ of a critical point to the region $\Delta(\delta) \times B$ a suitable critical point c_j such that the loss of hyperbolicity taking place at the return be, in some sense, proportional to the distance from $f_a^n(c_i)$ to c_j. Indeed, one key step in the proof is to *construct* a critical point c_j such that $f_a^n(c_i)$ is in tangential position to it: the horizontal distance is much larger than the vertical distance, as described in Figure 4.2. Then the same type of

Fig. 4.2. Tangential position

estimates as in the one-dimensional situation may be used to bound the loss of hyperbolicity. This point c_j is then used for the recovery argument, which is also similar to the one-dimensional case, although technically much harder.

These considerations also mean that the critical set $\mathcal{C}$ can not be chosen too small: it must contain enough points so that a convenient c_j in tangential position may be found at each return. But $\mathcal{C}$ should not be too large either: the more critical points there are, the more parameters must be excluded to impose recurrence conditions like (4.7) to each one of them. The following compromise is reached: at each step n one deals with not more than $e^{\theta n}$ critical points, where $\theta \approx 1/|\log b|$. In order to control the orbit of each one of them, a subset of parameters is excluded whose measure is bounded by $e^{-\gamma n}$, where the constant $\gamma > 0$ is independent of b. The assumption that b is small is crucial at this point: it ensures that $\sum_n e^{\theta n} e^{-\gamma n}$ is small, so that a positive Lebesgue measure subset of parameters survives all the exclusions.

But the main difficulty, that makes the proof so subtle and involved, lies in the precise definition of a critical point. The reason is that Pesin stable and unstable manifolds are not given a priori: one must assume some amount of hyperbolicity, which is precisely what one wants to prove using these arguments! The way to bypass this is to work with finite time *approximations to critical points,* defined by means of approximate stable directions. Roughly speaking, for a point $z \in W^u(P_a)$ such that the horizontal direction $(1,0)$ is uniformly expanded by the first n iterates of the derivative (a hyperbolicity hypothesis), one considers the direction $e^n(z)$ that is most contracted by $Df_a^n(z)$, and one says that z is a *critical value* of order n if $e^n(z)$ is tangent to the unstable manifold of P_a. Then the pre-image of z is a critical point of order n. Such critical approximations of order n are used at the n:th step of the induction to obtain the hyperbolicity hypothesis at time $n + 1$, after parameter exclusions. At this stage, critical approximations of order $n + 1$ may be defined, to replace those of order n, and the whole procedure may be repeated.

In view of the *ad hoc* nature of these constructions, the following fundamental problem arises naturally. Some progress in this direction has been reported by Pujals recently.

Problem 4.5. Give an intrinsic definition of critical point for a (dissipative) surface diffeomorphism.

4.2.3 Some conjectures and open questions

In the next sections we shall see that there is now a good description of the dynamics of Hénon-like maps for the parameter values in the proofs of Theorems 4.3 and 4.4. On the other hand, several fundamental questions remain open, concerning a deeper understanding of the dynamics for those chaotic parameters, the behavior of these families over larger parameter ranges, or the insertion of this theory in the broader context of Dynamics. Some of these questions are listed here, several others are scattered throughout this chapter.

On the existence of attractors and their basins:

For simplicity, the questions in this paragraph are stated for the Hénon family (4.1), but there are clear reformulations in the general Hénon-like setting.

Let us consider an (open) parameter domain $\mathcal{U} \subset \mathbb{R}^2$ for which $H_{a,b}$ has a topological attractor: there is an open set $U \subset \mathbb{R}^2$ that contains the closure of $H_{a,b}(U)$. The (non-transitive) topological attractor is $\Lambda_{a,b} = \cap_n H_{a,b}^n(U)$.

Conjecture 4.6. There exists an open dense subset of $\mathcal{U}$ for which $H_{a,b}$ has at least one periodic attractor in U.

This would correspond to density of hyperbolicity among interval maps; recall Theorems 2.9 and 2.10. The next question has a similar flavor:

Problem 4.7. For every $\varepsilon > 0$ there exists an open dense subset of parameters $(a, b) \in \mathcal{U}$ for which $H_{a,b}$ has a finite number of periodic attractors such that the complement of the union of their basins inside U has Lebesgue measure less than ε.

Let $\mathcal{T}$ be the closure of the subset of parameters $(a, b) \in \mathcal{U}$ for which $H_{a,b}$ has some homoclinic tangency.

Conjecture 4.8. Let $(a, b) \in \mathcal{U} \setminus \mathcal{T}$ be a parameter value far from homoclinic tangencies. Then $H_{a,b}$ has a finite family of periodic attractors such that the union of their basins has full Lebesgue measure in U.

Jakobson's Theorem 2.11 is a clear predecessor to the results in this chapter. Pushing the parallel to one-dimensional dynamics a bit further, one may ask whether an analogue of Lyubich' Theorem 2.13 holds in the present context:

Problem 4.9. For a full Lebesgue measure subset of $\mathcal{U}$ the map $H_{a,b}$ has a finite number of transitive attractors such that the union of their basins has full Lebesgue measure in U.

We mentioned in Section 4.1.1 that for many parameter values the topological attractor coincides with the closure of the unstable manifold of the fixed point $P_{a,b}$ (presently, the best results in this direction are in [439]). Moreover, the boundary of the basin of attraction coincides the stable manifold of the other fixed point $Q_{a,b}$ (see the last section in [54]).

Problem 4.10. Does $\Lambda_{a,b}$ coincide with the closure of $W^u(P_{a,b})$, and is its basin bounded by the stable manifold of $Q_{a,b}$ for all parameters $(a,b) \in \mathcal{U}$?

Hyperbolic attractors vs. *Hénon-like attractors:*

Let us call an attractor *Hénon-like* if it coincides with the closure of the unstable manifold of some periodic point, and contains dense critical orbits (critical means that there is a tangent vector which exponentially contracted by both forward and negative iterates of the derivative) such that some other tangent vector is exponentially expanded by all forward iterates of the map. For surface diffeomorphisms, Hénon-like attractors are the only known class of persistent non-hyperbolic attractors.

Problem 4.11. For generic 1-parameter families of surface diffeomorphisms, do almost all parameters correspond to diffeomorphisms whose attractors are either hyperbolic (periodic or not) or Hénon-like? Moreover, are the attractors finitely many?

We have already mentioned that, since Hénon-like diffeomorphisms on surfaces are area dissipative, Plykin [360] immediately gives that they can not have hyperbolic non-periodic attractors. It is natural to ask whether the same remains true in other situations where Hénon-like behavior arises but one does not have definite area dissipation:

Problem 4.12. Let g be a diffeomorphism on a compact surface, having two fixed saddle points that are heteroclinically related (transverse intersections between their stable and unstable manifolds) and such that the Jacobian of g is larger than 1 at one of the points and is smaller than 1 at the other. Consider the generic unfolding of a homoclinic tangency associated to some of the fixed points. Are there open sets in parameter space such that (i) the corresponding diffeomorphisms have some non-periodic hyperbolic attractor? (ii) for a residual subset of parameters there exist infinitely many non-periodic hyperbolic attractors?

In the same spirit, can homoclinic tangencies in higher dimensions, dissipative or not, produce non-periodic hyperbolic attractors?

Problem 4.13. Let $(g_\mu)_\mu$ be a family of diffeomorphisms in dimension larger than 2, unfolding a homoclinic tangency associated to some fixed point of g_0 (possibly sectionally dissipative). Are there open sets in parameter space such that (i) the corresponding diffeomorphisms have some non-periodic hyperbolic attractor? (ii) for a residual subset of parameters there exist infinitely many non-periodic hyperbolic attractors?

Examples of infinitely many coexisting hyperbolic attractors appear in [68]. These examples are C^1 locally generic, but involve infinitely many parameters.

Many of the results in this chapter hold for Hénon-like maps and attractors in manifolds of any dimension. Note, however, that by definition Hénon-like maps are "almost one-dimensional" and, thus, sectionally dissipative. In particular, they can exhibit at most one expanding direction. In Section 11.1.3 will discuss higher dimensional models, with several unstable directions, which turn out to have a much more robust character. However, these models are partially hyperbolic, in the sense that all but one expanding direction are actually uniform.

Problem 4.14. Construct persistent, or even robust, Hénon-like attractors with two (or more) strictly non-uniformly expanding directions.

It is natural to look for such examples amongst singular perturbations of surface maps, for instance of the form $(x, y) \mapsto (1 - a(x, y)x^2, 1 - b(x, y)y^2)$ exhibiting two positive Lyapunov exponents. Thus, the first key step should be some version of Jakobson's Theorem 2.11 on persistence of non-uniform hyperbolicity for such maps. The results on maps with critical points in [12] should be useful in this direction.

The geometry and topology of Hénon-like attractors:

Over the last decade, a rather complete ergodic theory of Hénon-like strange attractors was built, for the parameter values in the proof of Theorem 4.4. This we shall review in Sections 4.3 through 4.5. On the other hand, topological and geometric aspects of the attractor are still much less understood. Let us focus on the 2-dimensional case, for the sake of precision.

Conjecture 4.15. The Hausdorff dimension of the strange attractor coincides with its hyperbolic Hausdorff dimension, that is, the supremum of the dimensions of the (uniformly) hyperbolic sets contained in it. Moreover, this Hausdorff dimension varies continuously with a inside the set of good parameters.

The classical Bowen-Manning formula [89, 285] for surface diffeomorphisms says that the the stable dimension of a hyperbolic basic set (that is, the Hausdorff dimension of the intersection of the basic set with any stable manifold) is the unique solution $t > 0$ of

$$P(t \log |\det Df|E^u|) = 0$$

where P is the topological pressure. One important consequence is that in this context the Hausdorff dimension varies continuously with the dynamics.

Problem 4.16. Is there a version of the Bowen-Manning formula for Hénon-like strange attractors?

Another interesting question is the variational principle for the dimension, that we state next. The dimension of a measure is the infimum of the Hausdorff dimensions of all full measure subsets. See [48, 456] and also Section 12.6 below.

Problem 4.17. Is the supremum of the dimensions over all probability invariant measures supported on the attractor attained? Does it coincide with the Hausdorff dimension of the attractor?

It would also be nice to know how the Hausdorff dimension of the topological attractor varies on larger (open) parameter ranges:

Problem 4.18. For an open set $\mathcal{U}$ of parameters as in the first paragraph, such that a topological attractor exists, does the Hausdorff dimension of this attractor vary (semi)continuously with the parameters?

A very complete combinatorial description of the dynamics is now available for one-dimensional transformations, relying on such invariants as the rotation number of Poincaré, Denjoy, and the kneading sequence of Milnor, Thurston. See Chapter 2 and [147]. In contrast, the combinatorial theory of surface maps is very much incomplete, including the Hénon family (for the uniformly hyperbolic case, see [80]):

Problem 4.19. Develop a combinatorial model for the dynamics on the topological attractor of (4.1), and of its bifurcations as the parameters vary.

A general strategy for describing the dynamics of parametrized families of surface homeomorphisms in combinatorial terms has been proposed by Cvitanovic, Gunaratne, Procaccia [144]. A central open problem, sometimes referred to as the pruning front conjecture, states that every Hénon map may be viewed as what is left of a horseshoe map after part of the dynamics has been "pruned away". In this direction, Carvalho, Hall [145] study the class of maps that can be obtained from a horseshoe through this pruning procedure. More precisely, they show how, for certain topological disks (*pruning disks*), one can deform the homeomorphism continuously so as to make every orbit that enters the disk a wandering orbit, while leaving all the other orbits unaffected. The paper [145] also contains a precise statement of the pruning front conjecture and an updated discussion of this and related issues.

In the specific context of Hénon-like strange attractors (including the parameter values in the proof of Theorem 4.4) Wang, Young [447] give a kneading-type description of the dynamics on the attractor, and deduce some very nice consequences. Roughly speaking, their kneading invariant is an itinerary describing the position of successive iterates relative to the critical points of the Hénon-like map (recall Section 4.2.2). Using it, they prove that the dynamics on the strange attractor is semi-conjugate to a full shift with finitely many symbols, by a continuous surjection which fails to be injective only over the orbit of the critical set. They deduce that

- the topological entropy coincides with the exponential growth rate of the number of periodic points of period n;
- every continuous potential admits some equilibrium state (see [86] and Section 11.5.2).

In particular, the topological entropy is realized by the entropy of some invariant probability (variational principle for the entropy) .

We close with another long term problem inspired by the theory of one-dimensional maps.

Problem 4.20.
Develop a renormalization theory for Hénon-like maps. What does universality mean in this context?

Some progress in this direction has been recently announced by de Carvalho, Lyubich, Martens.

4.3 Sinai-Ruelle-Bowen measures

A very complete ergodic theory of Hénon-like attractors has been developed over the last ten years or so. We discuss some main results in Sections 4.3 to 4.5. Throughout, we let a be any fixed parameter value in the positive Lebesgue measure set exhibited in the proof of Theorem 4.4, and we consider $f = f_a$ and $\Lambda = \Lambda_a$.

4.3.1 Existence and uniqueness

The first main step was to prove that Hénon-like maps admit an SRB measure:

Theorem 4.21 (Benedicks, Young [56]). *Every Hénon-like attractor Λ has an SRB measure μ, which is ergodic, supported on the whole attractor, has one positive Lyapunov exponent, and is absolutely continuous with respect to arc-length along the corresponding unstable manifolds.*

In fact, they show that (f^n, μ) is ergodic for every $n \geq 1$. By results of Pesin [356] and Ledrappier [253], it follows that (f, μ) is equivalent to a

Bernoulli shift. In addition, they prove that μ is the unique f-invariant measure absolutely continuous along unstable manifolds [1]. See [56, Section 4.3]. It will follow from Theorem 4.24 that μ is also the unique physical (SRB) measure in Λ. Uniqueness is specific to (4.2) and other perturbations of unimodal maps: in the general setting of conditions (α), (β), (γ) the number of SRB measures is bounded by the number of critical points of the one-dimensional maps ϕ_a. See Wang, Young [447].

Global dynamics on the unstable manifold:

Let us comment on the proof of Theorem 4.21. The crucial ingredient is an extension, for (almost) all orbits in the unstable manifold $W^u(P)$, of the kind of control that [52, 297] provided for critical orbits. Indeed, the Main Proposition of Benedicks, Young [56] says

Proposition 4.22. *For every point $z \in W^u(P)$ in $\Delta \times B$ which is in a free state, there exists a critical point $\zeta \in \mathcal{C}$ such that z is in tangential position to ζ.*

Recall that $\Delta = \Delta(\delta)$ is the δ-neighborhood of the critical set. Let us explain the contents of this proposition, starting by saying what we mean by *free state*. By definition, every point w in a small segment of the unstable manifold around the fixed point P is in a free state. The first forward iterates $f^n(w)$ are also in a free state, up until the first time $f^n(w)$ hits $\Delta \times B$. This is called a *free return*. The proposition applies at that time, and asserts that there exists a critical point ζ such that (recall Figure 4.2)

- the horizontal distance between z and ζ is much larger than the vertical distance between z and ζ, and
- that horizontal distance is also much larger than the angle between the tangent directions $t(z)$ and $t(\zeta)$ to the unstable manifolds at both points.

Then one defines a time interval $[n, n+p]$ during which z is bound to ζ (recovery period): iterates in this time interval are *not* in a free state. If z coincides with ζ then the bound period is infinite. Otherwise, at time $n+p+1$ the orbit goes back to being in a free state, and remains in a free state until the next return to $\Delta \times B$. At that point the proposition can again be applied, and the whole procedure can be repeated all over again.

In this way, Proposition 4.22 allows for any orbit in the unstable manifold to be organized into bound periods and free periods, just as [52, 297] did for critical orbits. Most important, the fact that the bound periods start at tangential positions means that estimates similar to the one-dimensional case apply also in this generality: the loss of expansion at the return to $\Delta \times B$ is proportional to the distance to the binding critical point, and it is completely

---

[1] In their terminology: μ is the unique SBR measure. The notion of measure absolutely continuous along a foliation is recalled in Section C.4.

compensated for at the end of the bound period. In this way one finds $\sigma > 1$ and, for any point z whose orbit avoids the critical set, a constant $c > 0$ such that

$$\|Df^n(z)\,t(z)\| \geq c\sigma^n \quad \text{for all } n \geq 1. \tag{4.8}$$

Building on this expanding behavior, one then constructs an ergodic invariant measure μ with one positive Lyapunov exponent and absolutely continuous conditional measures along the associated Pesin unstable manifolds. Essentially, μ is obtained as a weak* Cesaro accumulation point

$$\lim_k \frac{1}{n_k} \sum_{j=0}^{n-1} f_*^j(m_{\bar{\gamma}})$$

of normalized arc-length along some segment $\bar{\gamma}$ of $W^u(P)$. Since these maps are area dissipative, the other Lyapunov exponent of μ must be negative. Consequently, almost every point has a Pesin stable manifold. By ergodicity, μ-almost every point is in the basin of μ. These points contain a positive (arc-length) measure subset in many unstable segments. Therefore, using also the fact that the Pesin stable lamination is absolutely continuous (see Theorem C.4), the union of the stable manifolds through those points has positive Lebesgue measure (area). Moreover, this union is contained in $B(\mu)$, since the basin always consists of entire stable manifolds. Thus, μ is an SRB measure.

We are going to detail the previous paragraph a bit more. Since the original proof alternative arguments have been proposed, in [53, 57, 447]. Our explanations follow more closely the proof in [53], since that approach will also be useful in Section 4.5 for treating random perturbations.

Inducing in Hénon-like attractors:

The main technical step is to construct a compact subset X of the attractor, consisting of long unstable leaves (approximately horizontal curves with length uniformly bounded from zero and which are uniformly contracted by every negative iterate of f) and such that almost every orbit goes across X. See Figure 4.3. Then we consider the first-return map $F : X \to X$ (F is defined by $F(z) = f^{n(z)}(z)$, where $n(z)$ is the smallest positive integer for which the iterate of z is in X) and show that it is uniformly expanding and Markov along the unstable leaves. More precisely,

(i) Markov structure) every unstable leaf γ is a union, up to a subset with zero arc-length measure, of segments γ_i each of which is mapped by F diffeomorphically onto some unstable leaf;

(ii) (expansion and distortion) the norm $|F'|$ of the derivative along every γ_i is larger than some uniform constant $\sigma > 1$, and the distortion $|(F^n)'(x)|/|(F^n)'(y)|$ is bounded over all $n \geq 1$, all $x, y \in \gamma_i$, and γ_i;

(iii) (fast returns) for each unstable leaf and $k \geq 1$, the set of points in the leaf with $n(\cdot) > k$ has arc-length measure $\leq C\lambda^n$, where the constants $C > 0$ and $\lambda < 1$ are uniform;

(iv) (saturation) almost every point (in the sense of arc-length) in the unstable manifold of P has some forward iterate in X.

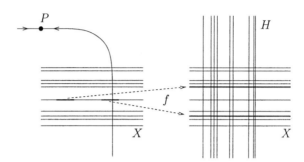

Fig. 4.3. Inducing in Hénon-like attractors.

We only say a few words about the construction of X, referring the reader to [53, Section 4] for the complete arguments. Proposition 4.22 permits to introduce a combinatorial description of the dynamics on the unstable manifold: to each free return n of the orbit of z to $\Delta \times B$ one associates a critical point ζ_n such that $f^n(z)$ is in tangential position to ζ_n; the *itinerary* of z collects the sequence of these *binding critical points*, together with the relative position

$$r_n \approx \left[\pm \log \mathrm{dist}(f^n(z), \zeta) \right] \tag{4.9}$$

(brackets mean integer part, and the sign describes whether $f^n(z)$ is to the left or to the right of ζ_n). Then the unstable manifold may be split into segments ω_n consisting of points with the same itinerary up to time n. One says that ω_n *escapes* at time n (a variation of a notion introduced in [52]) if n is a free return, yet $f^n(\omega_n)$ is relatively far from the binding critical point: essentially, one asks that $r_n = [-\log \delta]$. The set X is the closure of the union of these escaping curves $f^n(\omega_n)$.

Constructing the SRB measure:

At this point the situation is similar to the corresponding stage in the proof of Jakobson's Theorem 2.11. Let $m_{\bar{\gamma}}$ denote the normalized arc-length on some unstable leaf $\bar{\gamma} \subset X$. Using property (i) and, especially, property (ii) above, one gets that each forward iterate $F_*^j(m_\gamma)$ is absolutely continuous along the unstable leaves (see Appendix C for the definition), with uniformly bounded density. The density is also uniformly bounded from zero on the support, which consists of entire long unstable leaves. It follows that every Cesaro accumulation point μ_F of the sequence $F_*^j(m_{\bar{\gamma}})$ is an F-invariant probability measure which is still absolutely continuous along unstable leaves, with the same upper and lower bounds for the density. Then the saturation of μ_F

$$\mu_f = \sum_{k=1}^{\infty} \sum_{i=0}^{k-1} f_*^i (\mu_F | \{n(\cdot) = k\}).$$

is an f-invariant measure supported on the attractor Λ. The fast returns property (iii), and the fact that the density of μ_F with respect to arc-length is bounded, ensure that the measure μ_f is finite. Each term $f_*^i(\mu_F | \{n(\cdot) = k\})$ is absolutely continuous along the unstable foliation of Λ, because f is a diffeomorphism. Since F is the *first* return map, these terms are supported on pairwise disjoint sets. It follows that μ_f is also absolutely continuous along the unstable foliation. Its normalization

$$\mu = \frac{\mu_f}{\mu_f(\Lambda)}$$

is the SRB measure of the Hénon-like attractor, as we are going to explain.

Note that supp μ = supp μ_f contains supp μ_F and, as we have seen, the latter consists of entire long unstable leaves of X. Using that the stable manifold of the fixed point P intersects these unstable leaves transversely, and supp μ is a closed invariant set, we conclude that the whole unstable manifold of P is contained in the support of μ, and so supp $\mu = \overline{W^u(P)} = \Lambda$. See Figure 4.3. For proving ergodicity and the SRB property we need some additional information on the behavior of f transversely to the unstable foliation.

Long stable leaves:

Since these maps are area dissipative, expanding behavior along the unstable manifold must be accompanied by contracting behavior in a transverse direction. In particular, it was shown in [52] that immediately expanding points have long stable manifolds:

Lemma 4.23. *Let $\rho_0 \gg b$ be fixed and z be such that $\|Df^n(z)v\| \geq \rho_0^n$ for all $n \geq 1$ and some unit vector v with slope less than 1. Then there exists a long stable leaf through z, that is, a nearly vertical curve with radius $1/2$ which is uniformly contracted by all positive iterates of f (at rate Cb/ρ_0).*

One can show that points z whose orbits approach the critical set slowly enough, for instance such that

$$\text{dist}(f^n(z), \zeta_n) \geq \rho^n \quad \text{at every free return } n,$$

with $\rho \gg \rho_0$, satisfy the hypotheses of the lemma, and they constitute a positive arc-length subset of any long unstable leaf of X. Moreover, the corresponding long stable leaves form a Lipschitz lamination. Consequently, their union H has positive Lebesgue measure (area). See also Figure 4.3.

To prove ergodicity, consider any continuous function $\varphi : \Lambda \to \mathbb{R}$ and let

$$\tilde{\varphi}(x) = \lim_{n \to +\infty} \frac{1}{n} \sum_{j=0}^{n-1} \varphi(f^j(x))$$

be its (forward) time average. Given any $\alpha \in \mathbb{R}$, suppose $A = \{\tilde{\varphi} > \alpha\}$ has $\mu(A) > 0$. Then $\mu_F(A) > 0$, and so A intersects some unstable leaf on a positive arc-length subset. Now notice that $A \cap X$ must coincide $\bmod 0$ with a union of entire unstable manifolds: that is because the backward time averages of continuous functions are constant on unstable manifolds, and they coincide $\bmod 0$ with the corresponding forward time averages. It follows that A intersects some unstable leaf γ_1 on a *full* arc-length subset. Suppose the complement $A^c = \{\tilde{\varphi} \leq \alpha\}$ also had $\mu(A^c) > 0$. Then, by the same argument, A^c would intersect some unstable leaf γ_2 on a full arc-length subset. However, the set H intersects both γ_1 and γ_2 on positive arc-length subsets. So, this would contradict the fact that time averages of continuous functions are constant on stable leaves. Therefore, $\mu(A^c)$ must be zero. This proves that μ is ergodic.

The rest of the argument is quite general. By ergodicity, the basin $B(\mu)$ has full μ-measure, and so it intersects some unstable leaf of X on a full arc-length set. Then this set intersects H on a positive arc-length subset. Considering the long stable leaves through the points in the latter subset, and using once more the fact that time averages of continuous functions are constant on stable leaves, one obtains a positive Lebesgue measure (area) set contained in the basin of μ. Thus, μ is an SRB measure, as claimed.

4.3.2 Solution of the basin problem

Of course, we also want to know (Problem 1.11) whether the basin $B(\mu)$ of the SRB measure contains almost every point whose orbit converges to the attractor Λ. The answer is affirmative, and was given by the following theorem, proved a few years later: *the basins of Hénon-like attractors have no holes.*

Theorem 4.24 (Benedicks, Viana [54]). *Let Λ be the attractor of a Hénon-like diffeomorphism f and μ be its SRB measure. Then Lebesgue almost every point z in the topological basin $B(\Lambda)$ is contained in the stable manifold of some point in the attractor. Moreover, the ergodic basin $B(\mu)$ coincides with $B(\Lambda)$ up to a zero Lebesgue measure set.*

That is, almost every point x whose orbit converges to the attractor satisfies

$$\lim_{n \to \infty} \frac{1}{n} \sum_{j=0}^{n-1} \delta_{f^j(x)} = \mu$$

where δ_z denotes the Dirac measure at a point z, and the limit is in the weak* topology in the space of probabilities. It follows, easily, that the iterates of Lebesgue measure m restricted to any neighborhood U of Λ inside the basin $B(\Lambda)$ satisfy

$$\lim_{n \to \infty} \frac{1}{n} \sum_{j=0}^{n-1} f_*^j(m \mid U) \to m(U)\, \mu.$$

The next question has been raised by Michael Benedicks. It is also related to the issues of decay of speed of mixing and convergence to equilibrium that we discuss in Section 4.4 and in Appendix E.

Problem 4.25. Does $f_*^j(m \mid U)$ converge to $m(U)\mu$ as $n \to \infty$? If so, how fast is the convergence?

Before we comment on the proof of Theorem 4.24, let us discuss a few main issues. First of all, attractors with holes in their basins do exist: a (non-generic) example was constructed in [41]. On the other hand, it is a classical fact that uniformly hyperbolic attractors have the no-holes property; see Bowen [86]. Indeed, that follows from the fact that local stable manifolds of points in the attractor Λ cover a whole neighborhood of Λ (a full Lebesgue measure subset consists of local stable manifolds of points in $\Lambda \cap B(\mu)$), together with the elementary observation that the ergodic basin consists of entire stable manifolds.

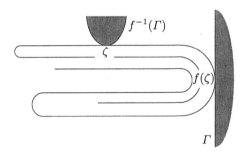

Fig. 4.4. Apparent holes in the basin of Hénon-like attractors

However, it is easy to see that Hénon-like attractors are very different from uniformly hyperbolic ones in this regard: *local stable manifolds miss whole open sets close to the attractor*. For instance, no local stable manifold of a point in the attractor can go through the shaded regions in Figure 4.4, where Γ represents the long stable leaf through the rightmost critical value $f(\zeta) \in \Lambda$. That is, simply, because distinct stable manifolds can not intersect each other.

So, what Benedicks, Viana [54] prove is that Lebesgue almost any point in a neighborhood of Λ is contained in some *global* stable manifold of a point inside $\Lambda \cap B(\mu)$. The main novelty in doing that is that one can not expect an analogue of Proposition 4.22 to hold in the basin of attraction $B(\Lambda)$. Indeed, there exist whole open sets close to $\{x = 0\}$ which are *not* in tangential position to any critical point. The most evident examples, like the shaded region close to ζ in Figure 4.4, are also domains missed by the local stable manifolds of points in Λ, which is perhaps not surprising. This means that the kind of control provided by Proposition 4.22 for orbits inside the attractor does not extend to the basin, especially not to those regions where it would be most needed.

Nevertheless, the core of the proof of Theorem 4.24 is a statistical argument showing that *for almost every point z in the basin of attraction, returns are eventually tangential and not too close to the critical set.* More precisely, for Lebesgue almost every $z \in B(\Lambda)$ there exists $\nu = \nu(z) \geq 1$ such that for every free return $n_k > \nu$ there exists a critical point ζ_k in tangential position and satisfying

$$d(f^{n_k}(z), \zeta_k) > ce^{-5(n_k-\nu)}. \tag{4.10}$$

This also implies that $f^{\nu+1}(z)$ is immediately expanding, in the sense of Lemma 4.23. Consequently, it has a long stable leaf, and this leaf is shown to intersect the attractor Λ at some point ξ. Then z belongs to the (global) stable manifold of $f^{-(\nu+1)}(\xi) \in \Lambda$. This gives the first claim in the theorem. Moreover, using the fact that the lamination by long stable leaves is Lipschitz continuous, and the SRB measure is ergodic and absolutely continuous along unstable manifolds, one gets that, for Lebesgue almost every z, the point ξ found in this way is in $B(\mu)$. Then z is also in $B(\mu)$, since the ergodic basin is invariant and consists of entire stable manifolds. That gives the second claim in the theorem.

In the sequel we outline this strategy in a bit more detail. Let us also mention that the conclusion of Theorem 4.24 has been extended to higher dimensions, by Muniz [317] and Wang, Young [447, 446].

Symbolic dynamics in the basin:

An important preparatory step is to construct, for each critical point ζ, a sequence $\Gamma_r(\zeta)$ of long stable leaves accumulating exponentially fast on the long stable leaf $\Gamma(\zeta)$ through the critical value $f(\zeta)$:

$$d(\Gamma_r(\zeta), \Gamma(\zeta)) \approx e^{-2r}.$$

This is used to give a symbolic description of the orbits in the basin, as we are going to describe. In fact, we outline a slightly simplified version, referring the reader to [54, Section 3] for the full details.

First, let ζ and ζ' be the critical points of lower generation, that is, the ones closest to the fixed point P inside $W^u(P)$. Lebesgue almost every orbit in $B(\Lambda)$ passes through the region close to $\{x = 0\}$ between the two segments of the unstable manifold containing ζ and ζ'; see [54, Section 5]. So, it suffices that we consider points in that region. For such points, we introduce the following symbol $i_0(z)$, that describes the position of $f(z)$ relative to the sequences of long stable leaves associated to those critical points:

$$i_0(z) = \begin{cases} (\zeta', r, \pm) & \text{if } f(z) \text{ is between } \Gamma_{r-1}(\zeta') \text{ and } \Gamma_r(\zeta') \\ (\zeta, r, 0) & \text{if } f(z) \text{ is to the right of } \Gamma(\zeta') \text{ between } \Gamma_{r-1}(\zeta) \text{ and } \Gamma_r(\zeta). \end{cases}$$

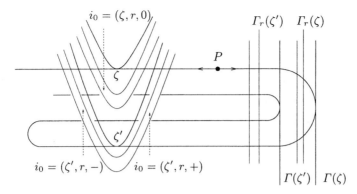

Fig. 4.5. Symbolic description: first step.

See Figure 4.5. The sign $\pm$ has the sole purpose of distinguishing connected components. For each possible value of i_0 let $R(i_0) = \{z : i_0(z) = i_0\}$. Note that every $R(i_0)$ is a quadrilateral, bounded by two segments of unstable manifold and two segments of stable manifold.

Next, we define $n_1 = n_1(i_0)$ to be the first free return for $R(i_0)$ that is, the first free return for the unstable segments on the boundary of $R(i_0)$. To it we associate binding critical points ζ_1 and ζ_1' for those two unstable segments. Figure 4.6 illustrates the case when the two points are distinct, but it may also happen that $\zeta_1 = \zeta_1'$. Then we define $i_1(z) = (\zeta_1(z), r_1(z), \varepsilon_1(z))$ in much the same way as before for i_0, with $\zeta_1, \zeta_1', f^{n_1}(z)$ in the place of ζ, ζ', z. Let $R(i_0, i_1) \subset R(i_0)$ denote the set of points sharing the same values for i_0 and i_1. Notice that $R(i_0, i_1)$ is also a quadrilateral, since two different stable, or unstable, leaves can never intersect each other.

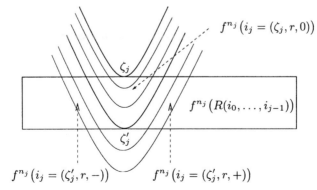

Fig. 4.6. Symbolic description: generic step.

Repeating this step, we obtain a sequence $i_k(z) = (\zeta_k(z), r_k(z), \varepsilon_k(z))$, $k \geq 0$, describing the positions of the iterates of z at free returns, in terms of binding critical points and corresponding long stable leaves. A very useful feature of this construction is that every set $R(i_0, i_1, \dots, i_{k-1})$ of points having the same values for the first $k \geq 1$ symbols is a quadrilateral. Together with bounded area distortion, this makes it feasible to estimate the relative probability of each value of the next symbol i_k inside $R(i_0, i_1, \dots, i_{k-1})$, which is crucial for the main estimate in (4.11) below.

The statistical argument:

Given a sequence $i_k = (\zeta_k, r_k, \varepsilon_k)$ with return times $0 < n_1 < \cdots < n_k < \cdots$, define the *close returns* to be the smallest subsequence $\nu_s = n_{k(s)}$ such that

$$r_j \leq 5(n_j - \nu_s) \quad \text{for all} \quad \nu_s < n_j < \nu_{s+1}$$

(by convention $\nu_0 = 0$). The point with this notion is that non-close returns are necessarily tangential, and points remain expanding for as long as they have no close returns. In particular, if $k(s+1) = \infty$ (in other words, if z has no other close return after ν_s), property (4.10) holds with $\nu = \nu_s$, and then Lemma 4.23 applies to the point $f^{\nu_s+1}(z)$.

The main technical result in the proof of Theorem 4.24 asserts that the conditional Lebesgue probability

$$\mathcal{P}\big(n_k \text{ is a close return } | i_0, i_1, \dots, i_{k-1} \big) \leq$$
$$\leq \min\{1 - \theta, \theta^{-1} e^{-(n_k - \nu_s)}\} \tag{4.11}$$

where $\theta > 0$ is a uniform constant and ν_s is the last close return prior to n_k. It follows that

$$\mathcal{P}\big(\text{all returns } n_j \geq n_k \text{ are non-close} | i_0, i_1, \dots, i_{k-1} \big) \geq \theta_0$$

for all $k, i_0, i_1, \dots, i_{k-1}$, where $\theta_0 > 0$ is another uniform constant. This implies that *Lebesgue almost every point z has only finitely many close returns.* The property in (4.10) is a consequence, with $\nu =$ last close return, as we explained already. This finishes our discussion of the proof of Theorem 4.24.

Some of the questions in Section 4.2.3 have direct reformulations at the ergodic level. For instance,

Problem 4.26. Let $\mathcal{U}$ be an (open) subset of parameters for which the Hénon map $H_{a,b}$ has some topological attractor inside some forward invariant open set U. For Lebesgue almost all $(a, b) \in \mathcal{U}$ is there a finite number of SRB measures supported inside the attractor, such that their basins cover a full Lebesgue measure subset of U?

As we mentioned before, [447] proves that every continuous potential admits some equilibrium state supported inside the attractor.

Problem 4.27. Under which conditions is the equilibrium state unique?

Another interesting problem, suggested by the proof of the basin Theorem 4.24:

Problem 4.28. Give a complete combinatorial model for the dynamics in the basin of the strange attractor.

4.4 Decay of correlations and central limit theorem

An ergodic system (f, μ) has *exponential decay of correlations* in a space $\mathcal{H}$ of integrable functions if iterates $f^n(\varphi\mu)$ of an initial mass distribution $\varphi\mu$ converge exponentially fast to the equilibrium $(\int \varphi \, d\mu)\mu$, in the following sense: there exists $\lambda < 1$ such that

$$\left| \int \varphi(\psi \circ f^n) \, d\mu - \int \varphi \, d\mu \int \psi \, d\mu \right| \leq C(\varphi, \psi)\lambda^n$$

for every $\varphi \in \mathcal{H}$, $\psi \in L^\infty(\mu)$, and $n \geq 1$. In Appendix E we give an overview of methods and results in the general study of decay of correlations and related problems. Here we are going discuss the following result.

Theorem 4.29 (Benedicks, Young [57]). *Let Λ be the attractor of a Hénon-like map f and μ be its SRB measure. Then (f, μ) has exponential decay of correlations and satisfies the central limit theorem in the space of Hölder continuous functions.*

The last part of the statement means that for any Hölder function φ, the deviation of time averages from the mean value

$$\frac{1}{n} \sum_{j=0}^{n-1} \varphi \circ f^j - \int \varphi \, d\mu,$$

scaled by $n^{-1/2}$, converges in distribution to a Gaussian law $\mathcal{N}(0, \sigma)$: when $\sigma > 0$ this means that

$$\mu\left\{ x : \frac{1}{\sqrt{n}} \sum_{j=0}^{n-1} \left(\varphi(f^j(x)) - \int \varphi \, d\mu \right) \leq z \right\} \to \frac{1}{\sqrt{2\pi}\sigma} \int_{-\infty}^{z} e^{-t^2/(2\sigma^2)} \, dt$$

for any $z \in \mathbb{R}$ (when $\sigma = 0$ the limit is the Dirac distribution). See Appendix E.

The starting point in the strategy to prove Theorem 4.29 is to identify a set $\Delta_0 \subset \Lambda$ and a return map $f^R : \Delta_0 \to \Delta_0$, $x \mapsto f^{R(x)}(x)$, such that

- Δ_0 has hyperbolic product structure and its intersection with every unstable leaf is a positive Lebesgue measure subset of the leaf,

- f^R preserves this product structure and admits a Markov partition with countably many rectangles.

These notions will be explained below. From $f^R : \Delta_0 \to \Delta_0$ one constructs a Markov tower extension $F : \Delta \to \Delta$ for f over $\cup_{n \geq 0} f^n(\Delta_0)$ which is essentially a uniformly hyperbolic transformation (on a non-compact space). One constructs an SRB measure μ_F for F and proves that (F, μ_F) has exponential decay of correlations. The corresponding statement for (f, μ) is a fairly easy consequence. Moreover, these arguments provide an L^2 control on the correlation sequence as in (E.14), so that Gordin's Theorem E.9 may be applied to get the central limit property.

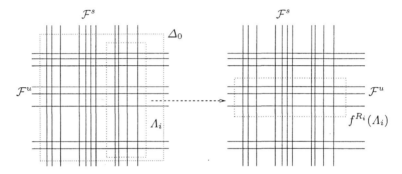

Fig. 4.7. Product structure and Markov property

Before giving a more detailed outline of the proof, let us mention that Young [460] has isolated a small set of properties under which these arguments go through, and observed that similar "horseshoes with positive measure" satisfying those properties can be found in other situations, so that exponential decay of correlations can then be proved along similar lines. See Appendix E.

We also mention that a different type of Markov extension, closer in spirit to the inducing method of Jakobson [227], has been introduced by Palis, Yoccoz [349] in a related situation, to study geometric and ergodic properties (including geometric invariant measures and speed of mixing) of non-uniformly hyperbolic horseshoes formed in the unfolding of homoclinic tangencies. See Section 4.6.

A "horseshoe" with positive measure:

Hyperbolic product structure (see Figure 4.7) means that Δ_0 coincides with the set of intersections between the leaves of a pair of transverse laminations, $\mathcal{F}^s$ and $\mathcal{F}^u$, such that

- the leaves of $\mathcal{F}^s$ are exponentially contracted by all forward iterates; analogously for $\mathcal{F}^u$ and backward iterates.

A subset of Δ_0 is an *s-subset* if it coincides with the union of entire sets $\gamma \cap \Delta_0$ over some subset of stable leaves $\gamma \in \mathcal{F}^s$. There is a dual notion of *u-subset*. By *Markov partition* we mean a family of pairwise disjoint s-subsets Λ_i covering almost all of Δ_0 in the sense that

$$m_\gamma\big(\Delta_0 \setminus \cup_i \Lambda_i\big) = 0 \quad \text{inside every unstable leaf } \gamma \in \mathcal{F}^u$$

(m_γ stands for arc-length), and satisfying also: there exist $R_i \in \mathbb{N}$ such that the return time $R(x) \equiv R_i$ on each Λ_i and the image $f^{R_i}(\Lambda_i)$ is a u-subset; the map f^{R_i} sends the stable leaves through Λ_i inside stable leaves, and its inverse has a dual property for the unstable leaves through the image. See Figure 4.7.

For the construction of Δ_0 (see [57, Section 3] for precise statements) one takes $\mathcal{F}^u$ consisting of escaping leaves (free segments of unstable manifold at distance $\approx \delta$ from their binding critical points), and obtains $\mathcal{F}^s$ through Lemma 4.23. That is, the orbits of points of Δ_0 approach the critical set at some bounded exponential rate:

$$\text{dist}(f^n(z), \zeta_n) \geq \rho^n \quad \text{at every free return,}$$

where ζ_n is the binding critical point and $\rho \gg b$. Notice that there is some resemblance to the construction of the X set in Section 4.3.1. One difference is that Δ_0 does not consist of entire unstable leaves, just positive measure subsets. Perhaps even more relevant, the return map f^R has a much more complicated definition presently and, in particular, it needs not be the *first* return map. However, it is important that it does share the fast returns property (iii) in Section 4.3.1:

- the return time R has exponential tail: there are $C > 0$ and $\lambda < 1$ such that

$$m_\gamma(\{x \in \gamma : R(x) > n\}) \leq C\lambda^n \quad \text{for every } n \geq 1 \text{ and } \gamma \in \mathcal{F}^u. \quad (4.12)$$

A few other properties follow from the construction and are needed for the sequel of the argument:

- backward iterates do not distort the Jacobian along unstable leaves $\gamma \in \mathcal{F}^u$ too much;
- the stable lamination $\mathcal{F}^s$ is absolutely continuous relative to the Lebesgue measure (arc-length) on unstable leaves;
- the greatest common divisor of the values taken by $R(x)$ is equal to 1.

The last property means that, if π denotes the projection along the stable lamination between two unstable leaves γ_1 and γ_2, then $\pi_* m_{\gamma_1}$ is absolutely continuous with respect to m_{γ_2}. A technical, yet relevant point, is that, rather than Riemannian distance, it is convenient to consider on each unstable leaf a dynamical distance $\beta^{s(x,y)}$ where $\beta < 1$ and $s(\cdot, \cdot)$ is the *separation time*:

$$s(x,y) = \min\{n \geq 0 : (f^R)^n(x) \text{ and } (f^R)^n(y) \text{ are in}$$

$$\text{different Markov rectangles}\}.$$

Markov extension and consequences:

The next step is to construct a Markov extension for f restricted to the invariant set $\Delta_* = \cup_{n \geq 0} f^n(\Delta_0)$. In general, by an *extension* of a given map $f : \Delta_* \to \Delta_*$ one means another map $F : \Delta \to \Delta$ together with a (surjective) projection $\pi : \Delta \to \Delta_*$ such that $\pi \circ F = f \circ \pi$. Benedicks, Young [57] propose a *tower extension*, that is, Δ is a disjoint union of copies of subsets of Δ_*, such that the return map to the ground level of the tower corresponds to the return map f^R. More precisely, they define

$$\Delta = \{(x, \ell) \in \Delta_0 \times \mathbb{N} : R(x) > \ell\} = \bigcup_{\ell=0}^{\infty} (\{x : R(x) > \ell\} \times \{\ell\}),$$

the projection $\pi : \Delta \to \Delta_*$ by $\pi(x, \ell) = f^\ell(x)$, and the map $F : \Delta \to \Delta$ by

$$F(x, \ell) = \begin{cases} (x, \ell + 1) & \text{if } R(x) > \ell + 1 \\ (f^R(x), 0) & \text{if } R(x) = \ell + 1 \end{cases}$$

The ground level is $\{(x, 0) : R(x) > 0\} = \Delta_0 \times \{0\}$. Observe that if f^R were the *first* return map to Δ_0 then π would be bijective, and so Δ would be naturally identified with Δ_*. This observation helps to understand why we need the extension in the first place: we want to be able to treat points $f^n(x)$ returning to Δ_0 prior to time $R(x)$ as "different points" (e.g. to consider different Riemannian structures on their tangent spaces and those of their iterates) even if they happen to coincide as points in $\Delta_* \subset M$. With this flexibility, one can introduce a Riemannian structure on Δ relative to which F is uniformly hyperbolic (with countably many Markov branches). The property of exponential decay of return times ensures that this structure has finite volume. Now classical arguments to prove exponential mixing (see Bowen [86]) may be adapted to F, as follows.

In simplified terms (see [57, Section 6], [460, Section 3], and Appendix E for more information), one considers the map $F^+ : \Delta^+ \to \Delta^+$ induced by F in the quotient space Δ^+ of Δ by the stable lamination. This F^+ is a uniformly expanding Markov map, and one proves that the corresponding transfer operator $\mathcal{L}$, relative to Lebesgue measure [2] on Δ^+, has a spectral gap. It follows that F^+ has an SRB measure μ^+ absolutely continuous with respect to Lebesgue measure, and the system (F^+, μ^+) has exponential decay of correlations in the space of Hölder functions. Then the same is true for $F : \Delta \to \Delta$, the corresponding SRB measure μ_F being absolutely continuous with respect to Lebesgue measure along unstable leaves. Then $\mu = \pi_* \mu_F$ is an f-invariant probability in Δ_* absolutely continuous with respect to Lebesgue measure along unstable manifolds. By uniqueness, μ must be the SRB measure of the Hénon-like map. This also proves that Δ_* is a full μ-measure subset of the

---

[2]There is a well defined notion of Lebesgue measure in the quotient space, up to equivalence, because the stable lamination is absolutely continuous.

attractor Λ. Now, every Hölder function $\varphi : \Delta_* \to \mathbb{R}$ lifts to a Hölder function on Δ, via $\varphi \mapsto \varphi \circ \pi$. It follows that the system (f, μ) has exponential decay of correlations in the space of Hölder functions, as claimed in Theorem 4.29.

4.5 Stochastic stability

Despite being metrically persistent in parameter space, Hénon-like attractors are very fragile under small perturbations of the map. A good measure of this is given by the result of Ures [434] stating that the parameter values exhibited in the proof of Theorem 4.3 are approximated by other parameters for which the map has a homoclinic tangency associated to the fixed point P. Consequently, under arbitrarily small perturbations one may cause the closure of $W^u(P)$ to contain infinitely many periodic attractors!

Quite in contrast, Theorem 4.30 below asserts that these systems are remarkably stable under small random noise. Let us explain this first in intuitive terms. Consider a point z close to the attractor, and suppose as we iterate it a mistake is made at each step: one uses some nearby diffeomorphism f_j in the place of f. Thus, instead of the orbit $f^n(z)$ one obtains a *random orbit*

$$z, \quad f_1(z), \quad (f_2 \circ f_1)(z), \quad \ldots, \quad (f_n \circ \cdots \circ f_1)(z), \quad \ldots.$$

The theorem says that, assuming the mistakes are small and independent, the statistical behavior of typical random orbits (that is, for almost all choices of f_j and almost every z) is essentially the same as that of typical genuine orbits: for any continuous function φ the time average

$$\lim_{n \to \infty} \frac{1}{n} \sum_{j=0}^{n-1} \varphi\big((f_j \circ \cdots \circ f_1)(z)\big) \quad \text{exists and is close to} \quad \int \varphi \, d\mu. \qquad (4.13)$$

In a more pictorial language: a computer screen plot of a typical random orbit looks essentially the same [3] as that of a typical orbit of the original map f. Next we give the precise statement. Appendix D contains general background information on random perturbations of dynamical systems.

Let us fix a bounded open neighborhood U of the attractor Λ, contained in the basin of attraction $B(\Lambda)$ and such that the closure of $f(U)$ is contained in U. We consider random perturbations scheme $\{\nu_\varepsilon : \varepsilon > 0\}$ where each ν_ε is supported in the ε-neighborhood of f relative to the C^2 topology. This means that we consider iterates

$$U \ni z \mapsto (f_n \circ \cdots \circ f_1)(z)$$

---

[3]This is sometimes seen as evidence that computer pictures of Hénon attractors are accurate, despite calculation errors and the chaotic character of such systems (but round-off errors and the like are not strictly random).

where each f_j is picked at random, independently of all previous choices, according to the probability distribution ν_ε. Throughout, ε will be smaller than the distance from $f(U)$ to the complement of U, so that every random orbit starting in U remains there for all times. The process is described by a Markov chain in U, with *transition probabilities* $p_\varepsilon(\,\cdot\mid z)$ defined by

$$p_\varepsilon(E\mid z) = \nu_\varepsilon\big(\{g : g(z) \in E\}\big)$$

for every $z \in U$ and any measurable set $E \subset U$. A probability measure μ_ε is *stationary* if, for every measurable set $E \subset U$,

$$\mu_\varepsilon(E) = \int p_\varepsilon(E\mid z)\, d\mu_\varepsilon(z).$$

Here we assume that the transition probabilities are uniformly distributed on the ball of radius ε around $f(z)$, up to constant factors. More precisely, they are absolutely continuous with respect to Lebesgue measure m on the ball of radius ε around $f(z)$, and there are $c_2 > c_1 > 0$ such that

$$c_1\, \varepsilon^{-2} \le \frac{dp_\varepsilon(\cdot, z)}{dm} \le c_2\, \varepsilon^{-2} \quad \text{for all } z \text{ and } \varepsilon \tag{4.14}$$

(in dimension d, replace ε^{-2} by ε^{-d}). In fact, the next theorem is valid for much more general noise (see [53, Section 1.5]).

Theorem 4.30 (Benedicks, Viana [53]). *Let f be a Hénon-like map, μ be its SRB measure, and $\{\nu_\varepsilon : \varepsilon > 0\}$ be a random perturbations scheme whose transition probabilities satisfy (4.14). Then there is a unique stationary measure μ_ε supported in the basin $B(\Lambda)$, and it is ergodic. Moreover, as $\varepsilon \to 0$ the measure μ_ε converges to μ in the weak* topology: $\int \varphi\, d\mu_\varepsilon \to \int \varphi\, d\mu$ for every continuous $\varphi : U \to \mathbb{R}$.*

Uniqueness also implies that, for any continuous function $\varphi : U \to \mathbb{R}$ and almost every random orbit,

$$\lim_{n\to\infty} \frac{1}{n} \sum_{j=0}^{n-1} \varphi\big((f_j \circ \cdots \circ f_1)(z)\big) = \int \varphi\, d\mu_\varepsilon\,.$$

So, the conclusion of the theorem does give (4.13) for small ε. In what follows we outline the main steps in the proof of Theorem 4.30.

An upper bound on random iterates of measures:

The claim that the stationary measure μ_ε is unique is relatively easy, relying on the fact that $f \mid \Lambda$ is transitive and the transition probabilities $p_\varepsilon(\cdot \mid z)$ are supported on a whole neighborhood of the corresponding $f(z)$. In order to prove the main statement, $\mu_\varepsilon \to \mu$ as $\varepsilon \to 0$, let us introduce the *random transfer operator* $\mathcal{T}_\varepsilon$, defined in the space of probabilities in U by

$$\mathcal{T}_\varepsilon(\eta)(E) = \int p_\varepsilon(E \mid z)\, d\eta(z) = \int \eta(g^{-1}(E))\, d\nu_\varepsilon(g).$$

A measure is stationary if and only it is fixed by $\mathcal{T}_\varepsilon$. Moreover, every accumulation point of the sequence $n^{-1} \sum_{j=0}^{n-1} \mathcal{T}_\varepsilon^j(\eta)$ is a stationary measure, for any initial measure η. So, since the space of probabilities is compact for the weak* topology, uniqueness of the stationary measure proves that

$$\mu_{\varepsilon,n} = \frac{1}{n} \sum_{j=0}^{n-1} \mathcal{T}_\varepsilon^j(\eta) \tag{4.15}$$

converges to μ_ε as $n \to \infty$, regardless of the initial measure η one takes. Fix $\eta = $ normalized arc-length on some segment of $W^u(P)$.

The main technical step in the proof is the following upper bound:

$$\mu_{\varepsilon,n} \leq \sum_{s=0}^{\infty} f_*^s(\lambda_{\varepsilon,n} \mid \{e(\cdot) > s\}) + M_{\varepsilon,n,N} + R_{\varepsilon,N}(\cdot). \tag{4.16}$$

for every $\varepsilon > 0$, $n \geq 1$, $N \geq 1$, where $e(\cdot)$ denotes the escape time for the unperturbed map f (recall the explanations following (4.9) above, and see also [53, Section 4.3] for a formal definition) and

- the $\lambda_{\varepsilon,n}$ are measures on the X set (Section 4.3.1), absolutely continuous on unstable leaves and with density and total variation bounded by C;
- the $M_{\varepsilon,n,N}$ are measures on the attractor Λ, with total variation bounded by Ce^{-cN};
- the $R_{\varepsilon,N}$ are positive functionals in the space of C^1 functions which, for each fixed $N \geq 1$, converge to zero pointwise when $\varepsilon \to 0$.

Here c and C denote various constants, respectively small and large, independent of ε, n, N. Given measures α and β and a positive functional $r(\cdot)$ in the space of C^1 functions, we write $\alpha \leq \beta + r(\cdot)$ to mean that there exists a third measure γ such that $\gamma(E) \leq \beta(E)$ for all measurable set E and

$$\left| \int \varphi\, d\alpha - \int \varphi\, d\gamma \right| \leq r(\varphi) \quad \text{for every } \varphi \in C^1.$$

A simple example that illustrates the way we use this terminology: Let β be normalized arc-length on some line segment ℓ, and α be normalized area on the rectangle whose larger axis is ℓ and with smaller axis of radius ε. Then $\alpha \leq \beta + r_\varepsilon(\cdot)$ where $r_\varepsilon(\cdot)$ converges pointwise to zero when $\varepsilon \to 0$.

Shadowing of random orbits:

Let us explain briefly where (4.16) comes from. The basic idea is to introduce a notion of escape time for random orbits (see [53, Section 5.7]) and to split the expression (4.15) into three parts:

(i) The mass carried by random orbits $(z, f_1, \ldots, f_j)$ at escape times.

One key ingredient in the whole proof is to show that such orbits are shadowed by genuine orbits in the attractor which are also at escape times. Thus, one proves that this part of $\mu_{\varepsilon,n}$ is bounded by

$$\lambda_{\varepsilon,n} + r_\varepsilon(\cdot)$$

for some $\lambda_{\varepsilon,n}$ absolutely continuous along escaping leaves and positive functional $r_\varepsilon(\cdot)$ that converges pointwise to zero as $\varepsilon \to 0$. Notice that $\lambda_{\varepsilon,n}$ is supported inside Λ which, of course, is not the case for $\mu_{\varepsilon,n}$. The functional is there, precisely, to take care of that: the fact that $r_\varepsilon(\varphi)$ converges to zero, for every C^1 function φ, depends on the fact that the support of $\mu_{\varepsilon,n}$ converges to Λ (uniformly in n) when $\varepsilon \to 0$.

(ii) The mass carried by random iterates in between escape times, with a cut-off N from the previous escape time.

Since all the maps are ε-close to f, from (i) and a continuity argument we get that the sth iterate is bounded by

$$f_*^s\big(\lambda_{\varepsilon,n} \mid \{e(\cdot) > s\} + r_\varepsilon\big) + \rho_{\varepsilon,s}(\cdot) \le f_*^s\big(\lambda_{\varepsilon,n} \mid \{e(\cdot) > s\}\big) + r_{\varepsilon,s}$$

where, for each fixed $s \ge 1$, the functionals $\rho_{\varepsilon,s}$ and $r_{\varepsilon,s}$ converge pointwise to zero as $\varepsilon \to 0$.

(iii) The mass carried by random orbits at iterates $> N$ from the last previous escape time.

Another crucial fact is that the total mass of random orbits with escape time larger than N decays exponentially. So this remainder term in $\mu_{\varepsilon,N}$ has total mass less than Ce^{-cn}. Putting this together with the previous estimates one gets (4.16).

Proving stochastic stability:

Now we explain how Theorem 4.30 can be deduced from (4.16). Firstly, making $n \to \infty$ along a suitable subsequence, we get that $\mu_{\varepsilon,n}$ accumulates on the unique stationary measure μ_ε and

- $\lambda_{\varepsilon,n}$ accumulates on some measure λ_ε on X, absolutely continuous along unstable leaves with density and total bounded by some $C > 0$
- $M_{\varepsilon,n,N}$ accumulates on some measure $M_{\varepsilon,N}$ with total mass bounded by Ce^{-cN}.

Therefore, (4.16) leads to

$$\mu_\varepsilon \le \sum_{s=0}^{\infty} f_*^s \lambda_\varepsilon + M_{\varepsilon,N} + R_{N,\varepsilon}(\cdot) \quad \text{for all } \varepsilon > 0 \text{ and } N \ge 1.$$

Keeping N fixed and making $\varepsilon \to 0$ along a suitable subsequence of any given sequence, we get that μ_ε accumulates on some measure μ_0 and

- λ_ε accumulates on some measure λ on X, absolutely continuous along unstable leaves with density and total mass bounded by C;
- $M_{\varepsilon,N}$ accumulates on some measure M_N whose total mass is less than Ce^{-cN}.

We also have $R_{N,\varepsilon}(\cdot) \to 0$, pointwise. Therefore, the previous inequality gives

$$\mu_0 \leq \sum_{s=0}^{\infty} f_*^s(\lambda \mid \{e(\cdot) > s\}) + M_N \quad \text{for all } N \geq 1.$$

The limit measure μ_0 is necessarily invariant, see [242, Theorem 1.1]. Finally, making N go to infinity, we obtain that

$$\mu_0 \leq \lambda_0 \quad \text{where} \quad \lambda_0 = \sum_{s=0}^{\infty} f_*^s(\lambda \mid \{e(\cdot) > s\}).$$

The measure λ_0 needs not be invariant, of course. On the other hand, it is absolutely continuous along unstable manifolds in the attractor Λ (because λ is). It follows that the f-invariant measure μ_0 is absolutely continuous along unstable manifolds of Λ. Since there exists a unique such probability measure [57], this μ_0 must coincide with the SRB measure μ of the attractor. This finishes our outline of the proof of Theorem 4.30.

As in most other situations where stochastic stability could be established, the previous arguments require much more detailed information on the dynamics of the unperturbed system than seems reasonable in view of the conclusion. At this point one may hope to obtain much more abstract statements of the following form:

Problem 4.31. Is every Hénon-like attractor supporting a unique SRB measure, hyperbolic and with the no-holes property, always stochastic stable?

4.6 Chaotic dynamics near homoclinic tangencies

We conclude this chapter with a brief discussion on the occurrence of chaotic behavior and Hénon-like attractors within certain bifurcation mechanisms.

4.6.1 Tangencies and strange attractors

The behavior of Hénon-like maps is very much characterized by the presence of critical regions hosting homoclinic phenomena, as we have seen. The next result means that a kind of converse is also true: the unfolding of homoclinic tangencies is also accompanied by the creation of Hénon-like attractors. A periodic point is called *sectionally dissipative* if the product of any pair of eigenvalues is smaller than 1 in absolute value. This implies that the unstable manifold is 1-dimensional.

Theorem 4.32 (Mora, Viana [297, 438]). *For an open dense subset of C^3 families $(g_\mu)_\mu$ of surface diffeomorphisms going through a homoclinic tangency at $\mu = 0$, there exists a positive Lebesgue measure set of parameters μ close to zero for which g_μ has some Hénon-like attractor or repeller. The same is true in arbitrary dimension, assuming the periodic saddle involved in the tangency is sectionally dissipative.*

This is a corollary of Theorem 4.4 together with the observation that *any parametrized family of diffeomorphisms going through a homoclinic tangency associated to a dissipative saddle point "contains" a sequence of Hénon-like families*. More precisely, return maps of g_μ to appropriate domains close to the tangency, and for certain parameter intervals close to $\mu = 0$, are conjugate to Hénon-like maps as in (4.2). Let us explain this in some detail.

Let $g_\mu : M \to M$, $\mu \in \mathbb{R}$, be a smooth family of diffeomorphisms such that g_0 has a homoclinic tangency associated to some fixed (or periodic) saddle point p. We assume the family to be generic, in the sense that the tangency is quadratic and is generically unfolded as the parameter μ varies. For the time being we take M to be a surface and g_0 to be area-dissipative at p, meaning that $|\det Dg_0(p)|$ is strictly smaller than 1, but we shall comment on the general case near the end of this section.

Theorem 4.33. *There is $l \geq 1$ and for each large n there exist*

- *an affine reparametrization $\mu_n : [1/4, 4] \to \mathbb{R}$ and*
- *C^r systems of local coordinates $\theta_{n,a} : [-4, 4]^2 \to M$, $1/4 \leq a \leq 4$*

such that $\mu_n([1/4, 4]) \to 0$ and $\theta_{n,a}([-4, 4]^2)$ converges to the homoclinic tangency, and the sequence $f_{n,a}(x, y) = \theta_{n,a}^{-1} \circ g_{\mu_n(a)}^{n+l} \circ \theta_{n,a}(x, y)$ converges to the map $\psi_a : (x, y) \mapsto (1 - ax^2, 0)$ in the C^r topology as $n \to +\infty$.

Here $r \geq 2$ is a fixed integer and convergence is meant in the strongest sense: as functions of (a, x, y), uniformly on $[1/4, 4] \times [-4, 4]^2$. The choice of this domain is rather arbitrary: uniform C^r convergence holds on any compact subset of $\mathbb{R}^+ \times \mathbb{R}^2$. Figure 4.8 helps understand the contents of this result: the image of $\theta_{n,a}$ is a small domain close to the point of tangency, that returns (close) to itself after $l + n$ iterates, at least for the parameter values in the image of μ_n. Up to n-dependent blowing ups for the domain and the parameter range, given by $\theta_{n,a}$ and μ_n, the family of return maps converges to the quadratic map as $n \to \infty$.

Let us describe the definition of μ_n and $\theta_{n,a}$. For technical simplicity we assume that there exist C^r μ-dependent coordinates (ξ, η) on a neighborhood U of p linearizing g_μ for μ small (this is a generic condition, and [387] shows how it can be removed):

$$g_\mu(\xi, \eta) = (\sigma\xi, \lambda\eta)$$

where $\sigma = \sigma_\mu$ and $\lambda = \lambda_\mu$ satisfy $|\sigma| > 1 > |\lambda|$. The dissipativeness assumption means that $0 < |\sigma\lambda| < 1$ for μ close to zero. Up to a convenient rescaling

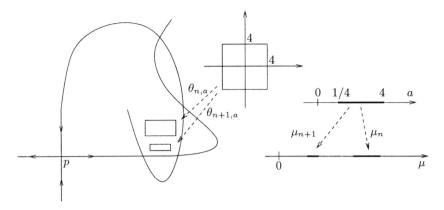

Fig. 4.8. Renormalization in homoclinic tangencies

of the coordinates, we may take the set $\{(\xi, \eta) : \|(\xi, \eta)\| \leq 2\}$ to be contained in U and $q = (1,0)$ to be a point in the orbit of tangency. Fix $l \geq 1$ such that $g_0^l(q) = (0, \eta_0) \in U$ and write $g_\mu^l(\xi, \eta)$ as

$$(\alpha(\xi - 1)^2 + \beta(\xi - 1)\mu + \gamma\mu^2 + b\eta + v\mu + r, \, \eta_0 + c(\xi - 1) + d\eta + w\mu + s)$$

where $\alpha, \beta, \gamma, b, c, d, v, w \in \mathbb{R}$ and $r = r(\mu, \xi, \eta)$, $s = s(\mu, \xi, \eta)$ are such that

$$r, s, Dr, Ds, \partial_{\xi\xi}r, \partial_{\mu\xi}r, \partial_{\mu\mu}r \quad \text{all vanish at } (0, 1, 0). \tag{4.17}$$

The hypotheses of nondegeneracy and generic unfolding of the tangency amount to having $\alpha \neq 0$ and $v \neq 0$ and, up to reparametrizing $(g_\mu)_\mu$, we may suppose $v = 1$. First we consider the n-dependent reparametrization

$$\nu = \nu_n(\mu) = \sigma^{2n}\mu + b\sigma^{2n}\lambda^n\eta_0 - \sigma^n. \tag{4.18}$$

It is easy to check that, given any large constant A, for n sufficiently large ν_n maps a small interval I_n close to $\mu = 0$ diffeomorphically onto $[-A, A]$. We let $\tilde{\mu}_n = (\nu_n|I_n)^{-1}$; in what follows we always take $\mu = \tilde{\mu}_n(\nu)$. Then we introduce (n, ν)-dependent coordinates $(\tilde{x}, \tilde{y})$ given by

$$(\xi, \eta) = \tilde{\theta}_{n,\nu}(\tilde{x}, \tilde{y}) = (1 + \sigma^{-n}\tilde{x}, \lambda^n\eta_0 + \sigma^{-2n}\rho^n\tilde{y}), \tag{4.19}$$

with $\rho = \sqrt{|\sigma\lambda|}$. Now we denote $\tilde{f}_{n,\nu} = \tilde{\theta}_{n,\nu}^{-1} \circ g_\mu^{n+l} \circ \tilde{\theta}_{n,\nu}$. A straightforward calculation gives

$$\tilde{f}_{n,\nu}(\tilde{x}, \tilde{y}) = (\alpha\tilde{x}^2 + \beta\tilde{x}(\sigma^n\mu) + \gamma(\sigma^n\mu)^2 + \nu + b\rho^n\tilde{y} + \sigma^{2n}r,$$
$$\pm c\rho^n\tilde{x} + d\rho^{3n}\eta_0 + d\lambda^n\tilde{y} \pm w\rho^n(\sigma^n\mu) \pm \rho^n\sigma^n s)$$

where $\pm$ is the sign of $\sigma\lambda$ and r, s are taken at $(\mu, \xi, \eta) = (\tilde{\mu}_n(\nu), \tilde{\theta}_{n,\mu}(\tilde{x}, \tilde{y}))$. Note that $\rho^n \to 0$ and

$$\sigma^n \mu = (1 + \sigma^{-n}\nu - b\sigma^n \lambda^n \eta_0) \to 1.$$

as $n \to +\infty$. On the other hand, $|\sigma^{2n}r|$, $|\rho^n \sigma^n s|$ also converge to zero, as a consequence of (4.17) and, recall (4.18) and (4.19),

$$|\mu| \leq C|\sigma|^{-n}, \quad |\xi - 1| \leq C|\sigma|^{-n} \quad \text{and} \quad |\eta| \leq C|\lambda|^n.$$

This proves that

$$\tilde{f}_{n,\nu}(\tilde{x}, \tilde{y}) \to \tilde{\psi}_\nu(\tilde{x}, \tilde{y}) = (\alpha\tilde{x}^2 + \beta\tilde{x} + \gamma + \nu, 0)$$

as $n \to +\infty$, uniformly on $[-A, A] \times [-A, A]^2$. Moreover, as the reader may check, the same type of estimations apply to all derivatives up to order r, proving that this convergence holds in the C^r topology.

Now, the affine diffeomorphism $h_a : (x, y) \mapsto (-ax/\alpha - \beta/(2\alpha), y)$ conjugates $\tilde{\psi}_\nu$ to $\psi_a : (x, y) \mapsto (1 - ax^2, 0)$, for

$$\nu = \hat{\nu}(a) = -a/\alpha - \beta/(2\alpha) + \beta^2/(4\alpha) - \gamma.$$

Thus, we may take $\mu_n(a) = \tilde{\mu}_n \circ \hat{\nu}$ and $\theta_{n,a} = \tilde{\theta}_{n,\hat{\nu}(a)} \circ h_a$. Clearly, the domain of definition contains $[1/4, 4] \times [-4, 4]^2$, if A is taken large enough.

Let us mention some extensions of this procedure. If $|\det Dg_0(p)| > 1$ then, obviously, g_0^{-1} is area-dissipative at p and so we get the same conclusions as before, just by iterating backwards. The area-conservative case $|\det Dg_\mu| \equiv 1$ is particularly interesting: a reasoning similar to the one above may still be applied, and the limit is the *conservative Hénon family* $\psi_a(x, y) = (1 - ax^2 + y, \pm x)$, see [171, 387]. Moreover, the arguments above also extend in a natural way to higher dimensions when g_0 is sectionally dissipative at p. In this case we get $\psi_a(x_1, x_2, \ldots, x_m) = (1 - ax_1^2, 0, \ldots, 0)$ as limit model, see [347, 438]. In fact, a more sophisticated restatement of this scheme is valid in any dimension for generic families unfolding a homoclinic tangency (with no dissipativeness conditions), see [387].

Using Theorem 4.32, Fornaess, Gavosto [184] concluded that the Hénon family (4.1) contains Hénon-like attractors for parameters close to the values $a = 1.4$ and $b = 0.3$ initially considered in [210]. These attractors have a local character, in the sense that they are fixed by some large iterate of the map. Hénon's original assertion remains open:

Problem 4.34. Is there a global strange attractor (containing the unstable manifold of the fixed point P) in the Hénon family, for parameter values close to $a = 1.4$ and $b = 0.3$?

4.6.2 Saddle-node cycles and strange attractors

Critical saddle-node cycles [325] are another main global bifurcation mechanism, closely related to homoclinic tangencies. By a *saddle-node cycle* of

a diffeomorphism we mean a finite family of periodic points p_0, p_1, ...,
$p_{\ell-1}$, $p_\ell = p_0$ such that p_0 is a saddle-node, p_i is a hyperbolic saddle for
$i = 1, \ldots, \ell-1$, and every $W^u(p_{i-1})$ has a transverse intersection with $W^s(p_i)$.
It is assumed that the saddle-node is sectionally dissipative, that is, all hyperbolic eigenvalues are less than 1 in absolute value. Then, its stable set
is a manifold with boundary (the strong-stable manifold of the saddle-node),
with the same dimension as the ambient space, and it admits a unique strong-
stable foliation of codimension-1 having the strong-stable manifold as one of
the leaves. The cycle is called *critical* if the unstable manifold of $W^u(p_{\ell-1})$ is
tangent to the strong-stable foliation of p_0 at some point (then the same is
true for $W^u(p_0)$). Figure 4.9 describes the formation of a 1-cycle, through the
collision of a saddle and a sink.

Fig. 4.9. A saddle-node cycle at the boundary of Morse-Smale diffeomorphisms

Building on Theorem 4.32, Díaz, Rocha, Viana [159] proved that Hénon-
like strange attractors are always a *prevalent* phenomenon in the generic unfolding of critical saddle-node cycles: this bifurcation always includes the unfolding of homoclinic tangencies and, hence, the creation of Hénon-like attractors, the key novelty being that these attractors occur for a set $\mathcal{S}$ of parameters
with positive Lebesgue *density* at the saddle-node cycle bifurcation. We shall
see in the next section that this is not true for homoclinic tangencies.

In the case of 1-cycles the results of [159] take a more global form: for
an open set of cases, one can give an explicit characterization of the strange
attractor, that ensures that its basin contains a fixed neighborhood of the
cycle:

Theorem 4.35 (Díaz, Rocha, Viana [159]). *There exists an open set of
1-parameter families $(g_\mu)_\mu$ unfolding a critical saddle-node 1-cycle at $\mu = 0$
such that there exists a set $\mathcal{S}$ of parameters with*

$$\liminf_{\varepsilon \to 0} \frac{m(\mathcal{S} \cap [0, \varepsilon])}{\varepsilon} > 0$$

and a neighborhood U of $W^u(p_0)$ such that the maximal invariant set $\Lambda_\mu = \bigcap_{n=0}^\infty g_\mu^n(U)$ is a Hénon-like attractor for every $\mu \in \mathcal{S}$.

We just make a few informal comments on the proof, to explain how prevalence comes about. As part of the definition of the open set in the statement, one requires that $W^u(p_0) \subset W^s(p_0)$. The starting point is to define a partition $I_k = (\mu_{k+1}, \mu_k]$ of some interval $(0, \varepsilon]$ in parameter space, a domain C in the ambient manifold, and a return map $R_\mu : C \to C$, given by some iterate g_μ^l with $l \approx k$ for all $\mu \in I_k$. This domain is bounded by two strong-stable leaves such that one is mapped to the other; by identifying points in the same orbit inside these leaves, the domain C becomes a cylinder and R_μ becomes a smooth map. Most iterates involved in R_μ take place near the saddle-node, and so this return map is strongly volume dissipative. In fact, up to a convenient k-dependent reparametrization $\mu = \psi_k(a)$, $a \in [0, 1)$, the families R_μ, $\mu \in I_k$, converge to a family of maps of the cylinder $C \approx S^1 \times B$, of the form

$$\Psi_a : S^1 \times B \to S^1 \times B, \quad \Psi_a(\theta, y) = (\phi(\theta) + a, 0),$$

where B is a ball in Euclidean space and $\phi : S^1 \to S^1$. The definition of the open set in the statement involves, mostly, conditions on the circle map ϕ to ensure that these $R_{\psi_k(a)}$ are Hénon-like families in the sense of Section 4.1. Now prevalence follows from two main observations. Firstly, there is a uniform lower bound for the Lebesgue measure of the sets of parameters a corresponding to Hénon-like attractors in each of these families: roughly, it corresponds to the measure of the set of parameters a for which $\theta \mapsto \phi(\theta) + a$ has chaotic behavior. Secondly, the parametrizations ψ_k also have uniformly bounded distortion. It follows that Hénon-like attractors correspond to a definite fraction of parameters μ in each I_k, and that implies the positive density claim.

In general, if the unstable manifold of p_0 is not contained in $W^s(p_0)$, it is not possible to find a convenient smooth return map R_μ as above. In fact, this is always the case if $\ell > 1$. However, Costa [136, 137] has been able to prove an analogue of Theorem 4.35 for *saddle-node horseshoes*, that is, saddle-node cycles where the saddle-node is part of the homoclinic class of some hyperbolic saddle. We refer the reader to [154] for a survey of this and other recent results on this topic.

4.6.3 Tangencies and non-uniform hyperbolicity

Concerning prevalence of strange attractors, the situation should be very different for homoclinic tangencies:

Problem 4.36. Let $(g_\mu)_\mu$ be a generic one-parameter family of diffeomorphisms unfolding a homoclinic tangency, as in Theorem 4.32. Does the set $\mathcal{S}$ of parameter values corresponding to Hénon-like attractors or repellers have zero Lebesgue density at $\mu = 0$?

The density does vanish in the context of Theorem 3.13 (Hausdorff dimension less than 1): the set $\mathcal{H}$ of parameters corresponding to hyperbolicity has full Lebesgue density at $\mu = 0$, and the two sets $\mathcal{S}$ and $\mathcal{H}$ are clearly disjoint.

A recent extension due to Palis, Yoccoz [349, 350] suggests that the answer to Problem 4.36 might be always positive. The setting is as in Section 3.3. Let $(g_\mu)_\mu$ unfold a homoclinic tangency associated to a periodic point p contained in a hyperbolic basic set H of g_0 (see Figure 1.2). Let

$$d_s(H) = \mathrm{HD}\big(W^u_{loc}(p) \cap H\big) \quad \text{and} \quad d_u(H) = \mathrm{HD}\big(W^s_{loc}(p) \cap H\big)$$

be the stable and unstable dimensions of H, as defined in (3.3). Palis, Yoccoz show that if the Hausdorff dimension $\mathrm{HD}(H) = d_s(H) + d_u(H)$ is not much larger than 1 then, for most parameter values close to $\mu = 0$, the maximal invariant set Σ_μ in a neighborhood of H union the orbit of tangency is a "non-uniformly hyperbolic horseshoe". We just quote the following consequence of their, much more detailed, statement:

Theorem 4.37 (Palis, Yoccoz [349, 350]). *There exists a neighborhood $\mathcal{D}$ of the set $\{(d_s, d_u) : d_s + d_u \leq 1\}$ inside $(0,1)^2$ such that if $(d_s(H), d_u(H)) \in \mathcal{D}$ then there exists a set $\mathcal{W}$ of parameters with full Lebesgue density at $\mu = 0$ such that for every $\mu \in \mathcal{W}$ the stable set and the unstable set of Σ_μ have zero Lebesgue measure in M and, consequently, Σ_μ contains neither attractors nor repellers.*

The strategy, inspired from Benedicks, Carleson [52], is to show that for most parameters μ close to zero the set Σ_μ is hyperbolic, in a delicate non-uniform sense. Essentially, although Σ_μ may contain tangencies, these correspond to very special points: at the majority of points there are transverse directions which are asymptotically contracted by forward and backward iterates, respectively. As in the Hénon-like case, the proof requires a careful analysis of how trajectories return close to the tangencies and, most importantly, a precise definition of what a "tangency" ("critical point") is. To ensure hyperbolicity, returns should not be too frequent nor too close. This is achieved through parameter exclusions. These exclusions are less and less significant near the original tangency parameter, and that is how full Lebesgue density of non-excluded parameters arises.

The rate of formation of tangencies as one iterates is a crucial ingredient in the exclusions estimates, and is closely related to the Hausdorff dimension of the original horseshoe. The assumption that this dimension is not much bigger than 1 ensures that the number of tangencies that must be considered at each stage grows fairly slowly (exponential growth with exponent not too big), so that a fairly small fraction of parameters needs to be excluded each time. Returns close to the tangencies yield quadratic type folds. The condition on the frequency and depth of returns is used to ensure that the folds always are "ironed-out" before a new very close return occurs. In this way, one never has to deal with folds of order larger than 2.

One can certainly expect to relax the hypothesis on the Hausdorff dimension, at the price of having to deal with higher order folds, and it may even be that the conclusion remain true in all cases:

Conjecture 4.38. The conclusion of Theorem 4.37 remains true for generic unfoldings of homoclinic tangencies by surface diffeomorphisms (with no assumption on the Hausdorff dimension).

On the horizon lies the case of area-preserving maps, like the conservative Hénon maps ($b = \pm 1$), or the family of standard maps on $\mathbb{T}^2$

$$g_\kappa(x, y) = (-y + 2x + \kappa \sin(2\pi x), x), \qquad (4.20)$$

for which the limit set (the whole ambient space) has dimension 2. The main dynamical issue in these maps is, once more, recurrence of criticalities. In fact, in this case one will probably have to deal with folds of all orders simultaneously. Notice that for $\kappa = 0$ the standard map is an integrable twist map: the circles in the direction $(1, 1)$ are invariant and rotated by the map, with rotation angles varying monotonically with the circle. At this point the topological entropy vanishes and, consequently, so does the entropy relative to Lebesgue measure. Moreover, g_κ has elliptic islands for all small κ. The most important problem in this domain is

Problem 4.39. Is there a positive Lebesgue measure set of values of κ for which the standard map g_κ (i) has positive entropy [4] relative to Lebesgue measure? (ii) has no Kolmogorov-Arnold-Moser elliptic islands? (iii) is ergodic and non-uniformly hyperbolic (non-zero Lyapunov exponents)?

In the direction of this problem it should be useful to handle first the following simplified model, inspired from [52] and [440], that we are going to describe. In order to motivate this model, let us informally recall a bit of the previous discussion. We have seen that control of the recurrence of tangencies (critical orbits) through parameter exclusions very much depends on the growth of the number of critical points one has to consider at each scale. The latter is related to fractal dimensions of the limit set, which is, itself related to the conservative/dissipative character of the dynamics: for strongly dissipative Hénon-like maps the dimension is $1 + \varepsilon$, whereas for the conservative standard map it is 2. Our model lives on the product of an interval by a Cantor set. Geometric aspects of the dynamics correspond to the interval direction. The fractal dimension of the Cantor set is a free parameter, that allow us to interpolate between strongly dissipative and conservative cases.

More precisely, let $M = [-1, 1] \times K(b)$ where $K(b)$ is the mid-$(1-b)$ Cantor set in $[0, 1]$. Let $a(\kappa) = a + \phi(\kappa)$ where a is a parameter and ϕ is some small Lipschitz function (e.g. $\phi \equiv 0$). Consider the transformation $f : M \to M$ given by

$$f(x, \kappa) = \left(1 - a(\kappa)x^2, \kappa_1\right), \quad \kappa_1 = \begin{cases} b\kappa/2 & \text{if } x < 0 \\ 1 - b\kappa/2 & \text{if } x \geq 0. \end{cases}$$

---

[4]The topological entropy is known to grow as $C\kappa^{1/3}$. See [171], for instance.

That is, f sends every half horizontal $[-1,0) \times \{\kappa\}$ and $[0,1] \times \{\kappa\}$ inside some horizontal $[-1,1] \times \{\kappa_1\}$ in a quadratic fashion, with a critical point at $x = 0$. The points $(1, \kappa)$, $\kappa \in K(b)$ play the role of critical values. Let f' denote the derivative along the horizontal. Let θ_b be the uniformly distributed probability on $K(b)$.

Problem 4.40. For any fixed $b \in (0,1)$, is there a positive Lebesgue measure set of values of a such that the derivative $(f^n)'(1, \kappa)$ grows exponentially with n for θ_b-almost every $\kappa \in K_b$? Is the growth rate uniform?

The answer is positive when $b \approx 0$, by a much simplified version of the arguments in [52].

5

Non-Critical Dynamics and Hyperbolicity

As we have seen in the previous chapters, much of the richness of the dynamics of surface diffeomorphisms is due to phenomena associated to the unfolding of homoclinic tangencies, as in the Hénon family. Moreover, these phenomena have many analogies with critical endomorphisms of the interval or the circle. For instance, homoclinic tangencies correspond in the one-dimensional setting to pre-periodic critical points.

For non-critical maps in dimension 1 which are sufficiently regular, one has good control of the distortion and, thus, a lot of rigidity: Theorem 2.5 asserts that non-critical maps are almost hyperbolic. Here we discuss some remarkable recent results of Pujals, Sambarino [371, 372, 370] extending this conclusion to *non-critical* surface diffeomorphisms, that is, such that one cannot create homoclinic tangencies by small C^1 perturbations of the diffeomorphism.

In this two-dimensional setting, the main tool involved in avoiding tangencies is the concept of dominated splitting of the tangent bundle. One key result states that an invariant compact set admitting a dominated splitting and whose periodic orbits are all hyperbolic coincides with the union of a hyperbolic set with finitely many periodic normally hyperbolic circles on which the dynamics is an irrational rotation. See Section 5.1 for the precise statement and Section 5.2 for an outline of the proof. The main result is deduced in Section 5.3, and Section 5.4 presents another very nice consequence: variation of entropy is always accompanied by homoclinic tangencies.

Finally, in Section 5.5 we discuss a few results and conjectures concerning non-critical behavior in higher dimensions.

5.1 Non-critical surface dynamics

Let M be a closed Riemannian surface, $f\colon M \to M$ be a diffeomorphism, and Λ be an f-invariant set. A Df-invariant splitting $T_x M = E(x) \oplus F(x)$, $x \in \Lambda$ of the tangent bundle over Λ is called *dominated* if there is $m \in \mathbb{N}$ such that

$$|Df^n \mid E(x)| \cdot |(Df^n \mid F(x))^{-1}| \le 1/2 \quad \text{for all } x \in \Lambda \text{ and } n \ge m.$$

See Appendix B for much more information about this notion.

In general terms, the results in this section show that surface diffeomorphisms far from homoclinic tangencies have dominated dynamics: the limit set admits a dominated splitting. Moreover, dominated diffeomorphisms on surfaces behave very much like non-critical one-dimensional maps.

Theorem 5.1 (Pujals, Sambarino [371]). *Let f be a C^2 surface diffeomorphism and $\Lambda \subset \Omega(f)$ be a compact f-invariant set admitting a dominated splitting $T_\Lambda M = E \oplus F$. Assume all periodic points in Λ are hyperbolic saddles. Then $\Lambda = \Lambda_1 \cup \Lambda_2$, where Λ_1 is a hyperbolic set and Λ_2 is the union of finitely many pairwise disjoint normally hyperbolic circles $C_1, \ldots, C_k$ such that $f^{m_i}(C_i) = C_i$ and $f^{m_i} : C_i \to C_i$ is an irrational rotation, for some $m_i \ge 1$.*

Here is a useful global reformulation of this theorem:

Theorem 5.2 (Pujals, Sambarino [372]). *Let f be a C^2 surface diffeomorphism whose periodic points are all hyperbolic and which can not be C^1 approximated by diffeomorphisms exhibiting homoclinic tangencies.*

Then the non-wandering set may be decomposed into compact f-invariant sets, $\Omega(f) = \Omega_1(f) \cup \Omega_2(f)$, such that $\Omega_1(f)$ is hyperbolic and $\Omega_2(f)$ is the union of finitely many pairwise disjoint normally hyperbolic circles $C_1, \ldots, C_k$ such that $f^{m_i}(C_i) = C_i$ and $f^{m_i} : C_i \to C_i$ is an irrational rotation, for some $m_i \ge 1$. In particular, Lebesgue almost every point of M is in the basin of attraction of either a hyperbolic attractor or a normally attracting periodic circle.

More recently, Pujals, Sambarino [370] extended these results by removing the assumption that all periodic points are hyperbolic: *If the limit set $L(f)$ admits a dominated splitting then the periods of the non-hyperbolic periodic orbits of f are bounded above. Moreover, $L(f) = L_1 \cup L_2 \cup L_3$, where L_1 is the union of finitely many pairwise disjoint homoclinic classes, each class containing at most finitely many non-hyperbolic periodic points; L_2 is the union of finitely many normally hyperbolic circles on which a power of f is a rotation; L_3 consists of periodic points contained in a finite union of normally hyperbolic periodic intervals.* Compare Mañé's Theorem 2.5.

A very interesting by-product of the arguments leading to the proof of this last result is the construction in [370] of surface diffeomorphisms having coexisting hyperbolic periodic orbits with normalized eigenvalues arbitrarily close to one, and such that there is no C^2 perturbations breaking the hyperbolicity of any of these points. By Frank's lemma (Theorem A.8), this phenomenon cannot occur in the C^1 topology.

5.2 Domination implies almost hyperbolicity

To prove Theorem 5.1 it is enough to see that, under the hypotheses, either Λ is hyperbolic or it contains a periodic circle where the dynamics of a power of f is conjugate to an irrational rotation. In fact, this implies the theorem, after noting that the number of periodic circles of f must be finite. The main step of the proof is

Proposition 5.3. *Consider a C^2 surface diffeomorphism f and a nontrivial transitive f-invariant compact set Σ admitting a dominated splitting $T_\Sigma M = E \oplus F$. Suppose that every proper compact invariant subset of Σ is hyperbolic, and Σ does not contain normally hyperbolic periodic circles C such that $f^m(C) = C$ and the restriction of f^m to C is conjugate to an irrational rotation, for some $m \geq 1$. Then Σ itself is hyperbolic.*

Assuming this proposition we now sketch the proof of Theorem 5.1. Suppose Λ is not hyperbolic. Consider the family $\mathcal{K}$ of compact f-invariant subsets of Λ which are not hyperbolic. The intersection of any decreasing sequence of non-hyperbolic invariant compact sets is also a non-hyperbolic invariant compact set. So, by Zorn's lemma, the family $\mathcal{K}$ has some minimal element Σ. By hypothesis, every periodic point of f in $\Lambda \supset \Sigma$ is hyperbolic. Hence, Σ cannot be trivial (a union of periodic orbits). We claim that Σ is a transitive set. Thus, since it is also minimal, Σ verifies all the hypotheses of Proposition 5.3. It follows that it is hyperbolic, which is a contradiction.

To see that Σ is a transitive set one uses the following useful alternative characterization of hyperbolicity, which is an analogue of Lemma 2.6 in higher dimensions:

Lemma 5.4. *Let Σ be an f-invariant compact set having a dominated splitting $T_\Sigma M = E \oplus F$ such that $|Df^n(x) \mid E(x)| \to 0$ and $|Df^{-n}(x) \mid F(x)| \to 0$ as $n \to \infty$ for all $x \in \Sigma$. Then Σ is a hyperbolic set: $E = E^s$ is uniformly contracting and $F = E^u$ is uniformly expanding.*

We are going to show that there exists $x \in \Sigma$ such that either $\Sigma = \omega(x)$ or $\Sigma = \alpha(x)$, which implies transitivity. Indeed, suppose the ω-limit set of a point $x \in \Sigma$ is properly contained in Σ. Then, by minimality, $\omega(x)$ is a hyperbolic set. It follows that $Df^n(x)(v) \to 0$ for every vector v of $E(x)$. In the same way, arguing with α-limit sets, $Df^{-n}(x)(v) \to 0$ for every vector v of $F(x)$. If this were valid for every $x \in \Sigma$, Lemma 5.4 would imply that Σ is hyperbolic, which is a contradiction. So, there is some $x \in \Sigma$ such that either $\omega(x)$ or $\alpha(x)$ is the whole Σ.

Let us now present some main ingredients in the proof of Proposition 5.3. A first one is a Denjoy type property about intervals transverse to the E-direction. Existence of a dominated splitting $T_\Sigma M = E \oplus F$ is equivalent to existence of forward invariant and backward invariant cone fields containing F and E, respectively, and defined on some neighborhood V of the set Σ (see

Appendix B). Then we say that a curve I contained in V is *transverse to the E-direction* if the tangent direction at every point $x \in I$ is contained in the corresponding forward invariant cone.

Fix an open neighborhood U of Σ whose closure is contained in V. Let Σ^+ be the maximal forward invariant set of f in the closure of U. By construction, Σ^+ has a dominated splitting and it contains Σ. We call a curve I a (δ, E)-*interval* if it is contained in Σ^+, all its forward iterates are transverse to the E-direction, and length $(f^n(I)) < \delta$ for every $n \geq 0$.

Proposition 5.5. *There is $\delta > 0$ such that the ω-limit set $\omega(I)$ of every (δ, E)-interval I satisfies one of the following two possibilities:*

- *either $\omega(I)$ is a normally hyperbolic periodic circle C where the restriction of f^m (m is the period of C) is conjugate to an irrational rotation,*
- *or $\omega(I)$ is contained in the set of periodic points f in the neighborhood V of Λ.*

By definition, $\omega(I)$ is the union of the sets $\omega(x)$ for $x \in I$. The Denjoy type statement implies that if the set Λ in Theorem 5.1 does not contain normally hyperbolic periodic circles then $\omega(I)$ is contained in the set of periodic points, for every (δ, E)-interval I.

Another key ingredient is to control center stable and center unstable manifolds. Existence of invariant families of curves $W^{cs}_\varepsilon(x)$ and $W^{cs}_\varepsilon(x)$ tangent to the subbundles E and F follows from general normal hyperbolicity theory [216]. In the present setting these curves are of class C^2. Most important, they are *dynamically defined*: for every $x \in \Sigma$ there exists $\varepsilon(x) > 0$ there is such that the lengths of $f^{-n}(W^{cs}_{\varepsilon(x)}(x))$ and $f^n(W^{cs}_{\varepsilon(x)}(x))$ go to zero as $n \to \infty$. In a second stage one proves that

$$\sum_{n=0}^{\infty} \text{length} \left(f^{-n}(W^{cu}_{\varepsilon(x)}(x)) \right) < \infty \quad \text{and} \quad \sum_{n=0}^{\infty} \text{length} \left(f^n(W^{cs}_{\varepsilon(x)}(x)) \right) < \infty$$

for every x in an open subset of Σ. One deduces that $|Df^{-n}(x) \mid F(x)| \to 0$ and $|Df^n(x) \mid E(x)| \to 0$ as $n \to \infty$, for every $x \in \Sigma$. In view of Lemma 5.4 this gives the hyperbolicity of Σ.

The proof of these facts uses ideas from the proof of Theorem 2.5, together the Denjoy-like Proposition 5.5 and the following property of the unstable, or stable, separatrices of hyperbolic points p of Σ: there is κ such that if a separatrix L of $W^u(p) \setminus \{p\}$ has length less than κ then the end-point of L other than p is either a sink or a non-hyperbolic saddle.

5.3 Homoclinic tangencies *vs.* Axiom A

An important consequence of Theorem 5.1 is that every diffeomorphism f that cannot be approximated by diffeomorphisms exhibiting homoclinic tangencies can be approximated by an Axiom A diffeomorphism:

Theorem 5.6 (Pujals-Sambarino [371]). *Let M be a closed surface. Then there is a dense subset $\mathcal{D}$ of* $\mathrm{Diff}^1(M)$ *of diffeomorphisms f such that either f has a homoclinic tangency or f satisfies the Axiom A.*

The proof has two main parts. The first one is the construction of a dominated splitting defined on the non-wandering set of every diffeomorphism g in an open and dense subset of a neighborhood of f. The second step is to prove that such a splitting is hyperbolic. The latter is done by approximating the diffeomorphism by a C^2 one, in the C^1 topology. This allows us to use Theorem 5.2 to decompose the dynamics into two parts, one corresponding to irrational rotations and another which is hyperbolic. Let us now comment both parts in more detail.

Consider the set $\mathcal{U}$ defined as the complement in $\mathrm{Diff}^1(M)$ of the closure of the diffeomorphisms exhibiting homoclinic tangencies. Theorem 5.6 just claims the density of the Axiom A diffeomorphisms in $\mathcal{U}$. Let $\mathcal{H}$ be the dense subset of Kupka-Smale diffeomorphisms $f \in \mathcal{U}$. In particular, every periodic point of $f \in \mathcal{H}$ is hyperbolic. For $f \in \mathcal{H}$, we let $\Sigma(f)$ be the set of (possibly infinitely many) sinks and sources, and $\Omega_0(f)$ be the complement of $\Sigma(f)$ inside the non-wandering set $\Omega(f)$. Clearly, $\Omega_0(f)$ is f-invariant and compact. The next step is to see that $\Omega_0(f)$ has a dominated splitting.

To prove this, one first observes that every point $x \in \Omega_0(f)$ is approximated by sequences $(p_n)_n$ such that each p_n is a hyperbolic periodic point of a diffeomorphism g_n arbitrarily C^1 close to f in the C^1 topology. This follows from the closing lemma, Theorem A.1. The key point here is that the natural splitting over the set of periodic points, given by the hyperbolic directions, is dominated uniformly in g_n, and so extends to the closure. The proof of this key fact is based on Franks' lemma (Theorem A.8) and arguments from the proof of the stability conjecture in [283]. In a few words, there are two cases. First, if the angles between the stable and unstable are of the periodic points p_n may be taken arbitrarily small, then we can obtain a homoclinic tangency after a small perturbation. So, this case is forbidden by the fact that $f \in \mathcal{U}$. Second, if the splitting is not dominated, a small perturbation allows us to recline the unstable direction against to the stable direction. Together with lack of domination, this leads to small angles between the two subspaces, and again we obtain a contradiction. See Section 7.2.1 for more details.

Now we move to the second and last part of the proof of Theorem 5.6. We only have to show that the diffeomorphisms in a dense subset of $\mathcal{H}$ verify the Axiom A. From the first part we know that, for every $f \in \mathcal{H}$, the set $\Omega_0(f)$ has a dominated splitting and so it verifies the hypotheses of Theorem 5.1. Since existence of normally hyperbolic curves supporting irrational rotations is a non-generic, Theorem 5.1 implies that $\Omega_0(f)$ is a hyperbolic set, for every f in a dense subset $\mathcal{D}$ of $\mathcal{H}$. Thus, the non-wandering set $\Omega(f)$ of $f \in \mathcal{D}$ is hyperbolic if, and only if, the set $\Sigma(f)$ of sinks and sources is hyperbolic.

The hyperbolicity of $\Sigma(f)$ is evident if this set is finite, and so we only have to consider the infinite case. Since $\Omega_0(f)$ is hyperbolic, it admit a neighborhood V such every f-invariant set in V is hyperbolic. Observe that every accumulation point of sinks or sources must be in $\Omega_0(f)$. Therefore, all but finitely many elements of $\Sigma(f)$ are contained in V, which is a contradiction. It follows that $\Sigma(f)$ is hyperbolic, and this concludes the proof of the hyperbolicity of $\Omega(g)$.

5.4 Entropy and homoclinic points on surfaces

The *topological entropy* of a transformation $f : M \to M$ is a number that, roughly speaking, measures how many distinct trajectories the transformation has. In precise terms, it is defined by (see [86])

$$h_{top}(f) = \lim_{\varepsilon \to 0} \limsup_{n \to \infty} \frac{1}{n} \log s(n, \varepsilon),$$

where $s(n, \varepsilon)$ is the maximum number of points with ε-distinct orbits of length n. We say that x and y have ε-distinct orbits of length n if $d(f^j(x), f^j(y)) > \varepsilon$ for some $0 \le j < n$.

Existence of transverse homoclinic points implies the existence of horseshoes (nontrivial hyperbolic sets), and so diffeomorphisms with transverse homoclinic points have strictly positive topological entropy. Conversely, the following deep result asserts that surface diffeomorphisms with positive topological entropy exhibit horseshoes [1]:

Theorem 5.7 (Katok [231]). *Let f be a $C^{1+\alpha}$ diffeomorphism, $\alpha > 0$, on a closed surface. If f has positive topological entropy then it has hyperbolic periodic saddles with transverse homoclinic points.*

In view of these results, it is a natural important problem to try and characterize the diffeomorphisms having zero topological entropy, as well as to understand the dynamical consequences of the variation of the topological entropy.

It is clear that Morse-Smale diffeomorphisms have zero entropy. The following consequence of Theorem 5.6 states that these are the only systems having zero topological entropy *robustly*. See also Gambaudo, Rocha [189].

Theorem 5.8 (Pujals, Sambarino [371]). *Let M be a closed surface and MS denote the C^1-closure of the set of Morse-Smale diffeomorphisms of $\mathrm{Diff}^1(M)$. Then there is an open and dense subset of $\mathrm{Diff}^1(M) \setminus \mathrm{MS}$ consisting of diffeomorphisms with transverse homoclinic orbits. Thus, MS coincides with the closure of the interior of the set of diffeomorphisms with zero topological entropy.*

---

[1] More generally, for diffeomorphisms in any dimension, horseshoes are dense in the support of any hyperbolic invariant measure. Positive topological entropy implies the existence of ergodic measures with positive *metric* entropy, by the variational principle [445]. In dimension 2 such measures are necessarily hyperbolic.

Since the set of diffeomorphisms with transverse homoclinic orbits is open, we only have to show that it is dense in $\text{Diff}^1(M) \setminus \text{MS}$. This is an easy consequence of Theorem 5.6. Indeed, the theorem gives that close to any $f \notin \text{MS}$ there is a diffeomorphism g either having a homoclinic tangency or verifying the Axiom A. In the first case, we may perturb g to create a transverse homoclinic orbit, thence proving the claim. In the second one, we may assume that the non-wandering set of g is infinite: otherwise g would be Morse-Smale, contradicting that $f \notin \text{MS}$. Then, since g is Axiom A, the non-wandering set of g contains nontrivial hyperbolic sets and, thus, transverse homoclinic orbits.

As an immediate consequence of Theorem 5.2 one gets the following result, that generalizes a theorem for diffeomorphisms of the 2-sphere in [189]:

Theorem 5.9. *Let f be a C^2 surface diffeomorphism with infinitely many periodic points, all of them hyperbolic. Then f is C^1 approximated by a diffeomorphism with a homoclinic point.*

This follows observing that if f can not be approximated by diffeomorphisms with homoclinic tangencies then it verifies Theorem 5.2. Since the non-wandering set is infinite, it must contain some nontrivial hyperbolic set and, hence, transverse homoclinic points.

We mention yet another important consequence of Theorem 5.2, concerning variation of the topological entropy. A C^k diffeomorphism f is a *point of entropy variation* if every C^k neighborhood $\mathcal{U}$ of f contains diffeomorphisms g such that the topological entropies of f and g are different. In what follows we consider only the case $k = \infty$, to benefit from the fact that the topological entropy depends continuously on the dynamics relative to the C^∞ topology [322, 455].

Theorem 5.10 (Pujals, Sambarino [372]). *Let f be a point of entropy variation. Then f is C^1 accumulated by diffeomorphisms exhibiting homoclinic tangencies.*

Suppose f is far from homoclinic tangencies. Let g be any diffeomorphism C^∞ close to f. Consider an arc $(f_t)_{t \in [0,1]}$ of C^∞ diffeomorphisms with $f_0 = f$ and $f_1 = g$, contained in a small neighborhood of f. By [421] we may take this arc such that the set S of parameter values t such that f_t is not Kupka-Smale is countable and $[0,1] \setminus S$ is a countable union of intervals. The assumption that f is far from homoclinic tangencies ensures that f_t satisfies the hypotheses of Theorem 5.2 for every $t \notin S$. Since invariant circles do not contribute for the entropy, one deduces that the entropy is constant on each interval in the complement of S. On the other hand, $h_{top}(f_t)$ is a continuous function of the parameter, because the topological entropy varies continuously with the dynamics in the C^∞ topology [322, 455]. As S is only countable, these two facts imply that the entropy is constant on the whole parameter interval $[0,1]$. This proves that $h_{top}(g) = h_{top}(f)$ for every g close to f, and so f is not a point of entropy variation.

5.5 Non-critical behavior in higher dimensions

Not much is known about the dynamics of invariant sets far from tangencies in higher dimensions. Recently, Wen has shown that such non-critical behavior implies the existence of a dominated splitting, thus extending the first step of the proof of Theorem 5.6 to any dimension. Let us call a point x i-*pseudo-periodic* for f if there exists y arbitrarily close to x and g arbitrarily close to f in the C^1 sense, such that y is a hyperbolic periodic point for g with stable manifold of dimension i. Then he proves that

Theorem 5.11 (Wen [449]). *If* $f : M \to M$ *is a diffeomorphism on a closed manifold of dimension d then f is C^1-far from diffeomorphism with homoclinic tangencies if and only if for every $0 < i < d$ the set of i-pseudo-periodic points admits a dominated splitting $E \oplus F$ with $\dim E = i$.*

This, and much of the progress in this direction is motivated by the following

Conjecture 5.12 (Palis [341]). In any dimension, the diffeomorphisms exhibiting either a homoclinic tangency or heterodimensional cycles are C^r dense in the complement of the closure of the hyperbolic ones, for any $r \geq 1$.

Of course, surface diffeomorphisms can not have heterodimensional cycles. So, in the 2-dimensional situation the case $r = 1$ of the conjecture reduces to Theorem 5.6, which had also been conjectured by Palis. A solution of Conjecture 5.12 has been announced very recently by Hayashi [206].

The following weaker version of Conjecture 5.12 has also been proposed:

Conjecture 5.13 (Palis [341]). The subset of dynamical systems that either have their limit set consisting of finitely many hyperbolic periodic orbits or else they have transverse homoclinic orbits, are C^r dense in the set of all dynamical systems, $r \geq 1$.

In this regard, Martin-Ribas, Mora [296] prove that every C^2 surface diffeomorphism that admits an *almost homoclinic sequence*, that is, a sequence of points that have iterates arbitrarily close to the stable manifold and to the unstable manifold of some saddle, either has transverse homoclinic points or can be C^1 approximated by diffeomorphisms with homoclinic tangencies. The proof goes as follows.

Under the hypotheses, Hayashi's connecting lemma, Theorem A.4, gives C^1 perturbations of the original map that have transverse homoclinic orbits. The closure of these homoclinic orbits contains a non-trivial invariant set with a dominated splitting. If the domination becomes weaker when the size of the perturbation goes to zero, methods of Mañé [278] permit to find a homoclinic tangency for a nearby system. Otherwise, the unperturbed map also has a non-trivial invariant set with dominated splitting. From Theorem 5.1 it follows that the original diffeomorphism has a nontrivial hyperbolic set, and so it has transverse homoclinic points.

Recently, Arroyo, Rodriguez-Hertz [33] extended some of the previous results to vector fields on 3-dimensional manifolds. Their main result says that *every C^1 vector field may be approximated by another which either is uniformly hyperbolic with no-cycles, has a homoclinic tangency, or has a singular cycle.* By cycle we mean a finite family of equilibrium points and periodic orbits cyclically related by intersections of their stable and unstable manifolds. The cycle is called singular if it does involve equilibria (not just regular periodic orbits). Singular cycles are studied in Section 9.2. This result is an important step towards the case $r = 1$ of the following reformulation of Conjecture 5.12 for flows:

Conjecture 5.14 (Palis [341]). Every vector field in a 3-dimensional manifold may be C^r approximated by another which either is uniformly hyperbolic, has a homoclinic tangency, or has a singular (Lorenz-like) attractor. More generally, in any dimension, the union of all vector fields which are uniformly hyperbolic or else have a homoclinic tangency, a heterodimensional cycle, or a singular cycle, should be C^r dense.

The basis for proving the main result is a continuous-time version of Theorem 5.1: Arroyo, Rodriguez-Hertz prove that *if a compact invariant set of a C^2 flow on a three-dimensional manifold has a dominated splitting, contains no equilibria, and all the periodic orbits in it are hyperbolic saddles, then it may be decomposed as the union of a hyperbolic set and finitely many tori supporting an irrational flow.* Let us point out that in this context, dominated splitting is defined in terms of the linear Poincaré flow. See Appendix B and [33, Proposition 3.3].

We close with a few more conjectures inspired by the previous results. Partially hyperbolic systems with invariant splitting $TM = E^u \oplus E^c \oplus E^s$ with one-dimensional central direction E^c are, in a sense, closest to being hyperbolic among all non-hyperbolic diffeomorphisms. Hence, one may expect a description of the dynamics to be easier and more complete in this case. In this spirit, we formulate

Conjecture 5.15. Generically, partially hyperbolic diffeomorphisms whose central direction has dimension 1 admit a filtration separating the non-wandering set into finitely many transitive pieces.

A more ambitious formulation would require the same conclusion for partially hyperbolic diffeomorphisms with splitting $TM = E^u \oplus E^1 \oplus \cdots \oplus E^l \oplus E^s$, where every E^i, $i = 1, \ldots, l$ is one-dimensional and, as before, E^u is uniformly expanding and E^s is uniformly contracting. On the other hand, Conjecture 5.15 might be easier when the invariant splitting contains only two sub-bundles:

Conjecture 5.16. Let f be a C^2 diffeomorphism such that the limit set $L(f)$ has a dominated splitting $E \oplus F$ where F has dimension 1 and E is uniformly contracting. Then the limit set $L(f)$ is the union of finitely many hyperbolic circles on which f is an irrational rotation, of (possibly infinitely many) periodic points contained in a finite union of periodic intervals, and finitely many pairwise disjoint homoclinic classes, each of them containing at most finitely many non-hyperbolic periodic points.

This corresponds to an extension of Theorems 2.5 and 5.6 to the general codimension-1 case. The definition of dominated splitting is in Definition B.1.

Crovisier [142] has obtained some progress in the direction of this conjecture, in the case when there is a unique non-hyperbolic periodic point.

6

Heterodimensional Cycles and Blenders

We call *heterodimensional cycle* the geometric configuration corresponding to two periodic points P and Q of Morse unstable index (dimension of the unstable bundle) $p < q$, respectively, such that the stable manifold of each point meets the unstable manifold of the other. In most cases we consider $q = p + 1$. Such cycles were first considered by Newhouse, Palis in [323], and were systematically studied in the series of papers [151, 152, 155, 156, 157, 158, 161]. This elementary phenomenon, which occurs for diffeomorphisms in any dimension larger than 2, is one of the main mechanisms for non-hyperbolicity as we have already mentioned in Section 1.4. For instance, the homoclinic classes of P and Q may coincide, for open sets in parameter space, in which case they can not be hyperbolic. See Section 6.1. In fact, a great variety of dynamical phenomena may arise in the unfolding of these cycles, depending in particular on the geometry of the heteroclinic intersections, as we also comment at the end of that section.

At the heart of heterodimensional cycles and the kinds of dynamics they yield, are certain transitive hyperbolic sets that we call *blenders*. This was introduced in [71], motivated by [152]. Topologically blenders are Cantor sets, but their distinctive geometric feature is that their embedding in the manifold is such that their stable set behaves as if it were of dimension higher than its actual topological dimension. More precisely, if p is the unstable index of the blender Γ, then the closure of the $(n - p)$−dimensional stable manifold of a periodic point $P \in \Gamma$ (in other words, the stable set of Γ) may intersect an unstable manifold of dimension $p - 1$ in a robust fashion. In Chapter 7 we shall use this property as a key ingredient in a semi-local argument for building robustly transitive dynamics.

In Section 6.2 we present an affine model of a blender, and then explore various generalizations of this model. In Section 6.3 we show that the unfolding of heterodimensional cycles yields blenders, leading to periodic orbits of different indices being heteroclinically related in a robust way.

6.1 Heterodimensional cycles

Consider a closed manifold M and a diffeomorphism f defined on M and having a pair of hyperbolic saddles P and Q with different indices, that is, different dimensions of their unstable subspaces. Assume $W^s(P)$ and $W^u(Q)$ have non-empty intersection, and the same for $W^u(P)$ and $W^s(Q)$. We say that f has a *heterodimensional cycle* associated to P and Q. Of course, heterodimensional cycles can only exist in dimension bigger or equal to 3.

6.1.1 Explosion of homoclinic classes

The main result in this section depicts heterodimensional cycles as a powerful mechanism for explosion of homoclinic classes of periodic points and creation of robust non-hyperbolic transitive sets. The assumptions are not minimal, we shall see later that this kind of behavior is typical of a large class of cycles.

For simplicity, we suppose that the periodic points P and Q in the cycle are actually fixed points. In addition, we always assume

- (codimension 1) the saddles P and Q have indices p and $q = p+1$, respectively,
- (quasi-transversality) the manifolds $W^s(P)$ and $W^u(Q)$ intersect transversely, and the intersection between $W^u(P)$ and $W^s(Q)$ is quasi-transverse: $T_x W^s(Q) \cap T_x W^u(P) = \{0\}$ for every intersection point x.

The next theorem asserts that the homoclinic classes of P and Q often explode, and become intermingled (non-empty intersection) when the cycle is unfolded.

A parametrized family $(f_t)_{t \in [-1,1]}$ of diffeomorphisms *unfolds generically* a heterodimensional cycle of $f = f_0$ if there are open disks $K_t^u \subset W^u(P_t)$ and $K_t^s \subset W^s(Q_t)$, depending continuously on t, such that $K_0^u \cap K_0^s$ contains a point of quasi-transverse intersection, and the distance between K_t^s and K_t^u increases with positive velocity when t increases. Here P_t and Q_t denote the continuations for f_t of the periodic points P and Q.

Theorem 6.1 (Díaz [152]). *There is a non-empty open set of parametrized C^∞ families of diffeomorphisms $(f_t)_{t \in [-1,1]}$ unfolding generically a heterodimensional cycle of $f = f_0$, such that for all small positive t*

1. *the transverse intersection between $W^s(P_t)$ and $W^u(Q_t)$ is contained in the homoclinic class of Q_t;*
2. *and the homoclinic class of P_t is contained in the homoclinic class of Q_t.*

Let d be the dimension of the ambient manifold M. The key fact behind Theorem 6.1 is that, for every small positive t, the $(d-p)$-dimensional manifold $W^s(P_t)$ is contained in the closure of $W^s(Q_t)$. This makes that the stable manifold of Q_t, which has dimension $(d-p-1)$, behaves in fact as a manifold of dimension $d-p$, that is one unit bigger. This type of phenomenon has been synthesized in the notion of blender, which we discuss in Section 6.2.

The proof of the theorem, which we are going to sketch next, relies on a reduction of the dynamics to a family of iterated functions systems on the interval. Besides the standing hypotheses of codimension 1 and quasi-transversality above, the main conditions defining the open set in the theorem are *connectedness* of the transverse intersection γ, *non-criticality* or partial hyperbolicity of the dynamics on γ, and *small distortion* of the transition from the neighborhood of Q to that of P. This will be explained in the sequel, but let us point out right away that these conditions may hold simultaneously for a parametrized family f_t and for its inverse f_t^{-1}: one gets open sets of families for which the homoclinic classes of P_t and Q_t coincide after the bifurcation. See Remark 6.3 below.

6.1.2 A simplified example

To explain the ideas of the proof of Theorem 6.1 while avoiding technical difficulties, we consider here a simple situation of a cycle in $\mathbb{R}^3$, which displays the main features of the general case we are dealing with.

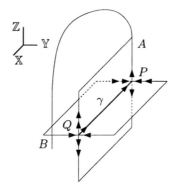

Fig. 6.1. Model heterodimensional cycle

We assume the restriction of f_t to a domain $C = [-1, 4] \times [-2, 2] \times [-2, 2]$ satisfies (see Figure 6.1):

Partial hyperbolicity and product structure along the transverse intersection:

(a) On some domain $[-1, 4] \times [-2, 2] \times [-a, a]$ whose image contains some $[0, 3] \times [-b, b] \times [-2, 2]$ the diffeomorphism f_t is independent of t and coincides with the product of its restriction to the $\mathbb{X}$-axis by a hyperbolic affine map, such that $\mathbb{Y}$ and $\mathbb{Z}$ are the strong stable and strong unstable directions: $\mathbb{Z} \oplus \mathbb{X} \oplus \mathbb{Y}$ is a dominated splitting for f_t.

(b) The points $Q_t = (0, 0, 0)$ and $P_t = (3, 0, 0)$ are hyperbolic saddles of indices 2 and 1, respectively. The *connection* $\gamma = (0, 3) \times \{(0, 0)\}$ is included in $W^u(Q_t) \cap W^s(P_t)$.

(c) The restrictions of f_t to the neighborhoods $U_Q = [-1,1] \times [-2,2] \times [-a,a]$ and $U_P = [2,4] \times [-2,2] \times [-a,a]$ are affine maps, and the eigenvalues of f_t at Q_t and P_t in the X-direction are $\beta > 1$ and $0 < \lambda < 1$, respectively.

Product structure along the quasi-transverse orbit and generic unfolding of the cycle:

(d) Consider the points $A = (3,0,1) \in W^u(P_0)$ and $B = (0,-1,0) \in W^s(Q_0)$. There is $\ell > 0$ such that $f_0^\ell(A) = B$, and so A and B are heteroclinic points for $f = f_0$. Moreover, there is a small neighborhood U_A of A where f_0^ℓ coincides with the translation $(x,y,z) \mapsto (x-3, y-1, z-1)$.

(e) The restriction of f_t^ℓ to U_A coincides with the translation

$$(x,y,z) \mapsto (x-3+t, y-1, z-1) = f_0^\ell(x,y,z) + (t,0,0).$$

By the first two conditions, $(0,3) \times [-2,2] \times \{0\}$ is contained in the stable manifold of P_t, and $(0,3) \times \{0\} \times [-2,2]$ is contained in the unstable manifold of Q_t. Then A and $B_t = (t,-1,0)$ are transverse homoclinic points of P_t, and $A_t = (3-t,0,1)$ and B are transverse homoclinic points of Q_t, for every small positive t.

Key properties of the diffeomorphism f_t may be derived from properties of the dynamics it induces in the quotient space by the sum $\mathbb{Y} \oplus \mathbb{Z}$ of the strong stable and the strong unstable directions. This quotient dynamics consists of an iterated functions system with two generators: the restriction of f to the X-axis, also denoted by f, and the quotient

$$g_t \colon [3 - \delta, 3 + \delta] \to \mathbb{R}, \quad g_t(x) = x - 3 + t,$$

of the restriction of f_t^ℓ to U_A, defined for small $t > 0$. Since we are interested in the dynamics in a neighborhood of the cycle, it is natural to restrict f to $[0,3]$ and g_t to $[3-t,3]$. See Figure 6.2.

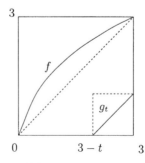

Fig. 6.2. Iterated functions system with 2 generators

The dynamics of the iterated functions system is, essentially, determined by the smooth conjugacy class and by the Mather invariant [288] of the interval map f. The former is described by the pair of eigenvalues (β, λ). Let us recall the definition of the latter.

By assumption, in a neighborhood of $Q = 0$ the transformation f is the time-1 map of the flow defined by the affine vector field $X_Q(x) = (\log \beta)\, x\, \frac{\partial}{\partial x}$. Similarly, in a neighborhood of $P = 3$ the transformation f is the time-1 of the vector field $X_P(x) = (\log \lambda)\,(x - 3)\,\frac{\partial}{\partial x}$. For $x \in (0, 3)$ close to Q, consider n big enough so that $f^n(x)$ is close to P, and then write

$$Df^n(x)(X_Q(x)) = \mu(x)\, X_P(f^n(x)),$$

for some $\mu(x) \in \mathbb{R}$. Since the vector fields X_Q and X_P are f-invariant, the function $\mu(\cdot)$ does not depend on n, and it satisfies $\mu(x) = \mu(f(x))$ for all x. Hence, the *Mather invariant* $\mu(\cdot)$ is a smooth function on the circle $(0, 3)/f$. It corresponds to what [152] calls the transition from Q to P.

As part of defining the open set in the statement of Theorem 6.1, we also assume that the transition has small distortion, in other words, that μ is close to 1. The precise condition appears in Lemma 6.2 below, involving also the eigenvalues β and λ. Observe that μ is constant equal to 1 if the restriction of f to $[0, 3]$ is the time-1 of a smooth vector field. We now sketch the proof of Theorem 6.1 for our simplified model.

Denote by $D_P^t = [3 - t, 3 - \lambda t]$ the fundamental domain of f at distance t from P. Let $s_t = f^{-n_t}(3 - t)$ be the unique backward iterate of $3 - t$ under f in the interval $(\beta^{-1} t, t]$. Then let $D_Q^t = [s_t, \beta s_t]$ be the corresponding fundamental domain of f close to Q. We are going to construct a return map to this last domain, for the iterated function system. See Figure 6.3.

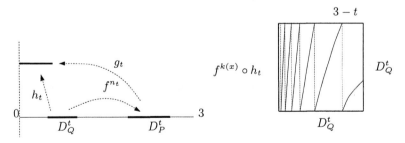

Fig. 6.3. A return map to D_Q^t

By construction, $f^{n_t}(D_Q^t) = D_P^t$. Consider the increasing surjective map

$$h_t = g_t \circ f^{n_t} : D_Q^t \to [0, (1 - \lambda)t].$$

Given a point $x \in D_Q^t$ there exists a unique integer $k(x) \geq 0$ such that $f^{k(x)} \circ h_t(x)$ is in D_Q^t. This defines a return map to the fundamental domain D_Q^t. Due to the jumps of $k(\cdot)$, the map $f^{k(x)} \circ h_t(x)$ has a sequence of discontinuities accumulating at the point $x = \beta^{-1} s_t$. We get rid of these discontinuities, though not the one at the accumulating point, by identifying the endpoints of the domain D_Q^t, so that it becomes a circle. To make this precise, consider the projection

$$\pi \colon (0,1] \to S^1 = \mathbb{R}/\mathbb{Z}, \qquad \pi(x) = \frac{\log x}{\log \beta} \pmod 1.$$

Let $c_t = \pi(s_t) = \pi(\beta\, s_t)$, and denote by π_t^{-1} the inverse of $\pi \colon (s_t, \beta\, s_t] \to S^1$. The *return map* is given by

$$R_t = \pi \circ h_t \circ \pi_t^{-1} \colon S^1 \to S^1.$$

This circle map R_t has a unique discontinuity, at c_t. Since $h_t(s_t) = 0$ and $k(x)$ goes to infinity when x goes to s_t inside D_Q^t, we have that the derivative $DR_t(z)$ goes to $+\infty$ when $z \to c_t$ from the right. Moreover, $R_t(z)$ "goes to $-\infty$", meaning that the graph winds infinitely many times around the circle.

Lemma 6.2. *Suppose the Mather function $\mu(\cdot)$ and the eigenvalues λ, β satisfy*

$$\frac{\mu(x)\, \lambda \,|\log \lambda|}{(1 - \lambda) \log \beta} > 1, \quad \text{for every } x.$$

Then, for every small $t > 0$, the map R_t is uniformly expanding: $|DR_t(z)| > 1$ for all $z \neq c_t$.

This lemma implies that for any open interval $I \subset [0,3]$ there is an element ϕ of the iterated functions system $\mathcal{G}(f, g_t)$ generated by f and g_t such that $\phi(I)$ contains $Q = 0$. Indeed, consider the projection $J = \pi(I)$. Since the map R_t is expanding, there is $n \geq 1$ such that $R_t^n(J)$ contains the discontinuity c_t. Then, if ψ is the element of $\mathcal{G}(f, g_t)$ corresponding to R_t^n, we have that $\psi(I)$ contains s_t. Then it suffices to take $\phi = h_t \circ \psi$ and to recall that $h_t(s_t) = 0$.

From the previous paragraph we get that *the stable manifold of Q intersects any disk Δ transverse to the stable manifold of P.* Indeed, suppose Δ is a vertical disk $I \times \{y\} \times [-2, 2]$ for some interval $I \subset [0, t]$ and some $y \in [-2, 2]$. Consider the image of Δ under the map ϕ (extended to the ambient space): the previous conclusion means that this image intersects $\{0\} \times [-2, 2] \times \{0\}$. Since this last segment is contained in the stable manifold of Q, it follows that $\Delta \cap W^s(Q)$ is nonempty. For a general disk Δ transverse to $W^s(P)$, the assertion follows after observing that the forward orbit of Δ contains disks arbitrarily close to vertical disks as before.

As an immediate consequence we get that the connection curve $\gamma = (0,3) \times \{(0,0)\}$ is contained in the homoclinic class of Q. The other claim in Theorem 6.1 follows from a variation of the same argument. To finish our sketch of the proof of the theorem, we give the proof of Lemma 6.2:

Proof. Consider $z \in S^1$ and let $x = \pi_t^{-1}(z)$. Observe that $DR_t(z)$ is defined by the relation

$$D(g_t \circ f^{n_t})(X_Q(x)) = DR_t(z) \cdot X_Q(g_t \circ f^{n_t}(x)).$$

Recalling that $Df^{n_t}(X_Q(x)) = \mu(x) X_P(f^{n_t}(x))$ and the expressions of X_Q and X_P, one has

$$\mu(x) \, |\log \lambda| \, (f^{n_t}(x) - 3) = DR_t(z) \, (\log \beta) \, g_t(f^{n_t}(x)).$$

Since, by definition, $f^{n_t}(x) \in D_P^t = [3 - t, 3 - \lambda t]$, one has $(f^{n_t}(x) - 3) \in [-t, -\lambda t]$. Thus $g_t(f^{n_t}(x)) = f^{n_t}(x) - 3 + t \in [0, t - \lambda t]$. Hence

$$DR_t(z) = \frac{\mu(x) \, |\log \lambda| \, (3 - f^{n_t}(x))}{(\log \beta) \, g_t(f^{n_t}(x))} \geq \frac{\mu(x) \, |\log \lambda| \, \lambda t}{(\log \beta) \, (1 - \lambda) \, t} \geq \frac{\mu(x) \, \lambda \, |\log \lambda|}{(\log \beta) \, (1 - \lambda)} > 1.$$

This ends the proof of the lemma.   □

Remark 6.3. When $\mu = 1$ and $\beta\lambda = 1$ the hypothesis of Lemma 6.2 becomes $\lambda > 1/2$ or, equivalently, $\beta < 2$. This condition does not change if we replace f_t by its inverse.

6.1.3 Unfolding heterodimensional cycles

The dynamics associated to the unfolding of general heterodimensional cycles is very rich, and still far from being completely understood. However, a good number of results are available, especially for the (codimension-1) quasi-transversal case. Here we review some of those results.

The example of heterodimensional cycle in the previous section is *connected*, meaning that the transverse intersection $W^s(P) \cap W^u(Q)$ contains a smooth invariant curve γ joining P to Q, and *non-critical*, meaning that the closed invariant set $\gamma \cup \{P, Q\}$ admits a Df-invariant dominated splitting of the form $E^u \oplus T\gamma \oplus E^s$. In addition we assumed that the restriction to the connection γ has *small distortion*. Due to non-criticality, no homoclinic tangencies related to either P or Q may occur in the unfolding of those cycles. As we are going to see, other types of cycles may exhibit very different dynamical behavior.

Firstly, [156] shows that the unfolding of *critical* cycles may lead to the phenomenon of persistence of tangencies. An interesting point is that this construction does not involve fractal dimensions, as was the case for the results in Section 3.2.

Next, in the same paper it is shown that the unfolding of *non-connected* cycles may lead to Morse-Smale dynamics after the bifurcation: essentially, one considers cycles such that the connected components of γ are small, compared to the length of fundamental domains of f near P and Q.

Finally, [157] exhibits connected non-critical cycles whose Mather invariant μ has *large distortion*, and for which there are sets of parameters t with positive Lebesgue density at $t = 0$ (the density can be made arbitrarily close to 1) such that the limit set of f_t in the neighborhood of the cycle is hyperbolic and coincides with the disjoint union of the homoclinic classes of P_t and Q_t.

A natural question in this context is whether having both homoclinic classes of P_t and Q_t hyperbolic suffices to guarantee hyperbolicity of the limit set in a neighborhood of the cycle. The answer turns out to be negative, as shown in [155]. In simple terms, the reason is that the unfolding of the cycle may contain the creation of secondary cycles, which are not associated neither to P_t nor to Q_t, and of new homoclinic classes disjoint from those of P_t and Q_t. Also, for any fixed k, it is possible to construct parametrized families generically unfolding a heterodimensional cycle such that, for a set of parameter values with positive density at $t = 0$, the limit set of f_t in a neighborhood of the cycle is hyperbolic and has at least k different homoclinic classes.

Problem 6.4. Given a parametrized family $(f_t)_{t \in [-1,1]}$ generically unfolding a connected non-critical cycle of f_0, is there a full Lebesgue measure subset of some neighborhood of $t = 0$ in parameter space, for which any homoclinic classes of f_t in a neighborhood of the cycle are pairwise disjoint (or equal) and their number is uniformly bounded?

Another pair of questions naturally raised by the previous remarks:

Problem 6.5. Are there examples of parametrized families of diffeomorphisms generically unfolding a heterodimensional cycle for which there are (infinitely many) parameter values t such that:

- The map f_t has infinitely many pairwise disjoint homoclinic classes in a neighborhood of the cycle?
- The homoclinic classes of P_t and Q_t are distinct but not disjoint?

If such examples do exist, generically, does the set of such parameter values have zero Lebesgue measure?

6.2 Blenders

While the arguments in Section 6.1.2 require a fair amount of differentiability, the conclusion of Theorem 6.1 does extend to the C^1 topology, as we shall see in Section 6.3. One way to prove this is using the notion of blender, which we discuss here.

Blenders were introduced by Bonatti, Díaz [72], and have been used to obtain persistence of cycles and transitivity, cf. Section 7.1.3. In this regard, they play a role similar to that of thick Cantor sets in the phenomenon of persistence of tangencies in Section 3.2. With the advantage that, unlike thick Cantor sets, blenders are robust already in the C^1 category.

Let us point out that what makes blenders so useful is not so much the dynamics they carry, but rather their geometry. Indeed, in simple terms, *a blender is a hyperbolic basic set such that convenient projections of its stable,*

*or unstable, set have larger topological dimension than the stable, or unstable,
set itself.*

One such model appeared already in Example 3.24 (discontinuity of fractal
dimensions). This was a hyperbolic Cantor $H \subset \mathbb{R}^3$, with 1-dimensional stable
manifolds and such that the projection of its stable set $W^s(H)$ onto the plane
generated by the stable and the weak unstable directions contained a whole
open set. In fact, the topologically 1-dimensional set $W^s(H)$ intersects any
curve roughly parallel to the strong unstable direction.

6.2.1 A simplified model

Let us start by describing a simple example that displays the main features of
blenders. Let $f\colon \mathbb{R}^2 \to \mathbb{R}^2$ be a diffeomorphism of $\mathbb{R}^2$ such that the rectangle
$R = [0,1]^2$ is a Smale horseshoe for f:

- $R \cap f(R)$ is the union of the vertical sub-rectangles $A = I_a \times [0,1]$ and
  $B = I_b \times [0,1]$, where I_a and I_b are closed intervals contained in $(0,1)$.
- $R \cap f^{-1}(R)$ is the union of the horizontal sub-rectangles $f^{-1}(A) = [0,1] \times J_a$
  and $f^{-1}(B) = [0,1] \times J_b$, where J_a and J_b are closed intervals contained
  in $(0,1)$.
- The restrictions of f to $f^{-1}(A)$ and to $f^{-1}(B)$ are affine maps with linear
  part $\begin{pmatrix} \pm\frac{1}{3} & 0 \\ 0 & \pm 3 \end{pmatrix}$. In particular, such restrictions preserve the horizontal
  and vertical directions.

Denote by $q = (x_0, y_0)$ the unique fixed point of f in A.

Consider now a diffeomorphism $F\colon \mathbb{R}^3 \to \mathbb{R}^3$ such that in $[-1,1] \times R$ is of
the form

$$F(t,x,y) = \begin{cases} (\frac{5}{4}t, f(x,y)) & \text{if } (x,y) \in f^{-1}(A) \text{ and } t \in [-1,1], \\ (\frac{5}{4}t - \frac{1}{2}, f(x,y)) & \text{if } (x,y) \in f^{-1}(B) \text{ and } t \in [-1,1]. \end{cases}$$

The dynamics of F in the cube $\Gamma = [-1,1] \times R$ is hyperbolic, with $\mathbb{T} \oplus \mathbb{Y}$
as unstable direction ($\mathbb{T}$ is weak unstable and $\mathbb{Y}$ is strong unstable), and $\mathbb{X}$
as stable direction. The *blender* is the maximal invariant set inside Γ. See
Figure 6.4.

The distinctive geometric feature of blenders is expressed in the next
lemma. For the statement we need some terminology. Observe first that
$\Gamma \cap F(\Gamma)$ contains the cubes $\mathbb{A} = [-1,1] \times A$ and $\mathbb{B} = [-1,\frac{3}{4}] \times B$ and
that $Q = (0, x_0, y_0) = (0, q) \in \mathbb{A}$ is a hyperbolic fixed saddle of F of index 2.
Let $W^s_{loc}(Q) = \{0\} \times [0,1] \times \{y_0\}$ be the connected component of $W^s(Q) \cap \Gamma$
that contains Q.

- A *vertical segment to the right of* $W^s_{loc}(Q)$ is a segment $\sigma_{t,x} = \{t\} \times \{x\} \times
  [0,1]$, where $x \in [0,1]$ and $0 < t \leq 1$.
- A *vertical strip to the right of* $W^s_{loc}(Q)$ is a rectangle $\Delta = [t_1, t_2] \times \{x\} \times
  [0,1]$, where $x \in [0,1]$ and $0 < t_1 < t_2 \leq 1$. The number $w(\Delta) = (t_2 - t_1)$
  is the *width of* Δ.

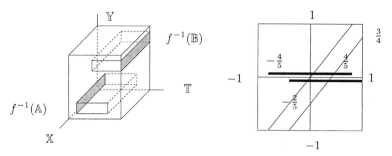

Fig. 6.4. Blender with one-dimensional reduction

Lemma 6.6. *(blender distinctive property) Every vertical strip Δ to the right of $W^s_{loc}(Q)$ intersects $W^s(Q)$.*

Proof. We are going to show that if Δ is a vertical strip to the right of $W^s_{loc}(Q)$ then either $F(\Delta)$ intersects $W^s_{loc}(Q)$, or else $F(\Delta) \cap \Gamma$ contains a vertical strip Δ_1 to the right of $W^s_{loc}(Q)$ with $w(\Delta_1) = (5/4)\, w(\Delta)$.

To deduce the lemma from this claim, start with Δ as in the statement and consider its forward iterates. For as long as these iterates do not intersect the local stable manifold of Q, they contain vertical strips whose widths grow exponentially fast. Since the width is bounded above by 1, some iterate $F^n(\Delta)$ must intersect $W^s_{loc}(Q)$. This implies that Δ intersects the (global) stable manifold of Q. Hence, we are left to prove the claim in the previous paragraph.

Consider any vertical strip $\Delta = [t_1, t_2] \times \{x\} \times [-1, 1]$ to the right of $W^s_{loc}(Q)$. Write $\Delta_A = F(\Delta) \cap \mathbb{A}$ and $\Delta_B = F(\Delta) \cap \mathbb{B}$.

- If $t_2 \leq \frac{4}{5}$ then $[\frac{5}{4} t_1, \frac{5}{4} t_2] \subset (0, 1]$. It follows that Δ_A is a vertical strip to the right of $W^s_{loc}(Q)$, with width $w(\Delta_A) = \frac{5}{4} w(\Delta)$. Thus the second possibility in the claim holds.
- If $t_2 \in (\frac{4}{5}, 1]$ then $\frac{5}{4} t_2 - \frac{1}{2} \in (\frac{1}{2}, \frac{3}{4}]$. By the definition of F, there is x' such that
$$\Delta_B = F(\Delta) \cap \mathbb{B} = \{[\frac{5}{4} t_1 - \frac{1}{2}, \frac{5}{4} t_2 - \frac{1}{2}]\} \times \{x'\} \times [0, 1].$$

There are now two subcases:

- If $\frac{5}{4} t_1 - \frac{1}{2} > 0$ then Δ_B is a vertical strip to the right of $W^s_{loc}(Q)$, with width $w(\Delta_B) = (5/4)\, w(\Delta)$. So, the second possibility in the claim holds in this case too.
- Otherwise $\frac{5}{4} t_1 - \frac{1}{2} \leq 0$ and $\frac{5}{4} t_2 - 1/2 \geq 0$. Then Δ_B meets $W^s_{loc}(Q) = \{0\} \times [0, 1] \times \{y_0\}$, which corresponds to the first possibility in the claim.

The proof is complete.   □

As we did in the previous section, we may consider the quotient of the dynamics by the sum of the strong stable and strong unstable directions. This yields an iterated functions system $\mathcal{G}(\phi_1, \phi_2)$ with two generators, as follows:

- for $t \in [-\frac{4}{5}, \frac{4}{5}]$ consider $\phi_1(t) = \frac{5}{4} t$, corresponding to the restriction of F to the domain $F^{-1}(\mathbb{A}) = [-\frac{4}{5}, \frac{4}{5}] \times f^{-1}(A)$,
- for $t \in [-\frac{2}{5}, 1]$ consider $\phi_2(t) = \frac{5}{4} t - \frac{1}{2}$, corresponding to the restriction of F to the domain $F^{-1}(\mathbb{B}) = [-\frac{2}{5}, 1] \times f^{-1}(B)$.

Notice that both maps ϕ_1 and ϕ_2 are expanding, and their domains of definition overlap on the interval $[-\frac{2}{5}, \frac{4}{5}]$. This reflects the fact that given any vertical segment $\sigma = \{t\} \times \{x\} \times [0,1]$ with t in the overlapping interval, there are x_1 and x_2 such that the vertical segments $\{\phi_1(t)\} \times \{x_1\} \times [0,1]$ and $\{\phi_2(t)\} \times \{x_2\} \times [0,1]$ are the two connected components of $F(\sigma) \cap \Gamma$.

Lemma 6.6 may now be reformulated for the iterated functions system, as follows. We also translate the proof, since it is very short.

Lemma 6.7. *Given any open interval $\alpha \subset [0,1]$ there is $\phi \in \mathcal{G}(\phi_1, \phi_2)$ such that $0 \in \phi(\alpha)$.*

Proof. Since $\phi_2(\frac{2}{5}) = 0$, it is enough to exhibit $\phi \in \mathcal{G}(\phi_1, \phi_2)$ such that $\phi(\alpha)$ contains $\frac{2}{5}$. Given $\alpha \subset [0,1]$ we let $\alpha_0 = \alpha$ and inductively define subintervals α_n as follows. Suppose α_n does not contain $\frac{2}{5}$. Then there are two possibilities. If $\alpha_n \subset (0, \frac{4}{5}]$, we let $\alpha_{n+1} = \phi_1(\alpha_n)$. Otherwise, $\alpha_n \subset (\frac{2}{5}, 1]$ and we let $\alpha_{n+1} = \phi_2(\alpha_n)$. In both cases, $\alpha_{n+1} \subset (0,1]$ and $\text{length}(\alpha_{n+1}) = \frac{5}{4} \text{length}(\alpha_n) = (\frac{5}{4})^n \text{length}(\alpha_0)$. This means that α_n must, eventually, contain $\frac{2}{5}$. At that point we are done. $\qquad\square$

To illustrate the dynamical usefulness of blenders, we close this section with the following direct consequence of Lemma 6.6:

Lemma 6.8. *Suppose that there is a hyperbolic periodic point P of F of index 1 whose one-dimensional unstable manifold crosses Γ along a vertical segment to the right of $W_{loc}^s(Q)$. Then $W^s(P) \subset \overline{W^s(Q)}$.*

So the one-dimensional stable manifold of Q looks like a 2-dimensional manifold, as its closure contains the 2-dimensional $W^s(P)$.

6.2.2 Relaxing the construction

So far we discussed a very simple affine model of blender. Of course, for most applications we need more flexible versions of the previous constructions, robust under C^1 small perturbations. Just as an example, in the context of Lemma 6.8 we would like to have the conclusion when the segment is just approximately vertical. We are going to explain how this flexibilization is carried out, starting with perturbations of the affine model, and to indicate other ways in which the construction may be generalized. Motivated by this, we discuss possible definitions of the notion of blender.

In the affine model F we have been considering, $\mathbb{X} \oplus \mathbb{T} \oplus \mathbb{Y}$ is a dominated splitting on Γ, with $\mathbb{X}$ as strong stable, $\mathbb{Y}$ as strong unstable, and $\mathbb{T}$ as central (weak unstable) direction. Hence, there are (thin) cone fields $\mathcal{C}^s$, $\mathcal{C}^{uu}$, $\mathcal{C}^u$

around the directions $\mathbb{X}$, $\mathbb{Y}$, $\mathbb{T} \oplus \mathbb{Y}$, respectively, strictly invariant under the derivative DF. These cone fields allow us to define *almost vertical segments (to the right of $W^s_{loc}(Q)$)*, as curve segments tangent to $\mathcal{C}^{uu}$ and crossing Γ (on the right hand side of the local stable manifold), as well as *almost vertical strips (to the right of $W^s_{loc}(Q)$)*, as 2-dimensional disks through Γ that are tangent to $\mathcal{C}^u$ and foliated by almost vertical segments (to the right of $W^s_{loc}(Q)$). The notion of width extends to these almost vertical strips: the length of the shortest curve transverse to $\mathcal{C}^{uu}$ joining the two boundary components inside the strip.

The families of almost vertical segments and strips are invariant under forward iteration, in the same sense as before for truly vertical segments and strips. Moreover, the widths of almost vertical strips not intersecting $W^s_{loc}(Q)$ increase exponentially under iteration, again in the same sense as before. This means that the proof of Lemma 6.6 carries through to this context, only with extra technical difficulties. We get that *every almost vertical strip at the right of $W^s_{loc}(Q)$ intersects $W^s(Q)$.*

Now, the cone fields remain invariant for any diffeomorphism in a C^1 neighborhood of F. Moreover, the width estimate is also robust under any small C^1 perturbation, because it depends only on the fact that the derivative expands all vectors in the cone $\mathcal{C}^u$. This means that the conclusion of the previous paragraph is true for any map near F: *the distinctive blender property is a robust property!*

Blenders from a constructive viewpoint. One way the construction of blenders may be relaxed, with respect to the affine model in Section 6.2.1, is allowing for more complicated combinatorics: any number of dynamical components, with variable return times to the ambient box.

Consider a C^1 diffeomorphism F and a box $\Gamma = D^k \times [-1, 1] \times D^l$, where $D^i = [-1, 1]^i$, in the ambient space, endowed with cone fields $\mathcal{C}^s$ and $\mathcal{C}^{uu} \subset \mathcal{C}^u$ around the directions of D^k, D^l, and $[-1, 1] \times D^l$, respectively. Assume that there are sub-boxes $(\Gamma_i)_{i=1}^k$ of Γ, and return times $(n_i)_{i=1}^k$ such that every return map $F^{n_i} : \Gamma_i \to \Gamma$ preserves both cone fields and expands uniformly all vectors in $\mathcal{C}^u$.

Define as above almost vertical (l-dimensional) disks, tangent to $\mathcal{C}^{uu}$, and almost vertical ($l+1$-dimensional) strips, tangent to $\mathcal{C}^u$ and foliated by almost vertical disks crossing Γ. Define the width $w(\mathcal{S})$ of an almost vertical strip $\mathcal{S}$, again as before. We turn the type of property in Lemma 6.6 into our first definition of a blender:

Definition 6.9. (Γ, F) is a *blender associated to a periodic saddle Q of index $l + 1$* if for every almost vertical strip $\mathcal{S}$ crossing Γ there is $i \in \{1, \ldots, k\}$ such that

- either $F^{n_i}(\mathcal{S} \cap \Gamma_i)$ contains an almost vertical strip $\mathcal{S}'$ intersecting $W^s_{loc}(Q)$
- or $F^{n_i}(\mathcal{S} \cap \Gamma_i)$ contains an almost vertical strip $\mathcal{S}'$ with $w(\mathcal{S}') \geq \lambda w(\mathcal{S})$, where $\lambda > 1$ is independent of the strip $\mathcal{S}$.

The periodic point Q in the definition does not need to belong to the cube Γ: our arguments just require that $W^s(Q)$ contains a disk Δ crossing Γ and tangent to the cone field $\mathcal{C}^s$. The reader may check that the unfolding of the heterodimensional cycle in Section 6.1.2 yields a blender inside a box close to the fixed point Q but not necessarily containing it.

In the particular case where the maps F^{n_i} preserve the D^k and D^l directions, as was the case for the affine blender, one may consider the quotient dynamics of F by the sum $D^k \oplus D^l$. In this way we obtain, for each i, a one-dimensional expanding map $\phi_i : I_i \to [0,1]$ induced by F^{n_i} on a compact $I_i \subset [0,1]$. (Γ, F) being a blender corresponds to the maps ϕ_i having the covering property: $\cup_i \text{int}(I_i) = (0,1)$. A point $x \in [0,1]$ is *periodic* for the iterated functions system $\mathcal{G}(\phi_1, \dots, \phi_k)$ if $\phi(x) = x$ for some $\phi \in \mathcal{G}$. Extending Lemma 6.7, we have

Proposition 6.10. *Given any system $\mathcal{G}(\phi_1, \dots, \phi_k)$ of iterated expanding functions with the covering property, there is a finite set $\mathcal{F} = \{x_1, \dots, x_s\}$, $1 \le s < k$, of periodic points such that, for any open interval $\alpha \subset [0,1]$, there is $\phi \in \mathcal{G}$ with $\phi(\alpha) \cap \mathcal{F} \ne \emptyset$.*

Blenders from an operational viewpoint. We now adopt a different point of view concerning blenders: we will think of them as tools for proving transitivity. Thus we propose another definition which takes into account their properties, rather than their construction.

Definition 6.11. Let $F : M \to M$ be a diffeomorphism and Q be a hyperbolic saddle of index $l + 1$. We say that F *has a blender associated to Q* if there is a C^1 neighborhood $\mathcal{U}$ of F, and a C^1 open set $\mathcal{D}$ of embeddings of an l-dimensional disk D^l into M, such for every $G \in \mathcal{U}$, every disk $D \in \mathcal{D}$ intersects the closure of $W^s(Q_G)$, where Q_G is the continuation of the periodic point Q for G.

The set $\mathcal{D}$ in the definition is the *superposition region* of the blender. For instance, in the simplified model considered before, $l = 1$ and the superposition region is the family of all almost vertical segments (that is, tangent to a thin vertical cone field and crossing the box Γ) to the right of $W^s_{loc}(Q)$. Bearing in mind the distinctive property of blenders, one says that the blender is *activated* by a saddle P of index l if the unstable manifold of P contains a disk of the superposition region.

Lemma 6.12. *Let F be a diffeomorphism having a blender associated to a saddle Q of index $l + 1$. Suppose that the blender is activated by a saddle P of index l. Then, for every diffeomorphism G in a C^1 neighborhood of F, the closure of $W^s(Q_G)$ contains $W^s(P_G)$, where P_G and Q_G are the continuations of P and Q for G.*

As noted before, the point with this result is that it means that the $(d - l - 1)$-dimensional manifold $W^s(Q_G)$ of Q_G looks like it contains a manifold with higher dimension, namely $W^s(P_G, G)$.

Proof. Let $\mathcal{D}$ the superposition region of the blender. As compact parts of invariant manifolds depend continuously on the diffeomorphism, $W^u(P_G)$ contains a disk $D_G \in \mathcal{D}$, for all G close to F. Let U be any open set intersecting $W^s(P_G)$. By the λ-lemma, for every n big enough, $G^n(U)$ contains a C^1-open set of disks arbitrarily close to D_G. Then these disks are in the superposition region. By definition of blender, it follows that $W^s(Q_G)$ intersects $G^n(U)$. As the stable manifold is invariant, it also intersects U. $\square$

6.3 Partially hyperbolic cycles

We are going to see that the formation of intermingled homoclinic classes is a frequent phenomenon in the unfolding of cycles, even in the C^1 topology.

Consider a diffeomorphism f with a cycle associated to saddles P and Q of indices p and $p+1$, respectively. We say that the cycle is C^k-*far from homoclinic tangencies* if every diffeomorphism g which is C^k close to f does not have homoclinic tangencies associated either of the continuations P_g and Q_g of P and Q.

Theorem 6.13 (Díaz, Rocha [158]). *Let f be a C^1 diffeomorphism with a heterodimensional cycle associated to saddles P and Q of indices p and $q = p+1$. Suppose that the cycle is C^1-far from homoclinic tangencies. Then there is an open set $\mathcal{U} \subset \mathrm{Diff}^1(M)$ whose closure contains f and such that the homoclinic classes of P_g and Q_g coincide for all $g \in \mathcal{U}$.*

The conclusion of this theorem also holds for non-critical connected cycles, without any assumption on the Mather invariant. In that case we even have the following version for one-parameter families generically unfolding the cycle: Given $t > 0$, let $B(t)$ be the interior of the set of parameters $s \in [-t, t]$ such that the homoclinic classes of P_s and Q_s coincide. Then $B(t)$ has positive Lebesgue measure for every $t > 0$. Moreover, if the cycle corresponds to a first bifurcation, the relative density of $B(t)$ at the bifurcation is positive:

$$\liminf_{t \to 0^+} \frac{1}{t} \mathrm{Leb}(B(t)) > 0.$$

Theorem 6.13 is a consequence of the following proposition, which uses blenders with several generators.

Proposition 6.14. *Let f be a diffeomorphism as in Theorem 6.13. Then there is an open set $\mathcal{V} \subset \mathrm{Diff}^1(M)$ whose closure contains f such that for every g in $\mathcal{V}$ there are blenders $\Gamma(g)$ and $\Gamma(g^{-1})$, defined for g and g^{-1}, respectively, such that:*

- $\Gamma(g)$ *is associated to a hyperbolic periodic point R_g homoclinically related to Q_g and is activated by P_g.*

- $\Gamma(g^{-1})$ *is associated to a hyperbolic periodic point* S_g *homoclinically related to* P_g *and is activated by* Q_g.

Observe that $\Gamma(g^{-1})$ is defined for g^{-1}: to be activated by Q_g means that the stable manifold of Q_g contains a (p-dimensional) disk of the superposition region. Since R_g and Q_g (respectively, S_g and P_g) are homoclinically related, Lemma 6.12 and Proposition 6.14 imply that, for every g in the open set $\mathcal{V}$,

$$W^s(P_g) \subset \overline{W^s(R_g)} = \overline{W^s(Q_g)} \quad \text{and} \quad W^u(Q_g) \subset \overline{W^u(S_g)} = \overline{W^u(P_g)}.$$

This is the main step towards showing that the homoclinic classes of P_g and Q_g coincide.

7

Robust Transitivity

A diffeomorphisms $f : M \to M$ is *robustly transitive* if every diffeomorphism in a C^1 neighborhood has dense orbits. More generally, we say that an invariant compact set Λ is robustly transitive if there is an open neighborhood U of Λ and a neighborhood $\mathcal{V}$ of the diffeomorphism f in $\mathrm{Diff}^1(M)$ such that for every $g \in \mathcal{V}$ the maximal invariant set Λ_g of g in $\bar{U}$ is a transitive set contained in U and $\Lambda_f = \Lambda$. The description of these systems is an important challenge: on the one hand, being robust, they can not be ignored in any global picture of dynamical systems; on the other hand, their dynamics is often not stable, indeed many different bifurcations may occur inside the set, without affecting transitivity. However, such bifurcations can never produce sinks nor sources. In the case of surface diffeomorphisms this implies that the periodic points inside a robustly transitive set are persistently hyperbolic, in a uniform fashion. In fact,

Theorem 7.1 (Mañé [278]). *Every robustly transitive set of a surface C^1 diffeomorphism is a hyperbolic set.*

Section 7.1 presents several constructions of robustly transitive diffeomorphisms. A point to be remarked about all these examples is that, while they need not be uniformly hyperbolic, they all have some weak form of hyperbolicity: at least, a dominated splitting of the tangent bundle. Indeed, the main conclusion in this chapter is that *existence of a dominated splitting is a necessary condition for robust transitivity*. See Theorem 7.5 in Section 7.2. The basic idea behind this statement is the observation that in the absence of a dominated splitting one can perturb the derivative at some periodic orbit, to turn that orbit into an attractor or a repeller, thus, breaking transitivity. A consequence of this main result is that homoclinic classes of C^1-generic diffeomorphisms either admit a dominated splitting or are contained in the closure of (infinitely many) periodic attractors or repellers.

The next crucial step is to try to describe the dynamics of diffeomorphisms and sets with a dominated splitting. This is yet far from being understood in general, but some results are reported in Section 7.2.3. See also Chapter 5 for the special case of surface diffeomorphisms.

In fact, in some situations (especially in low dimensions) one has the stronger property of partial hyperbolicity, and that is better understood. The main reason is that may use strong stable and strong unstable foliations as very useful tools for understanding topological and ergodic properties of the system. In Section 7.3 we introduce two useful properties of these invariant foliations, accessibility and minimality, and discuss how frequent they are among partially hyperbolic systems. We also discuss the behavior of central foliations, which is still much less understood.

7.1 Examples of robust transitivity

The best known examples of robustly transitive diffeomorphisms are the Anosov diffeomorphisms, especially the linear ones, that is the diffeomorphisms induced on the torus $\mathbb{T}^n$ by a hyperbolic map in $GL(n, \mathbb{Z})$. Several questions about Anosov diffeomorphisms remain open, in particular it is not known which manifolds support such diffeomorphisms. Here we are going to focus on *non-hyperbolic* robustly transitive diffeomorphisms.

The first examples of these diffeomorphisms were constructed in the seventies by Shub [410] and Mañé [277], respectively, on $\mathbb{T}^4$ and on $\mathbb{T}^3$. Their examples are partially hyperbolic with a dominated splitting of the tangent bundle $TM = E^s \oplus E^c \oplus E^u$ into 3 non-trivial subbundles, the central subbundle E^c being one-dimensional and non-hyperbolic. Much later, [71] and [84] gave new classes of robustly transitive diffeomorphisms, on much more general manifolds and also satisfying weaker hyperbolicity conditions.

Let us list the three classes of known examples, which we are going to discuss in detail in the next sections:

(A) Perturbations of the time-1 of any transitive Anosov flow.

These examples are isotopic to identity, and they exist in many manifolds, other than the tori, already in dimension 3. It is not known which 3-manifolds do support such diffeomorphisms. See Section 7.2.4

(B) Skew-products: $\mathbb{T}^n \times N \to \mathbb{T}^n \times N$, $(x, y) \mapsto (A(x), B(x, y))$ and their perturbations, where A is a transitive Anosov diffeomorphism, N is any compact manifold, and the behavior in the y direction is dominated by that of A along its stable and unstable directions.

In these examples the non-hyperbolic central direction may be chosen tangent to the fibers $\{x\} \times N$ and so $\dim E^c = \dim N$, providing examples of robustly transitive diffeomorphism having a central direction of any dimension.

(C) Robustly transitive diffeomorphisms derived from Anosov, that is, obtained by deforming Anosov torus diffeomorphisms by isotopy. They may be constructed

- (in dim ≥ 3) without any stable subbundle, that is, $TM = E^u \oplus E^{cs}$, where E^{cs} is indecomposable and non-hyperbolic. Analogously, without any unstable subbundle.
- (in dim ≥ 4) without any hyperbolic (stable or unstable) subbundle, that is, $TM = E^{cu} \oplus E^{cs}$ is a dominated splitting where E^{cs} and E^{cu} are indecomposable and non-hyperbolic.

As we shall see in Theorem 7.5, these two last constructions correspond to the weakest possible hyperbolicity for robustly transitive systems.

7.1.1 An example of Shub

Consider Anosov diffeomorphisms A and B defined on the torus $\mathbb{T}^2$ such that the product map $(A, B): \mathbb{T}^2 \times \mathbb{T}^2 \to \mathbb{T}^2 \times \mathbb{T}^2$ is partially hyperbolic with central bundle E^c tangent to the fibers $\{x\} \times \mathbb{T}^2$ (for that it is enough to choose $A = B^n$ for any $n > 1$). Assume that A has fixed points p and q and B has a fixed point r.

Consider a map $f: \mathbb{T}^2 \times \mathbb{T}^2 \to \mathbb{T}^2 \times \mathbb{T}^2$ of the form $f(x, y) = (A(x), F_x(y))$, partially hyperbolic with E^c tangent to the fibers $\{x\} \times \mathbb{T}^2$, and also that $F_p = B$.

Proposition 7.2. *The diffeomorphism f is robustly transitive.*

Proof. As the central foliation $\mathcal{F}^c$ (tangent to the subbundle E^c) is differentiable, Theorem B.10 implies that the pair $(\mathcal{F}^c, f)$ is structurally stable. It follows that the stable manifold and the unstable manifold of the continuation of the fiber $\{p\} \times \mathbb{T}^2$ are both dense in $\mathbb{T}^4$ for every C^1 perturbation of f. One deduces that the invariant manifolds of the fixed point (p, r) are both robustly dense in $\mathbb{T}^4$. This implies that f is robustly transitive: the stable manifold of the fixed point intersect any open set U of $\mathbb{T}^4$, thus the iterates $f^n(U)$, $n > 0$ accumulate arbitrarily large disks in the unstable manifold, which, on their turn, intersect any open set V. That is, $f^n(U) \cap V \neq \emptyset$. $\square$

To ensure that these examples are non-hyperbolic, it is enough to take $F_q: \mathbb{T}^2 \to \mathbb{T}^2$ having periodic points with different Morse indices (a saddle and a sink): then f also has points with different Morse indices, and transitivity implies that it can not be hyperbolic.

7.1.2 An example of Mañé

Consider a linear Anosov map $A: \mathbb{T}^3 \to \mathbb{T}^3$ having three real eigenvalues, $0 < \lambda_1 < 1 < \lambda_2 < \lambda_3$ and $3 < \lambda_3$. Let $\mathcal{F}^c$ be the linear foliation corresponding to the eigenvalue λ_2. Fix small domains $C' \subset C$ with C' much smaller than C. Consider a partially hyperbolic perturbation f of A preserving the foliation $\mathcal{F}^c$, having it as its central foliation, and coinciding with A outside C'.

Assume that the expansion of f in the strong-unstable direction remains bigger than 3. If C is small enough, there is a constant $L > 0$ such that, for every unstable segment σ of length $(\sigma) \geq L$, the set $(f(\sigma) \setminus C)$ contains an unstable segment of length bigger than L. On the other hand, given any unstable segment σ the lengths of its forward iterates $f^n(\sigma)$ are eventually bigger than L. This shows that every unstable segment contains some point x_0 whose forward orbit intersects C finitely many times only.

We claim that, given any small central segment α through such a point x_0, the lengths of its forward iterates increase exponentially. This is obvious while the iterates are shorter than the distance from C' to the complement of C, just because $f^n(\alpha)$ remains outside C' (discard finitely many iterates if necessary), where the central direction is actually expanding. When the segments are long just observe that the fraction of $f^n(\alpha)$ inside C' is very small: expansion of the big sub-segments outside C' suffices for expansion of the total length.

Now, given any open set $U \subset M$ consider an unstable segment $\sigma \subset U$, a point $x_0 \in \sigma$ visiting C finitely many times only, and a central segment $\alpha \subset U$ through x_0. By Theorem B.10 each leaf of $\mathcal{F}^c$ is dense in $\mathbb{T}^3$, in a robust fashion. So, the fact that the length of $f^n(\alpha)$ goes to infinity implies that the orbit of α is dense, and so the same is true for the orbit of U. This proves (robust) transitivity of f.

In order to guarantee that this construction yields non-hyperbolic maps, Mañé considered a fixed point p of A and performed a saddle-node bifurcation along a central leaf $\mathcal{F}^c$ inside a small neighborhood C' of p. In this way, f has periodic points with different Morse indices 2 and 1.

7.1.3 A local criterium for robust transitivity

In this section we explain the method used in [71] and [158] for proving robust transitivity. Let us start with the following simple criterium, which we used in Shub's example.

Lemma 7.3. *(Basic transitivity criterium) Suppose f has a hyperbolic periodic point p whose invariant manifolds are both (robustly) dense in M. Then f is (robustly) transitive.*

For simplicity, in what follows we restrict ourselves to 3-dimensional manifolds. See Section 6.2 for details and terminology concerning blenders. In the next statement note that $W^s(p)$ and $W^u(q)$ are two-dimensional.

Proposition 7.4. *(Blender criterium) Assume that*

1. *f is partially hyperbolic on the whole manifold and there is a dominated splitting $TM = E^u \oplus E^c \oplus E^s$ into three non-trivial subbundles such that E^u is uniformly expanding and E^s is uniformly contracting.*

2. *There are a constant $L > 0$ and hyperbolic periodic points p and q with Morse indices 1 and 2, respectively, such that the stable manifold $W^s(p)$ of p intersects every unstable segment of length bigger than L, and the unstable manifold $W^u(q)$ of q intersects every stable segment of length bigger than L.*

3. *The point q is homoclinically related to a blender Γ such that $E^c \mid \Gamma$ is uniformly expanding, and the one-dimensional unstable manifold $W^u(p)$ of p intersects every superposition region of Γ (we say that p activates Γ).*

Then f is robustly transitive.

Proof. From assumption 2 it follows that $W^s(p)$ intersects all unstable segments, since the length of forward iterates of unstable segments goes to infinity, and the stable manifold is invariant. This implies that the stable manifold of p is dense in M.

By compactness of the family of unstable segments of length $2L$, and the fact that the unstable foliation and stable manifold of p depend continuously on the dynamics, one easily gets that $W^s(p, g)$ intersects every unstable segment of length $2L$ for all g close to f. As before, this implies the density of $W^s(p, g)$. In other words, the stable manifold of p is robustly dense in M. The same arguments give robust density of the unstable manifold of q.

Using the principle of superposition of the blenders from Section 6.2, we deduce that the closure of the stable manifold of q robustly contains the stable manifold of p, thus it is robustly dense in M. At this point we know that both invariant manifolds of q are robustly dense in M, and so the basic criterium in Lemma 7.3 implies that f is robustly transitive.   □

The blender criterium is used in [71] to prove that both

- the time-1 map of a transitive Anosov flow and
- the product of an Anosov map by the identity on a compact manifold N

are approximated by (non-hyperbolic) robustly transitive diffeomorphisms. For the first situation one perturbs the dynamics along a closed orbit of the flow to get a pair of periodic orbits with different Morse index. In the second situation, one uses a chain of blenders to find a periodic point of index u (the dimension of the unstable manifold of the Anosov map) such that the closure of its unstable manifold contains the unstable manifold of a point of index $\dim N + u$.

7.1.4 Robust transitivity without hyperbolic directions

All robustly transitive diffeomorphisms we described so far admit hyperbolic directions, that is, either uniformly contracting or uniformly expanding. Here we describe the construction given in [84] of robustly transitive diffeomorphisms without any hyperbolic directions. This construction follows closely the one in Section 7.1.2, so we just emphasize the main differences.

Consider a linear Anosov map A of the torus $\mathbb{T}^4$ having 4 real eigenvalues, $0 < \lambda_1 < \lambda_2 < 1 < \lambda_3 < \lambda_4$. Fix small domains $C_1' \subset C_1$ and $C_2' \subset C_2$, with C_1 and C_2 disjoint. Fix an A-invariant cone field $\mathcal{C}^{cu}$ and an A^{-1}-invariant cone field $\mathcal{C}^{cs}$, around the planes corresponding to the expanding eigenvalues λ_3, λ_4 and the contracting eigenvalues λ_1, λ_2, respectively. Now consider diffeomorphisms f coinciding with A outside $C_1 \cup C_2$ and satisfying

1. Df preserves $\mathcal{C}^{cu}$ and expands area uniformly in this cone field.
2. Df^{-1} preserves $\mathcal{C}^{cs}$ and expands area uniformly in this cone field.
3. Df expands uniformly the norm of vectors in $\mathcal{C}^{cu}$ outside C_1'.
4. Df^{-1} expands uniformly the norm of vectors in $\mathcal{C}^{cs}$ outside C_2'.

By choosing sufficiently thin cone fields, we verify that there exists a constant $L > 0$ such that every center unstable disk D_1 (tangent to $\mathcal{C}^{cu}$) of radius bigger than L intersects any center stable disk D_2 (tangent to $\mathcal{C}^{cs}$) of radius larger than L.

Arguing as in Mañé's example, but using uniform expansion of area instead of uniform expansion of norms, we show that every center unstable disk D^1 contains a point whose forward orbit meets C_1 only finitely many times. This allows us to show that $f^n(D_1)$ contains a center unstable disk of radius bigger than L, for every large $n > 0$. The same argument shows that the large negative iterates of any center stable disk D_2 contains a center stable disk of radius bigger that L.

Finally, given any pair of open sets U and V, we may choose a center unstable disk $D_1 \subset U$ and a center stable disk $D_2 \subset V$. Thus, forward iterates of U and backward iterates of V intersect eventually: $f^n(U) \cap f^{-m}(V)$ for all large n and m. This implies that f is transitive.

In order to ensure that f does not preserve any hyperbolic subbundle, we choose the diffeomorphism such that C_1 contains a fixed point p_1 of f of index 2 with non-real contracting eigenvalues together with a fixed point q_1 of index 3. This prevents the existence of uniformly contracting subbundles: the presence of p_1 implies that a stable subbundle would have to be of dimension at least 2, while q_1 implies that the dimension is at most 1. We also choose f so that C_2 contains a fixed point p_2 of index 2 with non-real expanding eigenvalues together with a fixed point q_2 of index 1. Arguing as before, this prevents the existence of any uniformly expanding subbundle.

7.2 Consequences of robust transitivity

The main result in this chapter is the following generalization of Theorem 7.1 to arbitrary dimension:

Theorem 7.5 (Bonatti, Díaz, Pujals, Ures [73, 153]). *Every robustly transitive set of a C^1 diffeomorphism is volume hyperbolic.*

An invariant set Λ is called *volume hyperbolic* if there exists a splitting $T_\Lambda M = E^1 \oplus \cdots \oplus E^k$ into Df-invariant subbundles such that

1. the splitting is *dominated*: there exists $m \geq 1$ such that

$$\|Df^n \mid E^i(x)\| \, \|(Df^n \mid E^j(x))^{-1}\| \leq 1/2$$

   for every $n \geq m$, $i > j$, and $x \in \Lambda$.

2. the extreme subbundles E^1 and E^k are uniformly volume expanding and volume contracting: for all $n \geq m$ and $x \in \Lambda$,

$$|\det(Df^{-n} \mid E^1(x))| \leq 1/2 \quad \text{and} \quad |\det(Df^n \mid E^k(x))| \leq 1/2$$

In dimension 2, the only possibility is for the splitting to be into two 1-dimensional subbundles ($2 = 1+1$), and then volume hyperbolicity is the same as hyperbolicity. Therefore, Theorem 7.5 contains Theorem 7.1. Moreover, in dimension 3 there are two possibilities, $3 = 2 + 1$ and $3 = 1 + 1 + 1$. In either case, at least one of the subbundles is 1-dimensional and hyperbolic: either uniformly expanding or uniformly contracting. Thus, in this case robust transitivity implies partial hyperbolicity [153]: at least one of the invariant subbundles is hyperbolic. These results are best possible, by the examples in Section 7.1.

As mentioned before, the way to prove Theorem 7.5 is by showing that the lack of a dominated splitting leads to the creation of a sink or source, after a convenient perturbation. If the lack of dominated splitting is persistent, this yields infinitely many coexisting sinks or sources! We may apply this idea to a transitive set naturally associated to every hyperbolic saddle point p of a diffeomorphism f, its *homoclinic class*. This is defined as the closure of the transverse intersections between its invariant manifolds and denoted by $H(p, f)$. See Sections 10.4.1 and 10.4.2 for properties of homoclinic classes.

Theorem 7.6 (Bonatti, Díaz, Pujals [73])). *There exists a residual subset $\mathcal{R}$ of $\text{Diff}^1(M)$ such that for every diffeomorphism $f \in \mathcal{R}$ and periodic point p of f,*

- $H(p, f)$ *is contained in the closure of infinitely many sources or sinks,*
- *or else $H(p, f)$ admits a dominated splitting.*

In the rest of this section we explain the main ideas in the proofs of the theorems above. Roughly speaking, Theorem 7.5 follows from an idea in linear algebra: in the absence of domination, the eigenspaces of a linear map obtained by multiplication of several linear maps are very unstable: by small perturbation of each of the factors, one can mix the eigenvalues in order to get a homothetic transformation, which then leads to a new sink or a new source. The proof of volume expansion in the extremal subbundles uses the ergodic closing lemma [278] (see Appendix A).

7.2.1 Lack of domination and creation of sinks or sources

We begin the exposition with the, simpler, 2-dimensional case. The argument we are going to present, due to Mañé [278], shows the following: *if a set $\mathcal{P}$ of hyperbolic periodic saddles of a diffeomorphism of a compact surface does not admit a dominated splitting then it is possible to turn one of such periodic points to be a new sink or source, by a C^1 small perturbation.*

We begin with an elementary property of matrices in $GL(2, \mathbb{R})$. Denote by R_α the rotation of angle α in $\mathbb{R}^2$.

Lemma 7.7. *Consider an orientation preserving matrix $A \in GL(2, \mathbb{R})$ having two different real eigenvalues and whose eigenspaces form an angle less than $\varepsilon > 0$. Then there is $t \in [-\varepsilon, \varepsilon]$ such that $R_t \circ A$ has a pair of conjugate complex eigenvalues.*

A way to prove this is to consider the action h_t of $R_t \circ A$ on the projective space $\mathbb{RP}^1$, identified with the circle S^1. By hypothesis, the map h_0 has two hyperbolic fixed points, at distance less than ε. Moreover, outside the small interval between these fixed points, the sign of $h_0(x) - x$ is constant, say positive. So, $h_0(x) - x > -\varepsilon$ for every x in S^1. Consequently, $h_\varepsilon(x) - x > 0$ for all x in S^1. This implies that the rotation number of the circle map h_t is not constant in $[0, \varepsilon]$. Choosing t in this interval such that h_t has non-zero rotation number, we get that the eigenvalues of $R_t \circ A$ cannot be real, as claimed.

Now consider the natural splitting $T_p M = E^u(p) \oplus E^s(p)$ into eigenspaces, defined over $\mathcal{P}$.

Lemma 7.8. *Suppose that the angle $\mathrm{angle}(E^s(p), E^u(p))$ between $E^s(p)$ and $E^u(p)$ is not bounded away from 0 for $p \in \mathcal{P}$. Then there are g arbitrarily C^1-close to f and $p \in \mathcal{P}$ such that p is either a sink or a source of g.*

Theorem A.8 asserts that any small perturbation of the derivative of a diffeomorphism f along a periodic orbit p can be realized dynamically, that is, as the derivative of a C^1-nearby diffeomorphism g for which p is still a periodic point. Then, it is enough to see that there are arbitrarily small perturbations of the derivative of f along the orbit of some $p \in \mathcal{P}$ such that the resulting matrix has non-real eigenvalues of absolute value different from 1.

This is deduced from Lemma 7.7 as follows. Choose a point $p \in \mathcal{P}$ such that $\mathrm{angle}(E^s(p), E^u(p))$ is less than some small fixed ε and let n be the period of p. Then compose the derivative $Df(f^{n-1}(p))$ with a rotation $R_t \colon T_p(M) \to T_p(M)$ for an appropriate t, so that the resulting linear map

$$R_t \cdot Df(f^{n-1}(p)) \cdot Df(f^{n-2}(p)) \cdots Df(p)$$

has complex conjugate eigenvalues. Up to a final perturbation, the absolute value may be taken different from 1.

Lemma 7.8 completes the argument when the angle between E^s and E^u is not bounded away from 0. So, we are left to consider the case in which this angle is uniformly bounded from below on $\mathcal{P}$. In this case, up to a uniformly bounded change of the Riemannian metric, there are unitary vectors $e^s(p) \in E^s(p)$ and $e^u(p) \in E^u(p)$ forming an orthonormal basis of $T_p M$. In this basis, the derivative $Df(p)$ is given by a diagonal matrix. The set $\mathcal{P}$ is called *hyperbolic in the period* if there is $0 < \lambda < 1$ such that

$$\|Df^{n_p}(p)e^s(p)\| \le \lambda^{n_p} \quad \text{and} \quad \|Df^{-n_p}(p)e^u(p)\| \le \lambda^{n_p}$$

for every $p \in \mathcal{P}$, where n_p is the period of p.

Lemma 7.9. *Suppose $\mathcal{P}$ is not hyperbolic in the period. Then there exists a small C^1-perturbation g of f and a periodic point $p \in \mathcal{P}$ such that p is either a sink or a source of g.*

This follows by multiplying each matrix $Df(p)$ by a homothetic transformation of ratio $(1 + \varepsilon)$ or $(1 - \varepsilon)$, depending on which direction hyperbolicity fails, and then applying Theorem A.8.

Finally, consider the case where $\mathcal{P}$ is hyperbolic in the period. As we are assuming that $\mathcal{P}$ does not admit a dominated splitting, there are arbitrarily large n and $p \in \mathcal{P}$ such that $n < n_p$ and

$$\frac{\|Df^n(p)e^u(p)\|}{\|Df^n(p)e^s(p)\|} < 2.$$

Since $\mathcal{P}$ is hyperbolic in the period, $n_p - n$ must go to infinite as n goes to infinity.

Lemma 7.10. *Assume the splitting $E^u \oplus E^s$ over $\mathcal{P}$ is not dominated. Then there is an arbitrarily small perturbation of the derivative of f along the orbit of some point $p \in \mathcal{P}$ such that the angle between the eigenspaces of the corresponding matrix is arbitrarily small.*

This reduces the problem to the situation in Lemma 7.8, thus ending the proof of the statement made at the beginning of this section. So let us now explain the proof of this lemma.

Consider n and p as in the inequality above. Since n is large and the matrices of the derivatives are diagonal, by multiplying each $Df(f^i(p))$, $0 \le i < n$ by

$$C = \begin{pmatrix} \dfrac{1}{1+\varepsilon} & 0 \\ 0 & 1+\varepsilon \end{pmatrix}$$

we can make the quotient $\|Df^n(p)(e^u(p))\| / \|Df^n(p)(e^s(p))\|$ arbitrarily small, Denote $A(f^i(p)) = C \cdot Df(f^i(p))$. Then consider

$$B = \begin{pmatrix} 1 & 0 \\ \varepsilon & 1 \end{pmatrix}$$

and replace $A(p)$ by $A'(p) = A(p) \cdot B$. At this stage the direction $E^u(p)$ is no longer invariant, but our perturbations do keep $E^s(p)$ invariant. The linear map $A(f^{n-1}(p)) \circ \cdots \circ A(f(p)) \circ A'(p)$ sends $E^s(p)$ to $E^s(f^n(p))$, and $E^u(p)$ to some straight line $\tilde{E}(f^n(p))$ forming a small angle with $E^s(f^n(p))$, provided n is large enough. Through a final perturbation at times $> n$ we are going to recover the invariance of E^u, while keeping E^s invariant. This means they will again be the eigenspaces and so, since at time n their angle is small, the proof will be complete.

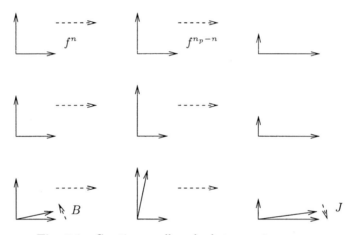

Fig. 7.1. Creating small angles between eigenspaces

So, let us explain this last step, with the aid of Figure 7.1. Since the periodic orbit is hyperbolic at the period, the image of $e^u(p) + \varepsilon e^s(p)$ under $A^{n_p}(p)$ makes a small angle with $E^u(p)$. In other words, $A(f^{n_p-1}(p)) \circ \cdots A(f(p)) \circ A'(p)$ maps $E^s(p)$ to itself and $E^u(p)$ to a subspace $\tilde{E}^u(p)$ forming a small angle with $E^u(p)$. Then there exists a matrix J close to the identity mapping $E^s(p)$ into $E^s(p)$ and $\tilde{E}^u(p)$ into $E^u(p)$. Thus, $E^s(p)$ and $E^u(p)$ are eigenspaces of

$$J \circ A(f^{n-1}(p))) \circ A(f^{n-2}(p))) \circ \cdot \circ A(f(p)) \circ A'(p).$$

This is the final perturbation mentioned above, and so the proof of the lemma is complete.

7.2.2 Dominated splittings *vs.* homothetic transformations

The previous ideas extend easily to arbitrary dimension. One gets that, given a set $\mathcal{P}$ of hyperbolic periodic saddles all with the same Morse index (the same dimension of the unstable direction), if the natural splitting induced by the stable and unstable directions is not dominated then it is possible to change the index of some saddle by a small C^1 perturbation.

However, in dimension larger than 2 such bifurcations do not break transitivity, and they do occur in robustly transitive sets. Towards proving Theorem 7.5, we are going to show that *in the absence of any dominated splitting it is possible to perturb the derivative of f along a periodic orbit in order to turn into a homothetic transformation and, after an additional small perturbation, to create a sink or a source for a diffeomorphism close to f*. An important tool is Franks' lemma (Theorem A.8), according to which C^1 small perturbations of the derivative can always be realized dynamically.

The main novelty with respect to the previous arguments is the introduction of *transitions* between different periodic orbits, which permit a kind of concatenation of such orbits and corresponding multiplication of eigenvalues. In a few words, while in the 2-dimensional situation presented above we deal with a linear cocycle defined over period points and their perturbations only, the arguments that follow use strongly the dynamics on a homoclinic class and the continuity of the derivative: the transitions correspond to heteroclinic orbits.

We begin with another argument of linear algebra which is crucial in the proof. Consider a basis $\beta = \{v_1, \ldots, v_n\}$ of $\mathbb{R}^n$ and let L_i be the subspace spanned by v_i. Consider a linear map $A \in GL(\mathbb{R}, n)$ which is diagonal relative to the basis β, and $(n-1)$ linear maps $A_1, A_2, \ldots, A_{n-1} \in GL(\mathbb{R}, n)$ such that each A_j

- leaves invariant the subspaces $L_1, \ldots, L_{j-1}, L_{j+2}, \ldots, L_n$ and
- permutes L_j and L_{j+1}, that is, $A_j(L_j) = L_{j+1}$ and $A_j(L_{j+1}) = L_j$.

For arbitrarily small neighborhoods $\mathcal{A}$ of A and $\mathcal{A}_i$ of each A_i, consider all possible products of matrices in $\mathcal{A} \cup (\cup_i \mathcal{A}_i)$. Then some of these products are homothetic transformations. The idea is that, as the matrices A_i generate all permutations of the eigenspaces of A, it is possible to mix the eigenvalues of A to obtaining a matrix having essentially the same rate of expansion in all directions.

In the model we have in mind, the matrices A and A_i correspond to the derivatives $Df^k(p)$, $k = \mathrm{per}(p)$ at periodic points p of a diffeomorphism f. Of course, to multiply derivatives corresponding to different points does not make dynamical sense. However, properties of homoclinic classes $H(p, f)$ allow us to give a sense to this procedure.

First of all, there is a dense subset K in $H(p, f)$ consisting of periodic points homoclinically related to p. As a dominated splitting extends to the closure of its domain (see Appendix B), absence of dominated splitting on $H(p, f)$ implies the same for K. Thus, it is enough to consider this set K of periodic orbits.

Next, given two fixed or periodic hyperbolic saddles x and y in K, and given any family $\{n_1, m_1, \ldots, n_k, m_k\}$ of arbitrarily big natural numbers, there are a natural number r (independent of n_i and m_i) and a saddle $z \in H(p, f)$

of period $t = kr + \sum_i (n_i + m_i)$ whose orbit spends alternately n_i consecutive iterates close to x and m_i consecutive iterates close to y. This means that the derivative of $Df^t(z)$ is, roughly, of the form

$$Df^t(z) = (T_2 \circ Df^{m_k}(q) \circ T_1 \circ Df^{n_k}(p)) \circ \cdots \circ (T_2 \circ Df^{m_1}(q) \circ T_1 \circ Df^{n_1}(p))$$

where T_1 and T_2 are called the *transitions* from x to y and from y to x, respectively. These transitions correspond to a bounded number of iterations, independent of n_i and m_i, so that their contribution to the product becomes negligible as n_i and m_i go to infinite.

Thus, it makes sense to multiply derivatives corresponding to different periodic points if they are homoclinically related. Besides the resulting linear maps correspond, roughly, to the derivative at another periodic saddle in the homoclinic class. Similar arguments show that in any homoclinic class there is a dense subset of periodic points whose derivative at the period can be made diagonal, with positive eigenvalues of multiplicity 1, by a small perturbation of the derivative along the orbit.

The construction above shows that a robust transitive set always admits a dominated splitting. Then the *finest* such splitting is well-defined. To get volume hyperbolicity it remains to prove uniform volume expansion and volume contraction along the two extreme subbundles. This is an application of Mañé's ergodic closing lemma (Theorem A.3), as we explain next.

Since E^1 does not admit any dominated sub-splitting, the previous arguments ensure that by a small perturbation we can produce a homothetic transformation along E^1. If E_1 is not uniformly volume expanding, by using the ergodic closing lemma we get a homothetic transformation along E^1 at a periodic point whose volume expansion rate is less than $1+\delta$. This is explained in more detail in Appendix A.2. So, maybe after a new perturbation, we get a homothetic transformation with contraction rate strictly smaller than 1. By domination, this implies that all the subbundles E_i are contracting. This means that the periodic point is really a source, which is a contradiction.

7.2.3 On the dynamics of robustly transitive sets

Here our goal is to deduce dynamical properties of robustly transitive sets, such as the kind of bifurcations they may exhibit, or the types of saddles they may contain, from the analysis of the finest dominated splitting. The arguments in the previous section imply that *if it is not possible to change, by perturbation, the index of a periodic point in a robustly transitive set Λ, then Λ is hyperbolic.* In the same spirit, the following result explains what happens when it is not possible to increase the indices of the periodic saddle points of Λ: it relates the maximal and minimal indices with the hyperbolicity of the subbundles in the finest dominated splitting.

Theorem 7.11 (Bonatti, Díaz, Pujals, Rocha [74]). *Consider $k \geq 1$ and an open subset U of a closed manifold. Let $\mathcal{U}$ be an open subset of* $\mathrm{Diff}^1(M)$

of diffeomorphisms f such that $\Lambda_f(U)$ is robustly transitive, has a dominated splitting $E_f \oplus F_f$ with $\dim E_f = k$, and does not contain saddles of Morse index less than k. Then E_f is uniformly expanding for every diffeomorphism f in $\mathcal{U}$.

If Λ is not hyperbolic then, up to a perturbation, it contains points of different indices. What can be said about the indices in between the maximal and minimal one?

Problem 7.12. Given a robustly transitive set Λ having points of indices k and m, does Λ contain a point of index l for each $l \in (k, m)$?

The answer is affirmative for most robustly transitive sets, as we are going to see. Given an open set $U \subset M$ denote by $\mathcal{R}(U)$ the subset of diffeomorphisms $f \in \mathrm{Diff}^1(M)$ for which $\Lambda_f(U) = \cap_{\mathbb{Z}} f^i(U)$ is a robustly transitive set.

Theorem 7.13 (Bonatti, Díaz, Pujals, Rocha [74]). *There is an open and dense subset $\mathcal{I}(U)$ of $\mathcal{R}(U)$ such that for all $g \in \mathcal{I}(U)$ the set of indices of the hyperbolic saddles of $\Lambda_g(U)$ is an interval in $\mathbb{N}$.*

The main ingredient in the proof is a notion of transitions associated to a heterodimensional cycle. These are similar to the transitions associated to homoclinically related periodic points in Section 7.2.2, but there are also some subtle differences that we now explain.

Given two periodic points x and y of Λ with different indices, using robust transitivity and the connecting lemma, Theorem A.5, we may create a heterodimensional cycle involving x and y. Then there is a perturbation of f having a periodic orbit which remains arbitrarily large periods close to x and close to y, with bounded transition times from one saddle point to the other. The derivative on this periodic orbit is a small perturbation of the product of arbitrarily large powers of the derivatives at x and y by a pair of bounded matrices, the *transitions*. One difference with respect to Section 7.2.2 is that here the new periodic orbits passes close to x and y only once, it closes the period already at the first return. A second difference is that in the homoclinic case the transitions are for the initial diffeomorphism, whereas here they are defined for perturbations of it.

Another important topic is the occurrence of homoclinic tangencies inside robustly transitive sets. A heuristic principle (see [74] for precise statements) is that *if the finest dominated splitting of $\Lambda_f(U)$ involves a non-hyperbolic subbundle of dimension bigger than 1 then there is g close to f such that $\Lambda_g(U)$ contains a homoclinic tangency*. Moreover, it is possible to determine the index of the hyperbolic saddle associated to the tangency. On the other hand, if all non-hyperbolic subbundles of the finest dominated splitting are 1-dimensional then it is not possible to create homoclinic tangencies inside the robustly transitive set by small perturbations of the system. In this case the prototypical bifurcations are heterodimensional cycles and saddle-node and period-doubling bifurcations.

Let us observe that homoclinic tangencies are indeed compatible with robust transitivity because, in higher dimensions, they need not lead to the creation of sinks or sources, unless the saddle point involved in the tangency is sectionally dissipative in the sense of Section 3.5. Indeed, it would important to know

Problem 7.14. What is the effect of homoclinic tangencies in the dynamics inside a robustly transitive set?

The same questions may, naturally, be asked for homoclinic classes.

7.2.4 Manifolds supporting robustly transitive maps

It is well-known that the existence of globally hyperbolic (Anosov) diffeomorphisms imposes very strong restrictions on the manifold; see [418]. In contrast, the corresponding question is wide open in the partially hyperbolic setting:

Problem 7.15. Which manifolds admit partially hyperbolic diffeomorphisms, respectively, diffeomorphisms admitting dominated splittings over the ambient space?

One obvious topological condition is that the tangent bundle should admit a decomposition into continuous sub-bundles. In the case of 3-dimensional manifolds this constitutes no obstruction at all, as the tangent bundle is always trivial. In fact, Mitsumatsu [295] proves that every 3-manifold supports a diffeomorphism with a dominated splitting into three sub-bundles. However, based on the fact that robustly transitive systems are partially hyperbolic, we expect that

Conjecture 7.16. There are no robustly transitive maps on the sphere S^3.

A partial proof was given by Díaz, Pujals, Ures [153], for the case of diffeomorphisms admitting either a center stable or a center unstable foliation. Their argument is based on Novikov's theorem [327, 420]: such a foliation must have a compact leaf separating the ambient space, and that would be an obstruction to transitivity. Also in this direction, we have

Theorem 7.17 (Brin, Burago, Ivanov [95]). *If $f : M \to M$ is a dynamically coherent partially hyperbolic diffeomorphism (with three invariant subbundles) on a 3-dimensional manifold M whose fundamental group is Abelian, then the action of f on the first homology $H_1(M, \mathbb{R})$ is also partially hyperbolic: it has some eigenvalue with norm larger than 1 and another eigenvalue with norm smaller than 1.*

Dynamical coherence (see [367]) means that the center stable bundle and the center unstable bundle are both integrable. The authors also observe that if the central direction E^c is uniquely integrable then f is dynamically coherent.

Notice that some hypothesis on the manifold is necessary for the conclusion, since the time-1 map of the geodesic flow on a hyperbolic surface is partially hyperbolic and dynamically coherent, yet acts trivially on the homology. On the other hand, one would like to remove all integrability conditions. As a direct corollary of the theorem one gets that *the sphere S^3 supports no partially hyperbolic dynamically coherent diffeomorphism with three invariant sub-bundles*, and the same is true for any other compact 3-manifold with finite fundamental group.

The proof of Theorem 7.17 is by contradiction, based on the following formulation of Novikov's theorem [327, 420]: if a continuous foliation of a 3-dimensional manifold admits a closed contractible curve transverse to the leaves then it has some compact leaf that bounds a solid torus. Let $\tilde{f}$ and $\tilde{W}^*$, $* \in \{u, s, c, cu, cs\}$ be the lifts of f and its invariant foliations to the universal covering $\tilde{M}$. Consider a small segment γ inside a leaf of $\tilde{W}^u$, and a small domain Γ around it. Assuming all eigenvalues of the action of f on the homology have norms less or equal than 1 (the case ≥ 1 is analogous, replacing f by its inverse) one gets that the diameter of $\tilde{f}^n(\Gamma)$ grows subexponentially with n. The condition on the fundamental group implies that the volume of balls in $\tilde{M}$ grows polynomially with the radius, and so the volume of $\tilde{f}^n(\Gamma)$ grows sub-exponentially as well. On the other hand, the length of $\tilde{f}^n(\gamma)$ grows exponentially fast, and so $\tilde{f}^n(\gamma)$ must be rather curly for large n. Thus, one finds unstable segments with length bounded from zero and endpoints arbitrarily close. Then there are small perturbations of such a segment which are closed and transverse to $\tilde{W}^{cs}$. In this one obtains closed contractible curves transverse to the center stable foliation W^{cs}, so that Novikov's theorem may be applied: there exists an f-invariant torus tangent to $E^c \oplus E^s$ at every point and bounding a solid torus in M. The sub-bundles E^s and E^c induce a pair of continuous invariant unit vector fields v^s and v^c on this torus, such that

$$|Df \cdot v^s| \leq \lambda \quad \text{and} \quad |Df \cdot v^c| \geq \theta, \qquad \exists \lambda < 1, \ \exists \theta > \lambda.$$

The key lemma asserts that any diffeomorphism of the 2-torus for which there exists such a pair of vector fields has non-trivial action on the fundamental group $\pi_1(\mathbb{T}^2)$ and, consequently, can not be extended to the solid torus. This gives the required contradiction.

Another very interesting set of results has been announced by Bennequin and by Herman (oral communications) concerning manifolds that support partially hyperbolic *symplectic* diffeomorphisms. In the symplectic category, partial hyperbolicity is equivalent to existence of a dominated splitting. Moreover, the stable and unstable sub-bundles have the same dimension and they are symplectic conjugate. See [65] for a proof. Corresponding to this, Bennequin proves that the Chern class of M must admit a factorization $c(M) = \alpha\beta$ where β involves only even terms. Then he checks that this condition excludes several manifolds, such as the complex projective spaces $\mathbb{C}P^n$ and their quotients by degree 2 equations (quadric manifolds), such as $S^2 \times S^2$.

7.3 Invariant foliations

Throughout this section $f : M \to M$ is a diffeomorphism admitting a partially hyperbolic invariant splitting $TM = E^u \oplus E^c \oplus E^s$ into three sub-bundles, all of them with positive dimension.

7.3.1 Pathological central foliations

Basic facts about invariant foliations of partially hyperbolic maps are recalled in Appendix B. In particular, the strong stable and the strong unstable foliations of C^2 diffeomorphisms are absolutely continuous with respect to Lebesgue measure [97]: the holonomy maps between any two transverse sections send zero Lebesgue measure sets to zero Lebesgue measure sets.

That is not true for the central foliation, in general:

Theorem 7.18 (Shub, Wilkinson [412]). *There is an open subset of C^1 volume preserving partially hyperbolic diffeomorphisms of the 3-torus admitting a unique central foliation which, nevertheless, is not absolutely continuous.*

Even more, Ruelle, Wilkinson [396] show that for these maps the conditional probabilities of the Lebesgue measure along central leaves are supported on a finite number of points, almost everywhere.

We prove the theorem using a variation of the construction in [412]. Start from the product $f = h \times \mathrm{id}$ of a linear area preserving Anosov diffeomorphism h in $\mathbb{T}^2$ by the identity on the circle $\mathbb{T}^1$. Let e_s, e_c, e_u be eigenvectors of Df corresponding to the eigenvalues $\lambda < 1 < \lambda^{-1}$, respectively. Now deform f by composing it with a diffeomorphism ϕ of $\mathbb{T}^3 = \mathbb{T}^2 \times \mathbb{T}^1$, of the form

$$\phi(\xi, z) = \big(\xi + \psi(z)(e_s + \alpha e_c), z\big)$$

for some $\alpha > 0$ and $\psi : \mathbb{T}^1 \to \mathbb{R}$ close to zero in the C^1 topology. Observe that ϕ preserves volume and the center unstable foliation of f. Since the central foliation is structurally stable, by Theorem B.10, the map $g = f \circ \phi$ (and any other map C^1 close to it) has a central foliation which is a topological fibration by circles.

Lemma 7.19. *If ψ is not constant then the integrated central Lyapunov exponent*

$$\lambda^c(g) = \int_{\mathbb{T}^3} \log |Dg \mid E^c_g| \, d\mathrm{Leb}$$

of $g = f \circ \phi$ along the corresponding central direction E^c_g is strictly negative.

In this setting the central Lyapunov exponent varies continuously with the map, because the central sub-bundle E^c does. In particular, λ^c remains negative for every map in a C^1 neighborhood of g. The following argument applies to any such map. Firstly, the central Lyapunov exponent must be strictly negative on a subset E of $\mathbb{T}^3$ with positive Lebesgue measure. Now, all central leaves are compact and so their forward and backward iterates have uniformly bounded length. Therefore, each leaf may intersect E only on a zero arc-length subset. However, if the central foliation were absolutely continuous then E would have to intersect many central leaves on subsets with positive arc-length. This shows that the central foliation is not absolutely continuous. To complete the proof of Theorem 7.18, we only have to give the proof of Lemma 7.19:

Proof. The matrices of Df, $D\phi$, $D(f \circ \phi)$ in the basis $\{e_s + \alpha e_c, e_c, e_u\}$ are

$$
Df = \begin{pmatrix} \lambda & 0 & 0 \\ a & 1 & 0 \\ 0 & 0 & \lambda^{-1} \end{pmatrix} \quad D\phi = \begin{pmatrix} 1 & \psi' & 0 \\ 0 & 1 & 0 \\ 0 & 0 & 1 \end{pmatrix} \quad D(f \circ \phi) = \begin{pmatrix} \lambda & \lambda\psi' & 0 \\ a & a\psi' + 1 & 0 \\ 0 & 0 & \lambda^{-1} \end{pmatrix},
$$

where $a = \alpha(1 - \lambda) > 0$. Since the perturbation preserves the original center stable bundle, we may represent the central direction E_g^c by $(\omega, 1, 0)$ with $\omega : \mathbb{T}^3 \to \mathbb{R}$ uniformly close to zero. By invariance, $Dg \cdot (\omega, 1, 0)$ may be written as $J^c g (\omega \circ g, 1, 0)$ where $J^c g$ takes values in $\mathbb{R}$. This translates into

$$
\omega \circ g = \frac{\lambda(\omega + \psi')}{1 + a(\omega + \psi')} \quad \text{and} \quad J^c g = 1 + a(\omega + \psi'). \tag{7.1}
$$

Now we use the elementary convexity fact that, for $t > -1/a$,

$$
\frac{\lambda t}{1 + at} \leq \lambda t \quad \text{and equality holds only if } t = 0.
$$

This gives that $\omega \circ g \leq \lambda(\omega + \psi')$, and equality holds only where $\omega + \psi'$ vanishes. Integrating the inequality over the 3-torus and using $\int \psi' \, d\text{Leb} = 0$ and invariance of Lebesgue measure,

$$
\int (\omega + \psi') \, d\text{Leb} = \int (\omega \circ g) \, d\text{Leb} \leq \lambda \int (\omega + \psi') \, d\text{Leb}
$$

and equality holds only if $\omega + \psi'$ is identically zero. In that case, using (7.1), both ω and ψ' would have to vanish identically. Thus, the assumption that ψ is not constant ensures a strict inequality:

$$
\int (\omega + \psi') \, d\text{Leb} < \lambda \int (\omega + \psi') \, d\text{Leb} \quad \text{and so} \quad \int (\omega + \psi') \, d\text{Leb} < 0,
$$

since $\lambda < 1$. Finally, $J^c g = |Dg \mid E_g^c|$ for appropriate choice of the Riemannian metric (the Lyapunov exponent is independent of this choice). Then

$$\lambda^c(g) = \int \log \left(1 + a(\omega + \psi')\right) d\mathrm{Leb} \le a \int (\omega + \psi') \, d\mathrm{Leb} < 0,$$

as claimed.   □

Applying a similar argument to the inverse, one proves that f is C^1 approximated by maps with central Lyapunov exponent strictly positive.

7.3.2 Density of accessibility

Due to the absolute continuity of the strong stable and the strong unstable foliations, one may propagate ergodic properties along their leaves. Thereafter, the topological features of these foliations are an important tool in the ergodic theory of partially hyperbolic systems, as we discuss in more detail in Section 8.2.

With this in mind we consider the following notion, first introduced in Dynamics by Brin, Pesin [97], and much developed in recent years, starting from the work of Grayson, Pugh, Shub [200]. Throughout, we assume that the diffeomorphism admits a partially hyperbolic splitting $E^u \oplus E^c \oplus E^s$ into three invariant subbundles.

Definition 7.20. A partially hyperbolic diffeomorphism $f : M \to M$ has the *accessibility property* if for any pair of points $x, y \in M$ there is a path joining x to y composed by finitely many segments inside strong stable or strong unstable leaves (a *su-path*) . More generally, f has the *essential accessibility property* if every set saturated by us-paths has either zero or full Lebesgue measure.

An important recent result of Dolgopyat, Wilkinson asserts that accessibility holds on an open and dense subset:

Theorem 7.21 (Dolgopyat, Wilkinson [170]). *There is an open and dense subset of C^1 partially hyperbolic diffeomorphisms on any compact manifold having the accessibility property. Moreover, this set intersects the spaces of volume preserving, or symplectic, partially hyperbolic diffeomorphisms on open and dense subsets.*

Combined with an older result of Brin [93], stated below as Theorem 8.13, this implies that among volume preserving partially hyperbolic diffeomorphisms the transitive ones contain a C^1 open and dense subset, and similarly in the symplectic case. See Section 8.5.

The proof of Theorem 7.21 has two main steps. Firstly, one proves *stable (robust) accessibility modulo a finite family of central disks*. That is, one chooses pairwise disjoint disks Δ_i, $i = 1, \ldots, k$ transverse to the sum $E^u \oplus E^s$ and such that

- given any points x and y in M there exists a path joining x to y and con- sisting of finitely many segments contained in strong stable leaves, strong unstable leaves, or the interior of some of the disks Δ_i ;
- there exists some large $N \geq 1$ such that the iterates $f^j (\cup_i \Delta_i)$, $j = 0, \ldots, N$ are pairwise disjoint.

Notice that this property is robust under small C^1 perturbations. Secondly, one carries out local perturbations close to each of the disks Δ_i in order to obtain that

- given any points x and y in Δ_i there exists a us-path joining x to y.

Combining the two steps one gets that any pair of points in M may be joined by a us-path, as claimed.

Let us outline the construction in this second step, in the case when the central sub-bundle is one-dimensional. See Figure 7.2.

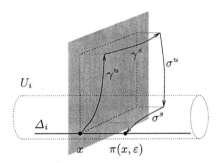

Fig. 7.2. Local perturbation to create accessibility

For each disk Δ_i one considers a continuous choice of two strong stable and two strong unstable arcs $\gamma^u(x, \varepsilon)$, $\gamma^s(x, \varepsilon)$, $\sigma^u(x, \varepsilon)$, $\sigma^s(x, \varepsilon)$, for $x \in \Delta_i$ and $0 < \varepsilon \leq \bar{\varepsilon}$, all with length $\approx \varepsilon$ and such that the concatenation

$$\gamma^u(x, \varepsilon) \cdot \gamma^s(x, \varepsilon) \cdot \sigma^u(x, \varepsilon) \cdot \sigma^s(x, \varepsilon)$$

links x to a point $\pi(x, \varepsilon) \in \Delta_i$. Next one considers a perturbation $g = h \circ f$ of the diffeomorphism f, where h is supported on a small box U_i close to Δ_i, in such a way as to change $\gamma^u(x, \bar{\varepsilon})$, uniformly over the whole Δ_i while keeping the other arcs essentially unchanged. To show that this is possible one uses the following observation: Due to the fact that the return time to U_i is very large (the choice of N in the first step), after the perturbation the strong unstable and strong stable sub-bundles on U_i are given by

$$E_g^u \approx Dh(E_f^u) \quad \text{and} \quad E_g^s \approx E_f^s.$$

In addition, we take U_i so that it avoids the other unstable segments $\sigma^u(x, \bar{\varepsilon})$. In this way one gets that, after perturbation,

$$\pi(x, \bar{\varepsilon}) \approx x + \bar{\varepsilon}\partial_i \quad \text{for every } x \in \Delta_i$$

where ∂_i is the tangent direction to Δ_i. By considering $\pi(x, \varepsilon)$, for the perturbed diffeomorphism g and for every $0 < \varepsilon \leq \bar{\varepsilon}$ we see that x may be joined to any points between x and $x + \bar{\varepsilon}\partial_i$ using a us-path formed by four arcs only. Since this holds uniformly for all x in the disk, we get that any two points in Δ_i are joined by a us-path, as claimed.

7.3.3 Minimality of the strong invariant foliations

A foliation $\mathcal{F}$ is *minimal* if every leaf is dense in the ambient manifold. The first result about minimality of strong invariant foliations applied to the geodesic flow of a compact hyperbolic surface: This is an Anosov flow, and its strong stable and strong unstable foliations are the horocyclic foliations. Furstenberg [187] proved that the horocyclic flow is uniquely ergodic.

More generally, we have

Theorem 7.22 (Plante [359]). *Let X be a transitive Anosov flow defined on a compact 3-manifold and $\mathcal{F}^u$ be its strong unstable foliation.*

1. *If there is a leaf of $\mathcal{F}^u$ which is not dense, then X is conjugate to the suspension of an Anosov diffeomorphism of the torus $\mathbb{T}^2$.*
2. *If there is a leaf of $\mathcal{F}^u$ which is dense then the unstable foliation is minimal.*

Minimality and unique ergodicity of the strong foliations were crucial tools to understand the ergodic properties of Anosov flows. Motivated by results such as Theorem 11.26, we expect the strong invariant foliations to play a similarly important role in the general partially hyperbolic setting.

So, here we are going to discuss some sufficient conditions for minimality of the strong foliations of diffeomorphisms. For simplicity, we always restrict ourselves to diffeomorphisms on 3-dimensional manifolds, partially hyperbolic with 3 invariant sub-bundles E^u, E^c, E^s.

Theorem 7.23 (Bonatti, Díaz, Ures [76]). *There is an open and dense subset of C^1 robustly transitive partially hyperbolic diffeomorphisms on any 3-dimensional manifold for which either the strong stable foliation or the strong unstable foliation is robustly minimal.*

The main step in the proof of this theorem is the construction of a *dynamical complete section* for the strong unstable (or the strong stable) foliation $\mathcal{F}^u$ (or $\mathcal{F}^s$). By such we mean a compact, possibly non-connected, surface S with boundary which is transverse to $\mathcal{F}^u$ and satisfies:

- $f(S)$ is contained in the interior of S,
- the restriction of f to S is a Morse-Smale diffeomorphism,
- every strong unstable leaf intersects transversely the interior of S.

Let us assume, for a moment, that a dynamical complete section S for $\mathcal{F}^u$ does exist. A simple argument, using the Morse-Smale condition and the λ-lemma, gives that every strong unstable leaf accumulates the leaf through some periodic point in S. So, it suffices to prove that the latter leaves are dense in the ambient space. Moreover, by a simple perturbation, it is enough to consider the leaves through the periodic points in S which are sinks for f restricted to S, that is, which have Morse index 1 in M. For these periodic points, the unstable manifold coincides with the strong unstable leaf. Hence, we are reduced to proving that the unstable manifolds of the periodic points of index 1 in S are dense in the ambient manifold. Now, a blender argument similar to the one in Section 7.1.3 implies that, for an open dense subset of C^1 partially hyperbolic diffeomorphisms, the closure of the unstable manifolds of any such periodic point contains the 2-dimensional unstable manifold of some periodic orbit of index 2. Such a 2-dimensional invariant manifold is always dense, just because it cuts the local stable leaf of some point with dense forward orbit. This proves that $\mathcal{F}^u$ is minimal, whenever f is in the intersection with that open dense subset.

Let us now explain how to construct a dynamical complete section for $\mathcal{F}^u$ of $\mathcal{F}^s$. It is clear that existence of such a section is an open property, so the real issue is density. So, we may suppose that all periodic points are hyperbolic. For simplicity, we consider only the case when the central bundle is uniquely integrable and orientable and f preserves this orientation. To each periodic saddle p of the diffeomorphism f we associate the half central leaf $L(p)$ to the right of p. This contains a finite number of periodic points, since they all have the same period as p. If $L(p)$ is not a circle, we say that p is of *type* 1 (respectively *type* 2) if the last periodic point p_* in $L(p)$ has Morse index 1 (respectively 2). If $L(p)$ is a circle then p is both of type 1 and type 2, and we set $p_* = p$.

Let Σ_1 be the set of type 1 and Σ_2 be the set of type 2 periodic points. Robust transitivity implies that, generically, periodic points are dense in M. It follows that the union of the two sets is dense in M, and so one of them must be dense in some open subset. Using transitivity once more, we may assume that at least of the two sets is dense in M. Let us suppose Σ_1 is dense: in this case we get a section for $\mathcal{F}^u$, allowing us to prove that the strong unstable foliation is minimal. A similar argument applies if Σ_2 is dense just replacing the diffeomorphism by its inverse: in this case we get a section for $\mathcal{F}^s$ and conclude that the strong stable foliation is minimal. In particular, if Σ_1 and Σ_2 are simultaneously dense then both strong foliations are minimal.

For every $p \in \Sigma_1$ there is a segment I_p contained in $L(p)$, with length bigger than some uniform constant and such that $f^n(I_p) \subset \text{int}(I_p)$ for some positive n: if $L(p)$ is a circle, just take $I_p = L_p$; otherwise let I_p be a segment starting close to p (to the left if p has index 1 and to the right if p has index 2) and extending long enough to the right of p_*. We construct a dynamical complete section of $\mathcal{F}^u$ by saturating finitely many (orbits of) segments I_p by local strong stable leaves. Since Σ_1 is dense, we may take sufficiently many such segments to ensure that this section does cut every strong unstable leaf.

For generic robustly transitive diffeomorphisms there is a unique homoclinic class, which coincides with the whole manifold. This is as immediate consequence of Theorem 10.26 below although, chronologically, it was first obtained in [72] as a consequence of Hayashi's connecting lemma [208]. As an application of Theorem 7.23, we can improve this result in dimension 3:

Proposition 7.24. *There is a C^1 open and dense subset of robustly transitive partially hyperbolic diffeomorphisms on any 3-dimensional manifold admitting some periodic point whose homoclinic class is the whole ambient manifold.*

To deduce this from Theorem 7.23, suppose for instance that the strong unstable foliation is robustly minimal. Then the stable and unstable manifolds of any hyperbolic periodic point p with Morse index 1 are both dense in M. This is immediate, by definition of minimality, for the unstable manifold. For the stable manifold it follows by observing that it intersects each sufficiently large strong unstable segment and so it is dense in the whole M; recall the arguments in Section 7.1.3. The density of the two invariant manifolds of p, together with partial hyperbolicity, imply that the homoclinic class of p is the whole manifold.

There are extensions of the previous results to higher dimensions, under the extra assumption that the central sub-bundle is robustly uniquely integrable [76]. Since it is known whether this property is dense, the following question is still open:

Problem 7.25. Let f be a robustly transitive partially hyperbolic diffeomorphism. Suppose that p is a periodic saddle of f whose continuation p_g is defined for every g in an open neighborhood $\mathcal{U}$ of f. Is $H(p_g, g) = M$ for all g in an open dense subset of $\mathcal{U}$?

A partial result is obtained in [74]: given any two hyperbolic periodic points p and q with different indices, there is an open set $\mathcal{U}$ of such that $H(p_g, g) = H(q_g, g)$ for all $g \in \mathcal{U}$.

7.3.4 Compact central leaves

The proof of the previous theorem leads to some natural questions about the dynamics in the central foliation. We continue to consider diffeomorphisms on 3-dimensional manifolds, partially hyperbolic with three invariant sub-bundles. Moreover, we assume that the periodic points are all hyperbolic. For simplicity, we consider only the case when the central sub-bundle is orientable and uniquely integrable, and its orientation is preserved by the diffeomorphism.

We say that a periodic central leaf L is of *stable type* if the first and the last periodic points in L have Morse index 1, and of *unstable type* if the first and the last periodic points of L have Morse index 2. The leaf L is of *saddle-node type* if the first saddle has index 1 and the last one has index 2, or vice-versa, and of *circle type* if it is compact.

The examples of robustly transitive diffeomorphisms listed in Section 7.1 follow into three classes, in what concerns their central leaves:

(1) all the central leaves are compact (perturbations of dominated products of Anosov diffeomorphism by other diffeomorphisms),
(2) there are both compact and non-compact central leaves, but those with periodic points are compact (perturbations of time-1 maps of transitive Anosov flows)
(3) all central leaves are non-compact and all are of the same (stable or unstable) type (diffeomorphisms derived from Anosov).

Problem 7.26. Are there partially hyperbolic transitive diffeomorphisms having

- simultaneously non-compact central leaves of stable and of unstable type?
- non-compact central leaves of saddle-node type?
- simultaneously non-compact and compact central leaves containing hyperbolic periodic points?

In a sense, all known robustly transitive diffeomorphisms are obtained by some sort of perturbation of hyperbolic systems. Examples as above, if they exist, should require essentially different constructions. For instance, we believe transitive diffeomorphisms with central leaves of both stable and unstable type can not be obtained via deformations of Anosov diffeomorphisms, or of skew-products based on Anosov diffeomorphisms. In this direction we have

Theorem 7.27 (Bonatti, Wilkinson [85]). *Let f be a transitive partially hyperbolic diffeomorphism on a compact 3-dimensional manifold M. Assume there exists a periodic circle γ such that $\mathcal{F}^u(\gamma) \cap \mathcal{F}^s(\gamma) \cap \gamma^c$ has some compact connected component. Then, f leaves invariant a Seifert circle bundle* [1]. *Consequently, the lift of some iterate of f to some finite covering of M is a dominated skew-product over an Anosov diffeomorphism of $\mathbb{T}^2$.*

Moreover, Pujals, Rodriguez-Hertz, Sambarino have recently announced that they can extend Theorem 7.17 to prove that every (non-hyperbolic) partially hyperbolic dynamically coherent diffeomorphism of $\mathbb{T}^3$ essentially belongs to one of two classes: deformations of Anosov or skew-products over Anosov diffeomorphisms of $\mathbb{T}^2$. A partially hyperbolic diffeomorphism f is *dynamically coherent* if its central foliation is uniquely integrable and it is the intersection of the center stable and center unstable foliations. Let us say that f is *robustly (stably) dynamically coherent* if any C^1 perturbation of it is dynamically coherent. Under this assumption, Theorem 7.23 can be improved as follows:

---

[1]Foliation of M by circles, which is trivial except for finitely many exceptional leaves.

Theorem 7.28 (Bonatti, Díaz, Ures [76]). *Let $\mathcal{D}$ be the subset of partially hyperbolic C^1 diffeomorphisms which are robustly dynamically coherent and have a periodic compact central leaf. There is an open and dense subset of $\mathcal{D}$ for which the strong stable and the strong unstable foliations are both minimal.*

In particular, we may apply Theorem 7.28 to perturbations of time-1 maps of transitive Anosov flows, and to perturbations of dominated skew products over Anosov diffeomorphisms, to conclude that their strong stable and the strong unstable foliations are both minimal. As for the other construction in Section 7.1, partially hyperbolic diffeomorphisms derived from Anosov, including Mañé's example in Section 7.1.2, we may apply Theorem 7.23 to conclude that, for a dense open subset, the strong stable foliation is minimal.

Problem 7.29. In Mañé's example is the strong unstable foliation also minimal (for a residual, or even open and dense subset)?

A crucial question, having in mind possible applications of these results to the study of ergodic properties:

Problem 7.30. Does robust minimality of a strong invariant foliation imply unique ergodicity of the diffeomorphism (at least for a residual subset of diffeomorphisms)?

One major challenge in this domain is to fill the gap between the C^1 and C^2 categories: while the generic statements above are proved for the C^1 topology, control of the ergodic behavior usually requires C^2 regularity at least.

8

Stable Ergodicity

A classical argument for proving ergodicity of conservative systems, going back to Hopf [218] and extended in a fundamental way by Anosov [16], may be summarized as follows. By Birkhoff's ergodic theorem, forward and time averages are well-defined and they coincide for every point on a full measure subset E. Clearly, forward time averages are constant on stable sets, and similarly for backward time averages and unstable sets. Therefore, to prove ergodicity it suffices to show that almost any pair of points in E may be joined by a path obtained by concatenation of finitely many arcs contained in stable sets and unstable sets, and *such that all endpoints are in E*.

This approach allowed Anosov [16] to prove that the geodesic flow of any compact Riemannian manifold with negative curvature is ergodic and, more generally, the same is true for any C^2 uniformly hyperbolic (Anosov) flow or diffeomorphism. The case of hyperbolic surfaces, as well as manifolds with constant negative curvature, had been dealt with by Hedlund and Hopf. This provided the first open sets of volume preserving systems which are ergodic: one speaks of stable ergodicity. Quite in contrast, KAM (Kolmogorov, Arnold, Moser) theory shows that *non-ergodicity* is very common among conservative systems: there often exist positive measure sets consisting of invariant tori; see for instance [454]. The material in this section is motivated by the fundamental problem of deciding which type of behavior is predominant.

A diffeomorphism f is called *stably ergodic* if it is in the C^2 interior of the set of all volume preserving ergodic diffeomorphisms [1]. There is a similar notion for volume preserving vector fields. This notion may be thought of as a kind of ergodic analogue, for conservative systems, of robust transitivity (see Section 8.5). However, the first examples of stably ergodic diffeomorphisms were discovered much later, when Grayson, Pugh, Shub [200] showed that

---

[1]Sometimes a stronger notion is used: f is in the C^1 interior of the set of all C^2 volume preserving ergodic diffeomorphisms. Moreover, C^2 is sometimes replaced by C^1 plus Hölder continuous derivative.

time-1 maps of the geodesic flow on surfaces with constant negative curvature are stably ergodic.

Since then, the theory has gone through rapid important progress, and the list of examples has also grown substantially. A broad survey can be found in Burns, Pugh, Shub, Wilkinson [105], and we also review the main examples in Section 8.1. Pugh, Shub conjectured that ergodicity should prevail among partially hyperbolic systems (an open dense subset), and that has been a central motivation for much work in this area. So far, the best characterization of stable ergodicity is Pugh, Shub's Theorem 8.6 in Section 8.3. The proofs of this and most other results in this area use sophisticated versions of the Hopf argument, relying on properties of the strong stable and strong unstable foliations. The most important property is accessibility, that we discuss in Section 8.2. Recently, [81] proved that any conservative partially hyperbolic diffeomorphism with 1-dimensional central direction is C^1-approximated by stably ergodic diffeomorphisms. See Section 8.5.

8.1 Examples of stably ergodic systems

The ergodicity problem was treated in the pioneer work of Brin, Pesin [97] in the context of partially hyperbolic systems. They needed very restrictive regularity assumptions on invariant foliations, and so could not exhibit new stable examples. In fact, the first non-hyperbolic stably ergodic systems were found only a few years ago, by Grayson Pugh, Shub [200]. The examples currently available run parallel to those of robustly transitive systems, reviewed in Section 7.1. The reasons for this are still not well understood.

8.1.1 Perturbations of time-1 maps of geodesic flows

Curiously, the first non-hyperbolic examples were found in a setting very close to the classical one:

Theorem 8.1 (Grayson Pugh, Shub [200]). *The time-1 map of the geodesic flow on the unit tangent bundle of a surface with constant negative curvature is stably ergodic.*

Theorem 8.1 was extended to surfaces with variable negative curvature by Wilkinson [451], and to compact manifolds with almost constant negative curvature in the work of Pugh, Shub [367]. In fact, the geodesic flow of any compact manifold with negative curvature and arbitrary dimension is stably ergodic, as we are going to see in Section 8.3.

8.1.2 Perturbations of skew-products

Stable ergodicity of compact Lie group extensions $F_\varphi : M \times G \to M \times G$, $F_\varphi(x, g) = (f(x), \varphi(x) \cdot g)$ over Anosov diffeomorphisms $f : M \to M$ has been investigated by several authors. We just quote

Theorem 8.2 (Burns, Wilkinson [107]). *Fix $r \in (1, \infty]$. Let $f : M \to M$ be a volume preserving Anosov diffeomorphism of class C^{r-} on a compact manifold, and let G be a compact connected Lie group. Then there exists a dense subset $\mathcal{E}$ of the space of C^{r-} functions $\varphi : M \to G$ such that the corresponding skew-product F_φ is stably ergodic*

When r is non-integer, C^{r-} means the same as C^r : the map is $[r]$-times differentiable and derivative of order $[r]$ is $(r - [r])$-Hölder . In the integer case it means that the map is $(r - 1)$-times differentiable and the derivative of order $r - 1$ is Lipschitz.

Theorem B in [107] also provides an explicit characterization of non-stably ergodic skew-products, in the case when M is an infranilmanifold. Recall that all known examples of Anosov diffeomorphisms live in this class of manifolds, which includes tori as a special case. In this setting, a skew-product is stably ergodic if and only if it is stably ergodic among skew-products.

8.1.3 Stable ergodicity without partial hyperbolicity

The previous constructions yield partially hyperbolic maps with invariant splittings $TM = E^u \oplus E^c \oplus E^s$, where all sub-bundles have positive dimensions. The first examples of robustly transitive diffeomorphisms without any invariant hyperbolic sub-bundle were given in [84]; recall Section 7.1.4. These examples contain a C^1 open subset of the space of volume preserving diffeomorphisms. Recently, Tahzibi has shown that such diffeomorphisms, and even a somewhat larger class, are stably ergodic:

Theorem 8.3 (Tahzibi [424]). *There exists a C^1 open set of volume preserving diffeomorphisms of the 4-dimensional torus which admit no uniformly contracting nor uniformly expanding invariant sub-bundle and such that every $C^{1+\alpha}$, $\alpha > 0$, diffeomorphism in this open set is ergodic.*

Conjecture 8.4. 1. Every stably ergodic diffeomorphism admits a dominated splitting and, in fact, is volume hyperbolic.
2. Every symplectic diffeomorphism stably ergodic (within symplectic diffeomorphisms) is partially hyperbolic.

When the ambient manifold has dimension 4, the second conjecture is contained in Arnaud [24, 26]. Observe that in the symplectic context, existence of a dominated splitting implies partial hyperbolicity, see [26, 65].

Concerning the first conjecture, [73] shows that in the absence of dominated splitting a completely elliptic periodic point may be created by a smooth C^1 small conservative perturbation. It remains to prove that such a periodic point may be used to break ergodicity, and perhaps even the transitivity, after another perturbation.

8.2 Accessibility and ergodicity

In this section we always consider diffeomorphisms admitting a partially hyperbolic splitting $TM = E^u \oplus E^c \oplus E^s$ where all three sub-bundles have positive dimensions.

The notion of accessibility, which we recalled in Definition 7.20, embodies the idea of joining different points in the manifold by paths formed by strong stable and strong unstable arcs (*su-paths*). However, by itself, accessibility is not enough to carry out the Hopf argument: what is missing is the crucial requirement that *all concatenation points of the su-path should belong to the full measure set E where the ergodic theorem holds*. This can obtained, for instance, for the geodesic flows on manifolds with constant negative curvature, using the fact that the strong stable and strong unstable foliations are smooth. Moreover, Hopf [218] proved that in the case of hyperbolic surfaces the two invariant foliations are always of class C^1, which was the crucial step in his proof of ergodicity of the corresponding geodesic flow.

A fundamental contribution of Anosov [16] was to realize that a much weaker regularity property, called absolute continuity, suffices for the Hopf argument, and that the stable and unstable foliations of geodesic flows of manifolds with negative curvature are always absolutely continuous. In fact, this is true for all C^2 uniformly hyperbolic systems. See Appendix C.

In the partially hyperbolic context, we still have absolute continuity, for the strong stable and the strong unstable foliations (Appendix B.2), but this alone is not sufficient to implement the Hopf argument: this time, stable/unstable leaves are not cross-sections for the stable/unstable foliation. Brin, Pesin [97] could bypass this difficulty, at the price of assuming Lipschitz regularity of the central foliation, which is not a robust property under small perturbations.

A breakthrough came with the recent work of Pugh, Shub, and their coauthors, see [105, 367] and references therein. Based on several partial results, they proposed

Conjecture 8.5 (Pugh, Shub [367]).

1. Stable accessibility is dense in the space of all partially hyperbolic C^2 diffeomorphisms (volume preserving or not).
2. A partially hyperbolic volume preserving C^2 diffeomorphism with the essential accessibility property is ergodic.
3. Thus, stable ergodicity is a dense property in the space of partially hyperbolic volume preserving C^2 diffeomorphisms.

Part (1) has been proved in the C^1 topology by Dolgopyat, Wilkinson, as we commented upon in Section 7.3.2. In the direction of part (2), the best current result is due to Pugh, Shub [367, 368]. This is the subject of the next section. Part (3) has been solved in the C^1 topology, when the central sub-bundle is 1-dimensional, as we shall see in Theorem 8.17 below.

8.3 The theorem of Pugh-Shub

Theorem 8.6 (Pugh, Shub [367, 368]). *Every C^2 volume preserving diffeomorphism which is partially hyperbolic, essentially accessible, dynamically coherent, and center bunched is ergodic.*

Recall that *essential accessibility* means that any subset saturated by *su*-paths has either zero of full Lebesgue measure, and *dynamical coherence* means existence and uniqueness of invariant center stable and center unstable foliations; see Appendix B.2. *Center bunching* is a technical condition which, roughly, means that the derivative is (uniformly) almost conformal along each of the three invariant sub-bundles, and almost an isometry along E^c. It ensures that the sub-bundles and the holonomies of their integral foliations are θ-Hölder for some θ not far from 1.

We only say a few words about the proof of Theorem 8.6, referring the reader to [105] for a detailed discussion and references. The basic strategy is to try and implement the Hopf argument by propagating *density points* of the set E, where the ergodic theorem holds, along the strong stable and strong unstable foliations. Density points are usually defined in terms of relative measure of the set inside small balls around the point. One important difficulty is that strong stable and strong unstable holonomies need not preserve the family of balls, nor any family of sets with bounded eccentricity. So, conceivably, they may not preserve the set of density points.

Instead, Pugh, Shub consider *julienne!density points*, defined using a different class of neighborhoods called juliennes such that, under the hypotheses of the theorem, their shape is not too distorted by the holonomies. For instance, center stable juliennes tend to be much longer in the center direction than in the stable direction (dually for center unstable juliennes), to compensate for poor distortion control along the central direction. Then the set of julienne density points is essentially preserved by the strong stable and the strong unstable holonomies, which is the key ingredient in setting the Hopf argument.

Based on Theorem 8.6, Burns, Pugh, Wilkinson [106] prove that the time-1 map of any volume preserving Anosov flow is stably ergodic, unless its stable and unstable sub-bundles are jointly integrable [2]. Joint integrability can not occur for contact Anosov flows, because they always satisfy accessibility [234]. Hence, in particular, *the time-1 map of the geodesic flow on any compact manifold with negative curvature is stably ergodic* (we learned this argument from Amie Wilkinson).

---

[2]That is, there exists some foliation tangent to the sum of the sub-bundles at every point.

8.4 Stable ergodicity of torus automorphisms

Let $A \in \mathrm{SL}(d, \mathbb{Z})$ and f_A be the automorphism induced by A on the d-dimensional torus. It is a classical fact that f_A ie ergodic if and only if no eigenvalue of A is a root of unity [205]. If A is hyperbolic then f_A is an Anosov diffeomorphism, and so it is stably ergodic. The following question was asked by Hirsch, Pugh, Shub [216]:

Problem 8.7. Is every ergodic linear automorphism of the torus stably ergodic?

A partial solution, and a very original approach to proving stable ergodicity, have been recently proposed by Rodriguez-Hertz [384]. In particular, he proves

Theorem 8.8 (Rodriguez-Hertz [384]). *Any ergodic linear automorphism of $\mathbb{T}^d$ is C^5 stably ergodic, for $d \leq 5$.*

For more precise statements we need the notion of pseudo-Anosov linear map. We say that A is *pseudo-Anosov* if no eigenvalue is a root of unity and the characteristic polynomial is irreducible over the integers and not a polynomial of a power x^k for any $k \geq 2$. This terminology is motivated by the following observation: Given a homeomorphism h on a surface S of genus g, if the action of h on the first homology group $H_1(S, \mathbb{Z}) = \mathbb{Z}^{2g}$ is pseudo-Anosov, in this sense, then h is isotopic to a pseudo-Anosov homeomorphism of S.

Theorem 8.8 is a direct consequence of the first claim in the statement that follows, together with the fact that for $d = 4$ any ergodic linear automorphism must be pseudo-Anosov, and for $d = 3$ or 5 it must even be Anosov. As a matter of fact, for odd dimensions pseudo-Anosov automorphisms are necessarily Anosov.

Theorem 8.9 (Rodriguez-Hertz [384]). *Let $\mathbb{T}^d$ denote the d-dimensional torus.*

1. *Any pseudo-Anosov linear automorphism of $\mathbb{T}^4$ is C^5 stably ergodic.*
2. *Any pseudo-Anosov linear automorphism of $\mathbb{T}^d$ with 2-dimensional central sub-bundle is C^{22} stably ergodic, for $d \geq 6$.*

The central sub-bundle corresponds to the eigenvalues of norm 1. We mention that Shub, Wilkinson [413] proved that f_A is *approximated* by stably ergodic diffeomorphisms, if A acts as an isometry on the central sub-bundle.

Let us comment on the proof of Theorem 8.9. The starting point is the criterium given by Theorem 8.6. Observe that the automorphism f_A is partially hyperbolic, dynamically coherent, and center bunched. Partial hyperbolicity and center bunching are stable properties under C^1 perturbations. So is dynamical coherence in this situation, because the unperturbed map has a C^1

central foliation; see [216, Proposition 2.3]. Therefore, the whole issue is accessibility or, more precisely, essential accessibility: the automorphism f_A itself is essentially accessible, but not accessible.

The main novelty in Rodriguez-Hertz's approach is that he is able to give a surprisingly strong description of the topology of accessibility classes. Let f be a small perturbation of f_A. The *accessibility class* of a point in $\mathbb{T}^d$ is the set of all points which may be joined to it by a su-path of f. It is useful to consider the lift $F : \mathbb{R}^d \to \mathbb{R}^d$ of f to the universal covering, which is also a partially hyperbolic map. Let us denote $C(x)$ the accessibility class of a point $x \in \mathbb{R}^d$ for the lift F. One key step is

Proposition 8.10. *Either* $C(0) = \mathbb{R}^d$, *in which case* f *has the accessibility property, or else* $\#\big(C(x) \cap W^c(0)\big) = 1$ *for all* $x \in \mathbb{R}^d$.

In fact, this is a consequence of the following, more detailed, statement:

Proposition 8.11. *For any* $x \in \mathbb{R}^d$ *one of the following holds:*

1. *either* $C(x)$ *is open*
2. *or* $C(x) \cap W^c(0)$ *is homeomorphic to the circle or to an interval*
3. *or* $C(x) \cap W^c(0)$ *contains exactly 1 point.*

Moreover, alternative (2) can not occur for the origin $x = 0$, *and the set of points* x *for which (3) occurs is either empty or the whole* $\mathbb{R}^d$.

To prove Theorem 8.9 from Proposition 8.10, there are two cases to consider. If $C(0) = \mathbb{R}^d$, the map f is accessible and thus ergodic. In the case where $C(x) \cap W^c(0)$ reduces to a single point, for all $x \in \mathbb{R}^d$, Rodriguez-Hertz ingeniously appeals to KAM arguments to prove that f is conjugate to f_A. It follows that f is essentially accessible and so, once more, ergodic.

8.5 Stable ergodicity and robust transitivity

The theories of stable ergodicity and robust transitivity had very parallel developments in these last few years and, as we mentioned before, all new classes of examples of robustly transitive systems may be adapted to get similar stably ergodic examples. It is then natural to ask

Problem 8.12. Does robust transitivity of conservative systems imply stable ergodicity?

Another fundamental question concerns the relations between accessibility and transitivity, namely in the dissipative context. The following very nice result of Brin gives a partial answer:

Theorem 8.13 (Brin [93]). *Let* f *be a partially hyperbolic diffeomorphism with the accessibility property and whose non-wandering set* $\Omega(f) = M$. *Then* f *is transitive.*

The hypothesis concerning the wandering set is automatic if the diffeomorphism is conservative. In that setting the hypotheses of Theorem 8.13 are satisfied on a C^1 open and dense subset, by a recent result of Dolgopyat, Wilkinson [170] quoted in Theorem 7.21. Thus, *robust transitivity is a dense property among conservative partially hyperbolic C^1 diffeomorphisms.* Even more recently, Bonatti, Crovisier used their closing lemma for pseudo orbits, Theorem A.7, to prove that

Theorem 8.14 (Bonatti, Crovisier [70, 69]). *Generic volume preserving C^1 diffeomorphisms are transitive.*

Theorem 8.13 is false outside the conservative world: Niţică, Török [326] construct an open set of non-transitive accessible diffeomorphisms. This is an application of the following result:

Theorem 8.15 (Niţică, Török [326]). *Stably accessible maps are C^2 dense in a neighborhood of the product of an Anosov volume preserving transitive diffeomorphism by the identity on the circle.*

Consider $f = h \times \phi$, where $h : M \to M$ is a transitive volume preserving Anosov diffeomorphism and ϕ is a circle diffeomorphism close to the identity and exhibiting some periodic attractor. Then there exists some open set $U \subset S^1$ such that $g(M \times \bar{U}) \subset M \times U$ for $g = f$ and, thus, for any g in a C^0 neighborhood. It follows that no such g is transitive, and we may take a C^2 open subset which are accessible.

Nevertheless, Niţică, Török [326] also show that stably ergodic diffeomorphisms are C^2 dense in a neighborhood of the product of a transitive volume preserving Anosov diffeomorphism by the identity on the circle, as well as in a neighborhood of the time-1 map of any transitive volume preserving Anosov flow, *inside volume preserving diffeomorphisms.*

8.6 Lyapunov exponents and stable ergodicity

Burns, Dolgopyat, Pesin [104] address Conjecture 8.5 within non-uniformly hyperbolic conservative systems, that is, such that all Lyapunov exponents are different from zero almost everywhere. The following theorem summarizes their main results:

Theorem 8.16 (Burns, Dolgopyat, Pesin [104]). *Let $f : M \to M$ be a partially hyperbolic volume preserving C^2 diffeomorphism on a compact manifold. Assume that the central Lyapunov exponents are negative on an invariant positive Lebesgue measure set A. Then*

1. *Every ergodic component of $f \mid A$ is an open set $\mod 0$. If f is transitive then A is dense and $f \mid A$ is ergodic.*

2. *If f is essentially accessible then it is ergodic. In particular, the central Lyapunov exponents are negative almost everywhere. If f is accessible then it is stably ergodic.*

The statement in (1) means that each ergodic component (more precisely, the set of points whose time averages are given by it) coincides with an open set up to zero Lebesgue measure. Let us sketch some main points in the proof of this theorem.

Using ideas from Alves, Bonatti, Viana [12, 84], together with the hypothesis on the Lyapunov exponents, one shows that almost every point in A is in the basin of some SRB measure ν which is an ergodic Gibbs u-state with negative central exponents. See Theorem 11.35 below. Then ν is the ergodic component of the Lebesgue measure that contains the point. For ν-almost every point, its (local) Pesin stable manifold is contained mod 0 in the basin of ν. Considering strong unstable leaves with uniform size through the points in that stable manifold, we get an open set which is mod 0 contained in the basin of ν. This gives the first half of (1). The second one is a direct consequence.

Next, essential accessibility implies that, given any ergodic component, almost every point may be connected to it by a *su*-path whose corner points are all regular, as in the Hopf argument. This gives ergodicity, as claimed in the first part of (2). The second one follows immediately. To get the last part, one begins by obtaining a uniform lower bound for the sizes of ergodic components of the perturbations of f, using the idea of hyperbolic times from [12]. Then one observes that if f is accessible then any perturbation is ε-accessible (the *su*-path saturation of any ε-ball is the whole manifold) which, in view of the previous observation, suffices for ergodicity.

In this context, Fayad [180] announces that essential accessibility alone does not ensure stable ergodicity.

Dolgopyat, Wilkinson [170] prove that stable accessibility is dense in the space of partially hyperbolic C^1 diffeomorphisms, as we have seen in Theorem 7.21. In another very recent paper, Baraviera, Bonatti [45] show how to approximate any volume preserving partially hyperbolic diffeomorphism by another such that the integrated sum of its central Lyapunov exponents is different from zero. This will be discussed in Section 12.5.1. In particular, in the case when the central sub-bundle is 1-dimensional the central Lyapunov exponent is non-zero for an open and dense subset of diffeomorphisms.

Combining these results, one gets a proof of Conjecture 8.5(3) in the C^1 topology, when the central sub-bundle is 1-dimensional:

Theorem 8.17 (Bonatti, Matheus, Viana, Wilkinson [81]). *Any partially hyperbolic volume preserving C^2 diffeomorphism on a compact manifold M having 1-dimensional central bundle is C^1 approximated by another which is stably ergodic.*

9

Robust Singular Dynamics

In this chapter we focus on the dynamics of flows on manifolds. While continuous time systems contain the main features of their discrete time counterparts, there is a very important phenomenon specific to the continuous case: *robust accumulation of singularities (equilibrium points) by regular orbits of the flow*. This is the main theme through the whole chapter. The first example of this phenomenon originated from the famous system of differential equations

$$\dot{x} = -\sigma x + \sigma y$$
$$\dot{y} = \rho x - y - xz \qquad (9.1)$$
$$\dot{z} = xy - \beta x,$$

proposed by Lorenz [267] as a simple model for sensitive dependence on initial conditions, loosely related to fluid convection and weather prediction. A detailed analysis of these equations may be found in the book of Sparrow [422]. Afraimovich, Bykov, Shil'nikov [5] and Guckenheimer, Williams [201, 452] proposed geometric models exhibiting a robust attractor with an equilibrium accumulated by regular orbits of the flow, for which the assertions made by Lorenz could be verified rigorously. An abstract class of *generalized hyperbolic sets* was then proposed by Pesin and Sataev [352, 405], extending those *geometric Lorenz models*.

However, it took yet another couple of decades before Tucker could prove that attractors similar to the geometric Lorenz models do occur in the original system (9.1):

Theorem 9.1 (Tucker [431, 432]). *The Lorenz equations support a robust strange attractor for the classical parameters, $\sigma = 10$, $\rho = 28$, $\beta = 8/3$.*

See Viana [443] for a historical account including an outline of the proof. By then, starting from the mid-nineties, a theory was being developed to explain the phenomenon of robust coexistence of singular and regular orbits in the same transitive set, and to describe the properties of such sets. A major progress was the work of Morales, Pacifico, Pujals [308, 306] showing that *robust singular sets of 3-dimensional flows are singular hyperbolic attractors or repellers* and, in fact, they share all the main properties of the geometric Lorenz attractors. An invariant set is called *singular hyperbolic* if it is partially hyperbolic with volume hyperbolic (expanding or contracting) central bundle and all singularities in it are hyperbolic. This remarkable result paved the way for a theory of generic flows unifying uniformly hyperbolic flows and Lorenz-like flows, which is already rather complete in dimension 3 as we shall see in Sections 9.3 through 9.5.

Before all that, cycles involving singularities have been studied by several authors, as part of the bifurcation theory of flows as well as a main mechanism for creating singular attractors. See Section 9.2. Finally, in Section 9.6 we briefly study flows combining critical behavior, due to homoclinic tangencies, with singular behavior, coming from the equilibrium points, and present some persistent non-robust singular attractors.

9.1 Singular invariant sets

In this section we describe the two main examples of singular hyperbolic sets, geometric Lorenz attractors and singular horseshoes, and also present a much more recent construction of multidimensional Lorenz-like attractors.

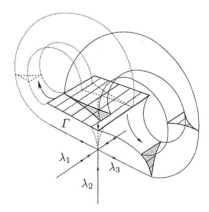

Fig. 9.1. Geometric Lorenz attractor

9.1.1 Geometric Lorenz attractors

An *attractor* for a flow X^t is a compact invariant set A containing a dense orbit and admitting an open neighborhood U such that

$$X^t(\overline{U}) \subset U \quad \text{for all} \quad t > 0 \quad \text{and} \quad A = \bigcap_{t \geq 0} X^t(U).$$

The attractor is called *robust* if $A_Y = \bigcap_{t \geq 0} Y^t(U)$ is also an attractor, that is, it is also transitive, for every vector field Y in a C^1 neighborhood of X. The attractor is *singular* if it contains equilibrium points of the vector field. Finally, we say that the attractor is *strange* if a dense orbit may be found exhibiting some positive Lyapunov exponent.

Historically, the first examples of robust singular strange attractors were the geometric Lorenz attractors, proposed in [5, 201, 452] as models for the behavior of the Lorenz equations (9.1) at parameter values near $\sigma = 10$, $\rho = 28$, $\beta = 8/3$. These are smooth flows in dimension 3, with the following main features (see Figures 9.1 and 9.2):

1. (existence of a singularity) The vector field X has a singularity θ such that $DX(\theta)$ has real eigenvalues $\lambda_3 > 0 > \lambda_2 > \lambda_1$ with $\lambda_3 + \lambda_2 > 0$.
2. (existence of a cross-section) There is a cross-section Σ intersecting the stable manifold of θ along a curve Γ that separates Σ into 2 connected components, and there is a smooth first return map $F : \Sigma \setminus \Gamma \to \text{int}(\Sigma)$ defined on the complement of Γ.
3. (invariant stable foliation) There is a smooth foliation of Σ into curves, having Γ as one of the leaves, invariant and uniformly contracted by forward iterates of the return map F. The quotient space of Σ by this foliation is a compact interval I.
4. (one-dimensional reduction) The quotient map f defined by F on the leaf space I is smooth and uniformly expanding on $I \setminus \{c\}$, with a discontinuity and infinite right and left derivatives at the point $c \in I$ that represents the leaf Γ. Moreover, f is topologically mixing (thus, transitive) in a strong sense: every interval is eventually mapped onto the whole I.

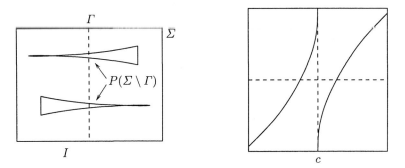

Fig. 9.2. Poincaré transformation and associated Lorenz map

For instance, one may ask that $|f'| \geq \sigma > \sqrt{2}$ and then topological mixing can be deduced as follows. The image of any interval J which does not contain

c is another interval with length $|f(J)| \geq \sigma |J|$. If J contains c then its image is the union of a pair intervals. Let J' be the largest of the two. Unless J' is fairly large (this case is treated at the end), it does not contain c, and so

$$|f(J')| \geq \sigma |J'| \geq (\sigma^2/2)|J| > |J|.$$

In this way one constructs a sequence of intervals with strictly increasing lengths contained in iterates of any given initial interval. This procedure stops when one encounters an interval J' large enough to contain c. Then, by construction, J' contains some connected component of $I \setminus \{c\}$. It follows that $f^2(J')$ is the whole I (see Figure 9.2), proving the strong mixing property.

The attractor:

The corresponding attractor Λ is characterized as follows. First, note that the restrictions of f to both $\{x < c\}$ and $\{x > c\}$ admit continuous extensions to the point $x = c$. Hence, f extends to a map $\tilde{f}$ that is 2-valued at $x = c$ and continuous on both $\{x \leq c\}$ and $\{x \geq c\}$. Correspondingly, the restriction of the Poincaré map F to each of the connected components of $\Sigma \setminus \Gamma$ admits a continuous extension to the closure, each one collapsing the curve Γ to a single point. Thus, F may be extended to a 2-valued transformation $\tilde{F} : \Sigma \to \Sigma$ defined on the whole cross-section and continuous on the closure of each of the connected components. Let

$$\Lambda_F = \bigcap_{n \geq 0} \tilde{F}^n(\Sigma) \subset \Sigma$$

and Λ be the saturation of the set Λ_F by the flow of X, that is, the union of the orbits of its points. Thus, orbits in Λ intersect the cross-section infinitely often, both forward and backward. This attractor has a complicated fractal structure that can been described as "a Cantor book with uncountably many pages" joined along a spine corresponding to the unstable manifold of the singularity. Notice that the unstable manifold accumulates on itself, and so the geometry is indeed very complex.

Dynamical properties of the attractor Λ may then be deduced from corresponding properties for the quotient map f. For instance, the fact that f is transitive implies that F is transitive on Λ_F and this implies that the flow itself is transitive on the attractor Λ. Similarly, the fact that f is sensitive to initial conditions (e.g. because $|f'| > 1$) implies sensitivity of the Poincaré map and of the flow on the basins of the corresponding attractors. Most important, a quotient map with similar properties exists for all nearby vector fields, and so properties such as transitivity and sensitivity are *robust* for these flows.

Yet, it should be noted that these systems are (structurally) *unstable*: the dynamics of a nearby vector field on the corresponding attractor is usually *not* equivalent [1] to the dynamics on the initial attractor. Once more, this

---

[1] Two vector fields are equivalent if there is a homeomorphism which maps the orbits of one to the orbits of the other, preserving the direction of time.

can be seen from the quotient map f, which is not necessarily conjugate to a perturbation. In fact, one can prove [201, 452] that the conjugacy classes are completely described by two parameters (the kneading sequences of the two singular values $f(c^+)$ and $f(c^-)$ relative to c: two maps are conjugate if and only if they have the same pair of kneading sequences), and then we have a corresponding statement for the flow itself.

9.1.2 Singular horseshoes

The primary motivation for this construction, due to Labarca, Pacifico [251], was to exhibit vector fields on compact manifolds *with boundary* that have a singularity accumulated by regular closed orbits and yet are structurally stable. As we pointed out before, geometric Lorenz attractors, although robust, are not stable (recall also Theorem 1.5).

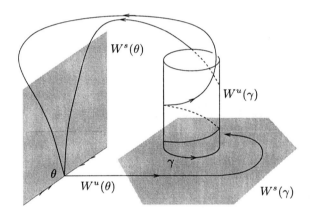

Fig. 9.3. Singular cycle giving rise to a singular horseshoe

Consider a vector field X on a 3-dimensional manifold having a singularity θ and a regular periodic orbit γ, both of saddle type, such that (see Figure 9.3)

1. $DX(\theta)$ has three real eigenvalues with $\lambda_1 < \lambda_2 < 0 < \lambda_3$ and $\lambda_3 + \lambda_2 > 0$.
2. There is a branch of $W^u(\theta)$ contained in $W^s(\gamma)$.
3. There are two orbits of transverse intersection between the invariant 2-dimensional manifolds $W^u(\gamma)$ and $W^s(\theta)$.

This geometric configuration is a codimension-1 bifurcation corresponding to a singular cycle formed by θ and γ (singular cycles will be defined and discussed in Section 9.2).

The singular horseshoe:

The dynamics of this vector field may be analyzed by reducing it to a one-dimensional map, in much the same way as in the case of the geometric Lorenz attractor. Consider a cross-section $S \simeq [0,1] \times [0,1]$ to the flow at some point p of the closed regular orbit γ. Let q be the first intersection with S of the branch of $W^u(\theta)$ contained in $W^s(\gamma)$. The Poincaré first return map F to the cross-section S is a modification of the standard horseshoe [418] with three symbols, as described in Figure 9.4. More precisely, the set of points of S for which F is defined consists of three strips S_i such that

- Each strip is mapped hyperbolically by F (roughly speaking, contracting the horizontal direction and expanding the vertical direction).
- S_0 contains the point $p \in \gamma$ (thus, p is a hyperbolic fixed point of F).
- The bottom sides of S_1 and S_2 are contained in the stable manifold of the singularity θ and so the return map F is not defined on them.

We call such a map a *singular horseshoe*. The last property is really the main difference with respect to the usual horseshoe. Note that F extends continuously (but not injectively) to the bottom sides of S_1 and S_2, as $F = q$.

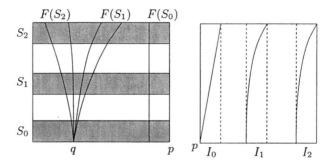

Fig. 9.4. Poincaré transformation and associated quotient map

One proves that F has an invariant stable foliation on Σ, whose leaves are roughly horizontal curves, and such that the quotient space is an interval I. The map f induced by F in this quotient space is illustrated on the right hand side of Figure 9.4. It is defined on the union of three compact subintervals I_i, corresponding to the three strips S_i, each of which is mapped onto the whole I. Moreover, f is uniformly expanding: $|f'(x)| \geq \sigma > 1$ at every point. The maximal invariant set Λ_f, consisting of the points such that the forward iterates are defined for all times, is a Cantor set. Moreover, $f \mid \Lambda_f$ is conjugate to the full one-sided shift with three symbols. Let Λ_F be the maximal invariant set for the return map F, that is, the set of points whose forward and backward iterates are all defined. The restriction $F \mid \Lambda_F$ is semi-conjugate to the full two-sided shift with three symbols: the semi-conjugacy fails to be a conjugacy (injective) only because of the pinching of stable directions caused by the lack of injectivity of F on the bottom sides of S_1 and S_2.

9.1.3 Multidimensional Lorenz attractors

Flows with robust singular attractors can be constructed in any dimension, in a sort of trivial way: the attractor lies inside a transversely contracting 3-dimensional submanifold. For a long time it was not known whether truly high dimensional Lorenz-like attractors, containing equilibrium points with more than one expanding direction, do exist. This problem was solved, affirmatively, by the following

Theorem 9.2 (Bonatti, Pumariño, Viana [82]). *For every $k \geq 2$ and $n \geq k+3$ (or for $k = 2$ and $n = 4$), there exist a manifold M of dimension n and a C^∞ flow on M exhibiting a robust attractor that contains a hyperbolic singularity with unstable dimension equal to k.*

Moreover, these attractors support a unique SRB measure, and they have the no-holes property: the ergodic basin coincides with the topological basin of attraction up to a set of Lebesgue measure zero.

For Lorenz-like attractors, and, as we shall see, for any robust singular attractor in dimension 3, the singularities have 1-dimensional unstable manifold. It is natural to expect that, in any dimension, the unstable manifolds of the singularities contained in robust attractors also have codimension 2 at least: this is the smallest codimension allowing the stable manifold of the singularity to intersect transversely the unstable manifold of some periodic orbit. In the construction of Theorem 9.2, the singularities have unstable manifolds of codimension 3, except in the case of 4-manifolds, where the codimension is indeed 2. This leads to the following question:

Problem 9.3. For any $n \geq 5$, does there exist a vector field on a manifold of dimension n with a robust singular attractor having a singularity whose unstable manifold has codimension 2?

Contrary to the classical geometric Lorenz models, these higher dimensional flows have *infinite modulus of stability*: the set of classes of topological equivalence in a neighborhood of the flow can not be parametrized using only finitely many parameters.

Let us comment on the proof of Theorem 9.2, in the most interesting case $n = k + 3$. In simple terms, one tries to construct the flow as a kind of suspension of a hyperbolic map defined on a manifold Σ of dimension $k+2$, in very much the same way as the classical geometric Lorenz models in dimension 3 may be thought of as suspensions of a piecewise smooth hyperbolic map on a rectangle. As it turns out, obstructions appear in higher dimensions that have no analog in the low-dimensional context. For instance, there are no smooth expanding maps from $D^k \setminus \{0\}$ to the unit k-dimensional disk D^k if $k \geq 2$ (clearly, there are plenty if $k = 1$). To overcome this sort of difficulties cross-sections Σ should have non-trivial topology: the construction in [82] uses $\Sigma = \mathbb{T}^k \times D^2$, where $\mathbb{T}^k$ is the k-dimensional torus.

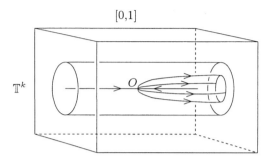

Fig. 9.5. Multidimensional singular attractors via suspension

The first step is to take some uniformly expanding map $g : \mathbb{T}^k \to \mathbb{T}^k$ and to consider its natural extension (or inverse limit), realized as a map $G : \mathbb{T}^k \times D^2 \to \mathbb{T}^k \times D^2$ on the product of the torus by the 2-dimensional disk D^2. For instance, if $k = 1$ then G would be the usual solenoid embedding [418]. Next, suspend G to a flow: this flow lives on the $(k+3)$-dimensional manifold $\mathbb{T}^k \times D^2 \times [0,1]/\sim$, where $\sim$ is the equivalence relation generated by $(z,1) \sim (G(z),0)$. Then deform the flow, so as to create an equilibrium point O with k expanding eigenvalues. This last step, resembling the classical construction of Cherry [128] flows, is illustrated by Figure 9.5 (for clearness, the D^2 coordinate is not represented).

9.2 Singular cycles

A *cycle* of a vector field X is a compact invariant subset $\Gamma = \Gamma_c \cup \Gamma_r$ where

- $\Gamma_c = \{\theta_1, \ldots, \theta_k\}$ is formed by finitely many critical elements (singularities or regular periodic orbits);
- Γ_r consists of non-periodic regular orbits whose α-limit set and ω-limit set are contained in Γ_c;
- for each $i \in \{0, \ldots k\}$ there is an orbit γ of Γ_r such that $\alpha(\gamma) = \theta_i$ and $\omega(\gamma) = \theta_{i+1}$ (respectively, $\omega(\gamma) = \theta_1$ if $i = k$).

The cycle Γ is called *singular* if it contains at least a singularity, and *hyperbolic* if all the critical elements involved in it are hyperbolic. The cycle in Figure 9.3 involves exactly one singularity and one regular periodic orbit.

The notion of singular cycle was introduced by Labarca, Pacifico [251], and a systematic study was initiated by Bamón, Labarca, Mañé, Pacifico [44], in the setting of vector fields on 3-dimensional manifolds. Singular cycles are a main mechanism for producing singular hyperbolic sets, and they may also be seen as a variant of heterodimensional cycles specific to continuous time systems.

9.2.1 Explosions of singular cycles

Let us say that a vector field is *simple* if its chain recurrent set [2] consists of finitely many hyperbolic critical elements. Bamón *et al* [44] considered 1-parameter families of vector fields $(X_\mu)_\mu$ on 3-dimensional manifolds, starting at a simple vector field X_{-1} and leaving the set of simple vector fields at a parameter $\mu = 0$ for which the vector field $X = X_0$ exhibits a singular cycle Γ. Their main assumptions are:

(a) The singular cycle Γ is hyperbolic and contains exactly one singularity θ.
(b) The eigenvalues of $DX(\theta)$ are real and satisfy $\lambda_1 < \lambda_2 < 0 < \lambda_3$ (up to time reversal).
(c) The intersections between the invariant manifolds of the elements of Γ_c are all transverse, except for the intersection between the unstable manifold of the singularity and the stable manifold of a periodic orbit $\gamma_1 \in \Gamma_c$.
(d) The latter consists of a unique orbit γ, which is tangent at infinity to the eigenspace of $DX(\theta)$ associated to the eigenvalue λ_2.
(e) The cycle has an *isolating block*, that is, an open neighborhood U such that Γ is the maximal invariant set of X in U.

The property that the eigenvalues be real, in (b), is a consequence of the assumption that the cycle at $\mu = 0$ corresponds to a first bifurcation. Concerning (c), a simple dimension counting argument shows that in a hyperbolic singular cycle not all intersections can be transverse.

Then one wants to describe the dynamics on the maximal invariant set

$$\Gamma(X_\mu) = \bigcap_{t\in\mathbb{R}} X_\mu^t(U)$$

of X_μ inside U, for small positive values of the parameter μ. Let $\mathcal{X}^r(M)$ denote the space of vector fields of class C^r. Throughout, we consider $r \geq 3$.

Theorem 9.4 (Bamón *et al* [44]). *There exists a neighborhood $\mathcal{U}$ of X in $\mathcal{X}^r(M)$ and a codimension-1 submanifold $\mathcal{N}$ of $\mathcal{U}$ containing X such that*

1. *for every $Y \in \mathcal{N}$ the maximal invariant set $\Gamma(Y)$ is a singular cycle, whose critical elements are the hyperbolic continuations of the critical elements in $\Gamma = \Gamma(X)$;*
2. *$\mathcal{U} \setminus \mathcal{N}$ has two connected components $\mathcal{U}^-$ and $\mathcal{U}^+$, such that every vector field Y in $\mathcal{N}^-$ is simple: the chain recurrent set inside $\Gamma(Y)$ consists of the continuations of the critical elements of X in Γ.*

---

[2]The chain recurrent set is formed by the points $p \in M$ such that for all $\varepsilon > 0$ there exist $(x_i)_{i=0}^m$ in M and positive numbers $(t_i)_{i=1}^{m-1}$ such that $x_0 = p = x_m$, and $X^{t_i}(x_i)$ is at distance less than ε from x_{i+1} for all $i = 0, \ldots, m-1$.

The main goal is to describe the behavior of vector fields in the component $\mathcal{U}^+$ corresponding to *a priori* non-simple vector fields. According to the next theorem, for the majority of such vector fields the dynamics related to the cycle is actually hyperbolic. Let $\mathcal{U}_h$ be the set of vector fields $Y \in \mathcal{U}^+$ such that the maximal invariant set $\Gamma(Y)$ is a (disjoint) union $\{\theta(Y)\} \cup \Sigma(Y)$ of the continuation $\theta(Y)$ of the initial singularity θ with a transitive hyperbolic set $\Sigma(Y)$. Let m denote Lebesgue measure in parameter space.

Theorem 9.5 (Bamón *et al* [44]). *There is $\delta > 1$ such that the set $B(t)$ of parameters $\mu \in [0, t]$ such that $X_\mu \notin \mathcal{U}_h$ satisfies*

$$\lim_{t \to 0} \frac{1}{t^\delta} m\big(B(t)\big) = 0,$$

for every parametrized family $(X_\mu)_\mu$ of vector fields starting in $\mathcal{U}^-$ and crossing $\mathcal{N}$ transversely at $\mu = 0$.

9.2.2 Expanding and contracting singular cycles

More precise information on the way $\mathcal{U}_h$ fills $\mathcal{U}^+$ depends on the eigenvalues at the singularity. The singular cycle is called *expanding* if $\lambda_3 + \lambda_2 > 0$ and *contracting* if $\lambda_3 + \lambda_2 < 0$. The next result illustrates the importance of this distinction.

Theorem 9.6 (Bamón *et al* [44]). *Under the hypotheses of Theorem 9.5,*

1. *If the cycle is expanding then the set $B(t)$ of parameters $\mu \in [0, t]$ such that $X_\mu \notin \mathcal{U}_h$ is a Cantor set, for any $t > 0$ such that $X_t \in \mathcal{U}_h$. Moreover, the limit capacity $c(B(t))$ goes to zero as $t \to 0$.*
2. *If the cycle is contracting then the set of parameters μ for which there exists a hyperbolic attracting orbit whose basin contains the unstable manifold of the singularity accumulates at $\mu = 0$.*

Part 1 implies, much in particular, that $m(B(t)) = 0$ for all small $t > 0$. In contrast, the parameters μ in part 2 form an open set with $X_\mu \notin \mathcal{U}_h$.

Let us explain what lies behind these differences. For simplicity, we focus on the case when the cycle involves only one regular periodic orbit and one singularity. The dynamics in the unfolding of such a singular cycle may be reduced to that of a 1-parameter family of maps of the interval. First, one reduces the flow to the Poincaré return map to some cross-section intersecting the regular orbit of the cycle. The assumptions imply that this return map admits an invariant stable foliation, for small values of the parameter $\mu > 0$. Next, taking the quotient with respect to this foliation, one reduces the return map to a one-dimensional transformation f_μ of the form

$$f_\mu \colon [0, \rho_\mu^{-1}] \cup [a_\mu, 1] \to [0, 1] \quad \begin{cases} f_\mu(x) = \rho_\mu x & \text{if } x \in [0, \rho_\mu^{-1}], \\ f_\mu(x) = \mu - K_\mu(x)(1-x)^{\alpha_\mu} & \text{if } x \in [a_\mu, 1], \end{cases}$$

where $a_\mu \to 1^-$ as $\mu \to 0^+$, $\rho_0 > 1$ is the expanding eigenvalue of the return map $F = F_0$ at its fixed point p, α_0 is the quotient $|\lambda_2|/|\lambda_3|$, and $K_\mu(1) > 1$ for all $x \in [a_\mu, 1]$. The constants ρ_μ and α_μ and the map K_μ depend continuously on the parameter μ.

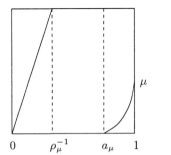 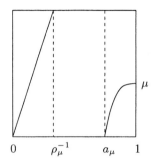

Fig. 9.6. One-dimensional reductions of expanding and contracting singular cycles

In the expanding case one has $\alpha_0 < 1$. Then, for all small μ, the derivative of the map f_μ in the interval $[a_\mu, 1]$ goes to infinity as μ goes to 0; notice that the interval also shrinks to the point 1. Since the derivative in $[0, \rho_\mu^{-1}]$ is always larger than 1, the map f_μ is uniformly expanding on its domain of definition. In particular, all periodic points are repelling. In fact, in this case bifurcations of periodic orbits may occur only when the point 1, which corresponds to the separatrix of the singularity involved in the cycle, becomes periodic: in terms of the flow this corresponds to the creation of a homoclinic connection due to collision between the singularity and a periodic orbit. In the contracting case one has $\alpha_0 > 1$. Then, $f'_\mu(1) = 0$ and so, for all small μ, the derivative is close to zero in absolute value at all points of $[a_\mu, 1]$. Then periodic attractors may arise, for instance, via saddle-node bifurcations and, in some cases, period-doubling bifurcations.

The next result completes a detailed description of the dynamics on $\mathcal{U}^+$ in the expanding case. Recall that the bifurcation set $\mathcal{U}^+ \setminus \mathcal{U}_h$ is extremely thin, by the first part of Theorem 9.6:

Theorem 9.7 (Bamón et al [44]). *Under the hypotheses of Theorem 9.5 and assuming that the singular cycle is expanding, the set $\mathcal{U}^+ \setminus \mathcal{U}_h$ is laminated by codimension-1 submanifolds such that for vector fields Y in the same lamina the dynamics on the maximal invariant set $\Gamma(Y)$ is the same up to topological equivalence. Moreover, $\Gamma(Y)$ is a chain recurrent expansive set for every $Y \in \mathcal{U}^+ \setminus \mathcal{U}_h$.*

We have seen in the second part of Theorem 9.6 that the unfolding of contracting cycles is comparatively richer, for it includes the presence of attracting periodic orbits. However, the result that we are going to state next means that, roughly speaking, that is the only main new phenomenon: if one takes the possible existence of a (unique) attracting periodic orbit in consideration, then it remains true that the bifurcation set has zero measure close to $\mu = 0$.

To be more precise, let $\mathcal{U}_{sh}$ be the subset of vector fields $Y \in \mathcal{U}^+$ whose chain recurrent set inside $\Gamma(Y)$ consists of the continuation of the initial singularity, a hyperbolic transitive set, and a unique attracting regular closed orbit. Given a family of vector fields $(X_\mu)_\mu$ crossing $\mathcal{N}$ transversely at $\mu = 0$, let $B_c(t) \subset B(t)$ be the set of parameters $\mu \in [0, t]$ such that $X_\mu \notin \mathcal{U}_h \cup \mathcal{U}_{sh}$.

Theorem 9.8 (Pacifico, Rovella [336], followed by [402, 250]). *Let* $(X_\mu)_\mu$ *be a one parameter family of vector fields crossing* $\mathcal{N}$ *transversely at* $\mu = 0$, *such that* X *has a contracting singular cycle and* $X_\mu \in \mathcal{U}^-$ *for* $\mu < 0$. *Then there exists* $t > 0$ *such that* $m(B_c(t)) = 0$.

The assumptions on the family of vector fields X_μ in this last theorem are similar to those in [44], that we recalled in Section 9.2.1; see the original papers [250, 336, 402] for precise formulations. In particular, it is assumed that the cycle is hyperbolic and contains a unique singularity. Theorem 9.8 was first proved by Pacifico, Rovella [336] under a stronger contractivity condition, and assuming also that the cycle contains a unique periodic orbit. The first condition was removed by San Martín [402], and then the second one was also removed by Labarca [250]. This latter paper also gives a detailed description of the bifurcation scenario for contracting cycles, in parallel with the one provided by Theorem 9.7 for the expanding case.

9.2.3 Singular attractors arising from singular cycles

Bifurcations leading to robust singular attractors:

Several authors have looked for general bifurcation mechanisms giving rise to robust singular attractors, as part of a strategy for proving that such attractors occur in specific dynamical systems.

Very successful attempts were due to Rychlik [399] and to Robinson [381], who showed that the unfolding of certain homoclinic connections in $\mathbb{R}^3$ yields robust singular attractors. They consider axially symmetric vector fields, that is, invariant under $(x, y, x) \mapsto (-x, -y, z)$. Observe that the Lorenz equations (9.1) are axially symmetric. By *homoclinic connection* we mean that the initial vector field has a hyperbolic singularity such that the unstable manifold is contained in the stable manifold. Clearly, this is the simplest kind of singular cycle.

Rychlik [399] assumes the singularity to be expanding Lorenz-like (eigenvalues $\lambda_3 > 0 > \lambda_2 > \lambda_1$ with $\lambda_2 + \lambda_3 > 0$) and the connection to be *inclination-flip*: the center unstable manifold and the strong stable manifold are tangent along the connection. Robinson [381] considers *resonant* connections instead: the eigenvalues satisfy $\lambda_2 + \lambda_3 = 0$ or, in other words, the singularity is neither expanding nor contracting. Both situations correspond to bifurcations of codimension-2 in the space of axially symmetric vector fields. Moreover, both are shown to occur in certain cubic differential equations, thus providing the first examples of algebraic differential equations with singular strange attractors.

In a similar spirit, Ushiki, Oka, Kokubu [435] and Dumortier, Kokubu, Oka [173] prove that Lorenz-like strange attractors occur in the unfolding of local bifurcations of certain degenerate singularities.

Singular attractors across the border of hyperbolic flows:

Another issue that has attracted a good deal of attention is the possibility to obtain Lorenz-like attractors through first bifurcations from hyperbolic systems, for instance, immediately after a parametrized family of vector fields leaves the region of simple vector fields. A good number of examples are now available.

Afraimovich, Chow, Liu [6] describe a codimension-1 bifurcation leading directly from Morse-Smale flows to Lorenz-like attractors, through a singular cycle containing a saddle-node periodic orbit and a hyperbolic singularity of expanding type. The unstable separatrices of the singularity are contained in the stable set of the saddle-node orbit and meet its strong stable leaves transversely.

At about the same time, Morales [300] provided a different construction of singular strange attractors from hyperbolic flows, via a saddle-node bifurcation. In this case the saddle-node sits at a singularity. The bifurcation leads directly from a flow with a Plykin (uniformly hyperbolic) attractor to another with a Lorenz-like attractor.

This was much extended by Morales, Pacifico, Pujals [307] (see also [311, 310, 301, 369]) who proposed various other constructions, where a main novelty is that the attractors are dynamically distinct from the geometric Lorenz models, in the sense that their flows are not equivalent. This provided a wealth of new equivalence classes of singular strange attractors.

Problem 9.9. Classify the singular attractors that can be obtained by deformation of uniformly hyperbolic (Axiom A) flows.

9.3 Robust transitivity and singular hyperbolicity

This section is the mathematical core of Chapter 9. We discuss the characterization of robust singular sets of vector fields in dimension 3 provided by the theorem of Morales, Pacifico, Pujals [308, 306] according to which *every robust transitive set containing singularities of the flow is a singular hyperbolic attractor or repeller*. The precise statement is given in Theorem 9.19. In Sections 9.4 and 9.5 we shall explore consequences and applications of this geometric property of singular hyperbolicity.

Theorem 9.19 was the first manifestation of the close link between robustness and weak hyperbolicity properties, at about the same time as the corresponding statement for diffeomorphisms in dimension 3 [153]. Thus, it generated a main theme that permeates also Chapter 7 and most of Chapter 10. Currently, most results for flows are proved for dimension 3 only. An obviously important program is

Problem 9.10. Extend the theory presented in Sections 9.3 to 9.5 to flows in any dimension.

Given a vector field X, we say that an invariant set Λ is *isolated* if there is a neighborhood U of Λ such that $\Lambda = \bigcap_{t \in \mathbb{R}} X^t(U)$. Then U is called an *isolating block* for Λ. An invariant set Λ is *transitive* if the forward orbit of some point is dense in Λ.

Definition 9.11. A compact isolated transitive set Λ is C^r-*robustly transitive* if there is an isolating block U and a neighborhood $\mathcal{U}$ of X in $\mathcal{X}^r(M)$ such that the set $\Lambda_Y = \bigcap_{t \in \mathbb{R}} Y^t(\overline{U}) \subset U$ is transitive for every Y in $\mathcal{U}$. In the case $r = 1$ we just say that Λ is *robustly transitive*.

9.3.1 Robust globally transitive flows

We begin by considering robustly transitive vector fields, that is, such that the whole ambient $\Lambda = M$ is a robustly transitive set. The following result is an important predecessor to Theorem 9.19:

Theorem 9.12 (Doering [162]). *Let X be a robustly transitive vector field on a closed manifold of dimension 3. Then X is uniformly hyperbolic (Anosov) and, in particular, it has no singularities.* •

The statement cannot be strictly true in higher dimensions: consider the suspension of any non-hyperbolic robustly transitive diffeomorphism such as exhibited in Chapter 7 for dimensions ≥ 3. However, an extension of Theorem 9.12 to arbitrary dimension, in the spirit of Theorem 7.5, was obtained recently:

Theorem 9.13 (Vivier [444]). *Let X be a robustly transitive vector field on any closed manifold. Then X has no singularities and it admits an invariant dominated splitting over the whole manifold.*

Here dominated splitting is for the linear Poincaré flow. See Appendix B. Let us say a few words about the proofs of these results. Theorem 9.12 is a consequence of the following

Proposition 9.14. *Let X be a vector field on a closed manifold M of dimension 3 whose non-wandering set $\Omega(X) = M$ and such that for every vector field in a C^1-neighborhood of X all the singularities are hyperbolic and all the regular periodic orbits are hyperbolic of saddle-type. Then X is Anosov and, hence, has no singularities.*

To see that the proposition does imply Theorem 9.12, consider any vector field Y in a small neighborhood of X. Transitivity implies that $\Omega(Y) = M$, and this implies that Y has no attracting nor repelling periodic orbits. Moreover, all regular periodic orbits must be hyperbolic: in this low dimensional setting a non-hyperbolic periodic orbit may always be perturbed to create a periodic attractor or repeller. This shows that all regular periodic orbits of Y are hyperbolic saddles. Suppose Y has some singularity. Then we can approximate it by vector fields Z with at least one singularity, and such that all singularities are hyperbolic (Kupka-Smale theorem, see [342]). The previous arguments ensure that Z satisfies the other assumptions of Proposition 9.14 as well. It follows that Z is Anosov and, consequently, has no singularities after all. This contradiction proves that Y has no singularity either. So, we may apply Proposition 9.14 to conclude that all vector fields Y in a neighborhood of X are Anosov.

Dominated splitting for the linear Poincaré flow:

Let us discuss the proof of Proposition 9.14. The first main step in the direction of proving hyperbolicity is to construct a dominated splitting for the linear Poincaré flow of X over the set $R(X) = \{x \in M : X(x) \neq 0\}$ of regular points. See Appendix B for definitions and comments on this notion. The starting point is the observation that there is a naturally defined splitting over periodic orbits, given by their eigenspaces. Crucial arguments of Pliss [358], Mañé [277], Liao [260] show that this decomposition is dominated, in a very uniform way: the domination constants may be taken the same for every periodic orbit and even for every nearby vector field. The reason is that, otherwise, one would be able to create a periodic attractor or repeller by a small perturbation of the system, which would contradict transitivity; see Section 7.2.1 for similar arguments. Roughly speaking, the next step is to extend this dominated splitting to the closure. Notice, however, that it is not known whether periodic orbits are necessarily dense in $\Omega(X) = M$. More generally,

Problem 9.15. Are periodic orbits dense in any robustly transitive set Λ of a vector field (or a diffeomorphism)?

This difficulty is bypassed by means of Pugh's closing lemma [364]: for every $x \in \Omega(X)$ there exist vector fields X_n converging to X and for which x is a periodic point. Consider $x \in R(X)$. Then, there is an X_n-dominated (even hyperbolic) splitting $N_n = N_n^1 \oplus N_n^2$ over the X_n-orbit of x. Using the aforementioned uniformity, one proves that N_n converges to an X-dominated splitting for the linear Poincaré flow over the X-orbit of x. Finally, taking the union over all x, one gets an invariant splitting for the linear Poincaré flow of X over the whole $R(X)$, and we are left to show that it is dominated. This amounts to proving uniqueness and continuity of the splitting, which is not quite as automatic as in the case of dominated splittings over compact invariant sets. The proof uses the fact that every singularity is hyperbolic.

Dominated splitting implies absence of singularities:

Now we explain that the dominated splitting for the linear Poincaré flow prevents the existence of equilibria: use the next proposition with $\Lambda = R(X)$.

Proposition 9.16. *Let X be a vector field on a closed 3-dimensional manifold, Λ be an invariant subset of $R(X)$ admitting a dominated splitting for the linear Poincaré flow, and σ be any hyperbolic singularity of saddle-type. Then σ does not belong to the interior of $\Lambda \cup \{\sigma\}$.*

The proof is a clever use of an elementary topological argument. Suppose the unstable manifold of σ has dimension 2; otherwise just consider the inverse flow. Consider a cylinder C transverse to $W^u_{loc}(\sigma)$ and orthogonal to the flow direction. Consider a small disk D through a regular point $y \in W^s(\sigma)$ and also orthogonal to the flow direction. Then there is a smooth Poincaré map $P : D \setminus \{y\} \to C^+$ where C^+ denotes one of the connected components of $C \setminus W^u_{loc}(\sigma)$. Let S_s, $s \in (0,1]$, be a partition of $D \setminus \{y\}$ into circles centered at y, with $S_1 = \partial D$ say. Then $C_s = P(S_s)$, $s \in (0,1]$, is a foliation by circles of a neighborhood of $C_0 = C \cap W^u_{loc}(\sigma)$ inside C^+. Let $N^1 \oplus N^2$ be the dominated splitting. It defines splittings on C and on D, since we took them orthogonal to the flow. One proves that for points in $W^u(\sigma)$ the (most expanded) subbundle N^1 must be contained in the tangent direction to the unstable manifold. It follows that N^2 is transverse to C_0 and, hence, to C_s for every small $s > 0$. Then, by invariance, N^2 is transverse to S_s for every small $s > 0$. This implies that the line bundle N^2 has no continuous extension to the point y, which is a contradiction.

Conclusion and higher dimensions:

The final step is a kind of continuous time version of Theorem 5.1: in the absence of singularities the dominated splitting of the linear Poincaré flow is actually hyperbolic. More precisely, see Liao [260] and [162, Theorem 3.4],

Proposition 9.17. *Let X be a vector field without singularities on a closed manifold M of dimension 3, whose non-wandering set $\Omega(X) = M$, and such that all the periodic orbits of any vector field in a C^1 neighborhood are hyperbolic. Then X is Anosov.*

The proof of Theorem 9.13 has a similar global structure. Firstly, one constructs a dominated splitting for the linear Poincaré flow, using methods from Theorem 7.5 together with the fact that all perturbations of the flow have neither attracting nor repelling periodic orbits. Secondly, one argues that a dominated splitting $N = N^1 \oplus N^2$ (where N^1 dominates N^2) prevents the existence of singularities. The key is to prove that no vector can exhibit, under the linear Poincaré, growth at an exponential rate close to that of a vector in N^1, for an arbitrarily large period of time, followed by decay at an exponential rate close to that of a vector in N^2, also for an arbitrarily large period of time. Such behavior always occurs close to a saddle-type singularity, and so this proves that there can be no singularity.

9.3.2 Robustness and singular hyperbolicity

In view of the previous results, it is tempting to look for some hyperbolicity for general robustly transitive sets, by building on the hyperbolic structure of their periodic orbits. The problem is that, in general, one can not avoid the presence of singularities, as shown by the examples in Section 9.1. Because of this fundamental difficulty [3], a conceptual understanding of robustness for flows with singularities stalled for several years, until the problem was solved by Morales, Pacifico, Pujals [308, 306] for vector fields in dimension 3, by means of the following notion:

Definition 9.18. A compact invariant set Λ of a vector field X on a closed manifold of dimension 3 is *singular hyperbolic* if Λ is partially hyperbolic with volume hyperbolic central direction, and every singularity in Λ is hyperbolic. A *singular basic set* is a singular hyperbolic set which is transitive and isolated.

Here the notion of partial hyperbolicity is relative to the flow X^t (not the linear Poincaré flow). See Appendix B. There are two possibilities for the splitting: either $T_\Lambda M = E^c \oplus E^s$ with 2-dimensional volume expanding E^c and 1-dimensional norm contracting E^s, or $T_\Lambda M = E^c \oplus E^u$ with 2-dimensional volume contracting E^c and 1-dimensional norm expanding E^u. Of course, the two possibilities are exchanged if one replaces the vector field X by its symmetric $-X$.

In either case, the flow direction is contained in the central bundle E^c. This and other basic properties of singular hyperbolic sets are collected in Proposition 9.25 below. Another important property, used in the proofs of the results that follow, is that *any subset of a singular hyperbolic set that contains no singularities is uniformly hyperbolic*. See Corollary B.13.

Examples of non-hyperbolic singular basic sets include the geometric Lorenz attractors and the singular horseshoes, presented in Section 9.1. The main result of the theory of singular hyperbolicity states that every robustly transitive set containing singularities looks very much like a geometric Lorenz attractor, either for the flow or for its inverse:

Theorem 9.19 (Morales, Pacifico, Pujals [308, 306]). *Let Λ be a robustly transitive set for a C^1 vector field X on a closed manifold of dimension 3. Then Λ is singular hyperbolic and all the singularities in it are Lorenz-like and have the same unstable index. If there are no singularities then Λ is a hyperbolic set. Otherwise, Λ is a non-hyperbolic attractor or a non-hyperbolic repeller, depending on whether the unstable index is 1 or 2.*

A singularity σ of a 3-dimensional vector field X is called *Lorenz-like* if it has eigenvalues $\lambda_1 < \lambda_2 < \lambda_3$ with $\lambda_1 < 0 < \lambda_3$ and either $\lambda_2 \in (-\lambda_3, 0)$

---

[3]It is clear that a hyperbolic splitting $E^u \oplus E^X \oplus E^s$, where $E^X = \mathbb{R}X$, can not extend continuously to singularities, and the linear Poincaré flow is not defined outside the set $R(X)$ of regular points.

or $\lambda_2 \in (0, -\lambda_1)$. These relations are inspired from the geometric Lorenz attractors. Lorenz-like singularities are hyperbolic, and so there are only finitely many of them in Λ. We shall see in Proposition 9.25 that singular hyperbolicity implies that non-isolated singularities are Lorenz-like. However, in the present context we must prove the Lorenz-like property first, on the way to proving singular hyperbolicity.

Indeed, the starting strategy for proving Theorem 9.19 is similar to that of Theorem 9.12, in that one builds the invariant dominated splitting from the eigenspaces of the periodic orbits. However, the real problem is to show that the splitting extends to the singularities. This is dealt with through a local analysis, which relies on the fact that the singularities are Lorenz-like. This last step does not make sense for the linear Poincaré flow P^t, of course, which is the reason why the theorem must be formulated in terms of an invariant splitting for the actual flow X^t.

In what follows we give an outline of the proof of Theorem 9.19, referring the reader to [308, 306] for the full details.

Characterizing attractors and repellers:

The first step in the proof of Theorem 9.19 is the following sufficient criterium for a robustly transitive set to be an attractor, valid in any dimension:

Proposition 9.20. *Let Λ be a robustly transitive set of a vector field X on a manifold of dimension $n \geq 3$ such that all critical elements in it are hyperbolic. Suppose Λ strictly contains a singularity σ with unstable dimension 1. Then Λ is an attractor.*

This is a consequence of the following, somewhat more general,

Proposition 9.21. *Let Λ be a transitive set admitting an isolating block U and such that every critical element in Λ is hyperbolic. Suppose there is a critical element $\theta \in \Lambda$ such that the unstable manifold of the continuation θ_Y is contained in U for every vector field Y in a C^1-neighborhood $\mathcal{U}$ of X. Then Λ is an attractor.*

To deduce Proposition 9.20 from this last result we only have to check that

Lemma 9.22. *The unstable manifold of the continuation σ_Y is contained in Λ_Y for every Y in a neighborhood of X.*

To see that this is so, choose points q_Y^+ and q_Y^- in distinct branches of $W^u(\sigma_Y)$, depending continuously on Y. Since Λ_Y is transitive, the singularity is accumulated by dense orbits and, hence, so is some of the branches of $W^u(\sigma_Y)$. This implies that at least one of the branches is contained in Λ_Y or, in other words, that the orbit of either q_Y^+ or q_Y^- is entirely contained in U. If the conclusion of the lemma were false then there would exist vector fields Z close to X such that, say, $q_Z^+ \in \Lambda$ and $Z^s(q_Z^-) \notin U$ for some s. We may take the isolating block to be compact, and then

$$Y^s(q_Y^-) \notin U \quad \text{for every } Y \text{ close to } Z. \tag{9.2}$$

This implies that q_Y^- is not contained in Λ_Y, for all these vector fields. Moreover, q_Z^+ is accumulated by some orbit dense in Λ_Z. Clearly, this dense orbit must also accumulate some point of $W^s(\sigma_Z) \setminus \{\sigma_Z\}$. Then, we may use the connecting lemma (Theorem A.4) to obtain a nearby vector field for which the orbit of q_Z^+ is a homoclinic connection associated to the continuation of σ_Z. After another small perturbation, to break the homoclinic connection, we obtain a vector field Y such that the forward orbit of q_Y^+ passes arbitrarily close to $Y^s(q_Y^-)$. In view of (9.2), this implies that the orbit of q_Y^+ leaves U, and so q_Y^+ is not contained in Λ_Y either. This contradiction proves the lemma, and that gives Proposition 9.20 from Proposition 9.21.

The proof of Proposition 9.21 is also based on the connecting lemma. To show that a transitive isolated set Λ is an attractor we only have to show that it is Lyapunov stable: there exists a neighborhood $U_1 \subset U$ such that $X^t(U_1) \subset U$ for all $t > 0$. Indeed, once that is done, the neighborhood $U_2 = \cup_{t>0} X^t(U_1)$ is an isolating block for Λ such that $X^t(U_2) \subset U_2$ for all $t > 0$ and, together with transitivity, that means Λ is an attractor.

So, suppose Λ is not Lyapunov stable. Then there exist points x_n converging to some $x \in \Lambda$ and times $t_n > 0$ such that $X^{t_n}(x_n)$ converges to some $q \notin U$. Since Λ is transitive, we may also find points y_n converging to some point $p \in W^u(\theta) \setminus \{\theta\}$ and times $s_n > 0$ such that $X_{s_n}(y_n)$ converges to x. Suppose that x is not a critical element of the flow. Then we are in a position to use the version of the connecting lemma in Theorem A.5, to obtain a vector field Y close-by such that $q \in W^u(\theta_Y)$. This contradicts the assumption that the unstable manifold of θ is robustly contained in Λ. The case when x is a critical element is treated along the same lines, using the fact that the critical element is hyperbolic. This shows that Λ is indeed Lyapunov stable.

Singularities of robustly transitive sets:

The next step is a local analysis of the singularities inside a robustly transitive set, to prove that they all have the same index and are Lorenz-like.

Proposition 9.23. *Let Λ be a robustly transitive set for a vector field on a 3-dimensional manifold. Then all the singularities in Λ are hyperbolic and have the same unstable index. Supposing there is some singularity, Λ is an attractor if the index is 1 and a repeller if the index is 2.*

Assume for a while that all the singularities are hyperbolic. Recall, from Corollary B.13, that all periodic orbits are hyperbolic as well. Thus, we may conclude from Proposition 9.20 that Λ is an attractor if it contains some singularity with unstable index 1 and, applying the proposition to the inverse flow, Λ is a repeller if it contains some singularity with unstable index 2. Suppose the two possibilities coexist. Then Λ is both an attractor and a repeller. It is not difficult to deduce that Λ is an open subset of M and so, by connectedness, it coincides with the whole ambient manifold. However, by Theorem 9.12, in that case the vector field is Anosov and has no singularities at all. This contradiction proves that the unstable index is the same for all singularities. Finally, suppose there is some non-hyperbolic singularity. Bifurcating this singularity we may construct two new ones, with different indices. Then, using the Kupka-Smale theorem [342], we may find a nearby vector field such that all the singularities are hyperbolic and the continuation of Λ contains two singularities with different indices. This contradicts the previous case, and that gives the conclusion of Proposition 9.23 in general.

Corollary 9.24. *Let Λ be a robustly transitive set for a vector field on a closed 3-dimensional manifold. Then all the singularities in Λ are Lorenzlike. Moreover, Λ intersects the strong invariant (strong stable or strong unstable)manifold of every singularity only at the singularity.*

Transitivity implies that all singularities in Λ are of saddle-type. Consider a singularity $\sigma \in \Lambda$ and let $\lambda_1, \lambda_2, \lambda_3$ be the eigenvalues of $DX(\sigma)$. Suppose there is a pair of non-real eigenvalues, for instance, with negative real part. Using the fact that σ is accumulated by some orbit dense in Λ, we may use the connecting lemma (Theorem A.4) to construct a nearby vector field with a homoclinic connection associated to σ. The unfolding of this saddle-focus connection leads to the creation of periodic attractors or periodic repellers (see [373, page 30]). In either case, this contradicts the assumption that Λ is robustly transitive. Thus, all eigenvalues must be real. A similar argument shows that all eigenvalues have multiplicity 1: eigenvalues with higher multiplicity can be turned into complex eigenvalues by a small perturbation.

Now let us prove that the singularities are Lorenz-like. To fix notations, let the eigenvalues be $\lambda_1 < \lambda_2 < \lambda_3$. We already know that $\lambda_1 < 0 < \lambda_3$. Suppose $\lambda_2 < 0$. Using the connecting lemma once more, we obtain a nearby vector field with a singular cycle associated to σ. Up to a small perturbation, we may assume that the homoclinic orbit γ is tangent to the weak stable direction at σ, that is, the eigendirection corresponding to λ_2. Then γ is normally contracting. Suppose $\lambda_2 + \lambda_3 \leq 0$. Up to another small perturbation we may suppose that the inequality is strict. Then the unfolding of the homoclinic connection gives rise to the formation of a periodic attractor, contradicting transitivity. Therefore, $\lambda_2 + \lambda_3 > 0$. Considering the inverse flow we also get that if $\lambda_2 > 0$ then $\lambda_1 + \lambda_2 < 0$.

To prove the last part one argues as follows (see also [306, Lemma 2.16]). On the one hand, using that the periodic orbits in the continuation of Λ are always of saddle-type, one obtains a dominated splitting for the linear Poincaré flow over the set of regular points in Λ_Y for any nearby vector field Y. This is analogous to the proof of Theorem 9.12. On the other hand, if there was a regular point in $\Lambda \cap W^{ss}(\sigma)$ then by a small perturbation one could create an inclination-flip [4] homoclinic connection associated to the continuation of σ.

---

[4]The stable manifold and the center unstable manifold of the singularity are tangent along the connection.

However, this configuration is incompatible with the existence of a dominated splitting. This contradiction proves that $\Lambda \cap W^{ss}(\sigma) = \{\sigma\}$ as claimed.

Proving singular hyperbolicity:

After these crucial preparations, one is now ready to prove that Λ is a singular hyperbolic set. Suppose the singularities in Λ have unstable index 1. Start from the splittings $\tilde{E}_p^u \oplus E_p^X \oplus \tilde{E}_p^s$ into eigenspaces on regular periodic points $p \in \Lambda$. Define $E_p^c = \tilde{E}_p^u \oplus E_p^X$ and $E_p^s = \tilde{E}_p^s$. The whole point is to show that these splittings are dominated, with domination constants uniform over all periodic orbits and all nearby vector fields. The proof of this key fact splits into two cases.

For periodic orbits far from singularities, the arguments go back to Pliss [358], Mañé [277], Liao [260]: Essentially, one shows that in the absence of domination one could perturb the vector field to create an attracting or a repelling periodic orbit, contradicting transitivity. For periodic orbits approaching some singularity σ one uses a local control of the eigenspaces, based on the information provided by Corollary 9.24: In a few words, one proves that $E_p^s \to E_\sigma^1$ and $E_p^c \to E_\sigma^{23}$ as $p \to \sigma$, where E_σ^1 is the most contracting eigenspace of $DX(\sigma)$ and E_σ^{23} is the sum of the other two eigenspaces. Thus, one gets uniform domination also in this case.

Now, given any regular point $x \in \Lambda \subset \Omega(X)$, use the closing lemma [364] to find vector fields $X_n \to X$ for which x is a periodic point. Let $E_{x,n}^c \oplus E_{x,n}^s$ be the corresponding X_n-dominated splittings, as in the previous paragraph. By uniformity, they converge to an X-dominated splitting $E_x^c \oplus E_x^s$ over the X-orbit of x. Extend this splitting to the singularities inside Λ (if there is any) by $E_\sigma^s = E_\sigma^1$ and $E_\sigma^c = E_\sigma^{23}$. Since the dimensions of the subbundles are constant, and we have uniform domination on every orbit, this does define an X-dominated splitting $T_\Lambda = E^c \oplus E^s$ over the compact invariant set Λ.

The claim that E^s is uniformly contracting is a consequence of robust transitivity and Mañé's [278] ergodic closing lemma. Indeed, if E^s is not uniformly contracting then, by the ergodic closing lemma, that can be seen on periodic orbits: There exist periodic points p_n for vector fields $X_n \to X$ such that $E_{p_n}^s$ is, at most, weakly contracting. Then by another small perturbation one can make E^s expanding over some of these periodic orbits. That means the orbit became a periodic repeller, contradicting transitivity. A similar argument proves that E^c is volume expanding: Otherwise one can find periodic orbits of nearby vector fields over which E^c is, at most, weakly volume expanding. By another small perturbation one can make E^c volume contracting, and that means the orbit became a periodic attractor. Once more, that contradicts transitivity. This completes our sketch of the proof of Theorem 9.19.

9.4 Consequences of singular hyperbolicity

In this section we investigate some basic features of singular hyperbolic sets
of vector fields on closed 3-dimensional manifolds.

9.4.1 Singularities attached to regular orbits

Given a compact invariant set Λ we say that a singularity $\sigma \in \Lambda$ is *attached* to
Λ if it is accumulated by regular orbits of Λ. Some of the properties we proved
before for singularities of robustly transitive sets are shared by all singularities
attached to singular hyperbolic sets:

Proposition 9.25 (Morales, Pacifico, Pujals [309]). *Let Λ be a singular
hyperbolic set for a vector field X on a closed 3-dimensional manifold M.
Then Λ is uniformly hyperbolic if and only if it has no attached singularities.
Supposing there are attached singularities,*

1. *All singularities attached to Λ have the same index and they are Lorenz-
   like.*
2. *Λ intersects the strong invariant manifold of any attached singularity only
   at the singularity.*
3. *The flow direction is contained in the central subbundle at every point.*

By strong invariant manifold we mean the strong stable manifold or the
strong unstable manifold, depending on whether the index is 1 or 2. Before
we proceed to outline the proof of this proposition, let us point out that
Morales [299] has recently shown that *there are singular basic (transitive isolated) sets with any given number of attached singularities.*

The first statement in the proposition is well-known to us by now: the
only if part is easy and the converse is contained in Corollary B.13. Let the
singular hyperbolic splitting be $T_\Lambda M = E^s \oplus E^c$ with $\dim E^s = 1$. We begin
by observing that X is never tangent to E^s, at any regular point. Indeed,
otherwise there would be some regular point x such that

$$\|X(X^{-t}(x))\| = \|DX^{-t} \cdot X(x)\| \geq c\lambda^{-t}\|X(x)\| \quad (\lambda < 1)$$

converges to infinity as $t \to +\infty$. This would contradict the fact that the
vector field is bounded.

Suppose there is an attached singularity $\sigma \in \Lambda$ with two expanding directions. Since σ is accumulated by orbits of Λ, at least one of the branches of
$W^s(\sigma)$ must be contained in Λ. Let q be a regular point in $\Lambda \cap W^s(\sigma)$. All
tangent vectors at q are expanded by the flow DX^t, except for the direction
tangent to $W^s(\sigma)$. Therefore, we must have $E^s_q = T_q W^s(\sigma)$. However, this is
also the direction of $X(q)$, so this conclusion contradicts the previous paragraph. Thus, every attached singularity σ has unstable index 1. It is easy to
deduce that σ is Lorenz-like. Indeed, we already know that the eigenvalues

satisfy $\lambda_1, \lambda_2 < 0 < \lambda_3$. The existence of the invariant splitting $E_\sigma^s \oplus E_\sigma^c$ implies that $\lambda_1 < \lambda_2$ and, in particular, they are both real. Moreover, the stable subspace E^s coincides with the eigenspace associated to λ_1, and the central subspace E^c coincides with the sum of the eigenspaces associated to λ_2 and λ_3. The assumption that E^c is volume expanding corresponds to $\lambda_2 + \lambda_3 > 0$. This completes the proof of part 1.

Part 2 is similar. Suppose there exists a regular point $q \in \Lambda \cap W^{ss}(\sigma)$. All tangent vectors are either expanded or contracted by DX^t at rate $\geq \lambda_2$, except for the direction of $T_q W^{ss}(\sigma)$ which is contracted at a faster rate λ_1. Therefore, $E_q^s = T_q W^{ss}(\sigma)$. However, this is also the direction of $X(q)$, and so this is a contradiction.

We already know that the flow direction never coincides with E^s. Then, by continuity, the angle between the two directions is bounded from zero outside any neighborhood of the singularities. Let V be a small neighborhood of any attached singularity σ. Taking V sufficiently small, we ensure that E_x^s is very close to $T_\sigma W^{ss}(\sigma)$ for every $x \in \Lambda \cap V$. By the previous paragraph, $\Lambda \cap V$ avoids a neighborhood of $W^{ss}(\sigma) \setminus \{\sigma\}$. Outside that neighborhood the flow direction is far from $T_\sigma W^{ss}(\sigma)$. This shows that the angle between X and E^s is also bounded from zero on $\Lambda \cap V \setminus \{\sigma\}$. This proves that the angle between E^s and X is uniformly bounded from zero on the whole set of regular points in Λ. Since the splitting $E^c \oplus E^s$ is dominated, it follows that the forward iterates of the line bundle defined by X converge exponentially fast to E^c. Since that line bundle is obviously invariant, this means that X is contained in E^c at every point. That was claim 3 in Proposition 9.25.

Problem 9.26. For vector fields in dimension larger than 3, can there be robustly transitive sets containing singularities with different unstable indices?

9.4.2 Ergodic properties of singular hyperbolic attractors

The ergodic theory of singular hyperbolic attractors has just been initiated. Colmenárez [134] proves that these attractors carry a unique SRB measure, whose basin has full Lebesgue measure inside the topological basin of attraction. Existence of an SRB measure in the particular case of the geometric Lorenz attractors had been proved much earlier by Bunimovich, Sinai [103], and Metzger [291] obtained the corresponding result for Lorenz-Rovella attractors (see Section 9.6). Recall that the multidimensional Lorenz-like attractors of Bonatti, Pumariño, Viana [82] also carry an SRB measure with the no-holes property.

In Appendix E we discuss the issue of rates of mixing and, in particular, we ask (Problem E.4) whether singular hyperbolic attractors have exponential decay of correlations. The proof of this and other properties should require a more detailed knowledge of the structure of these systems. Indeed, the ergodic theory of uniformly hyperbolic flows carried out by Bowen, Ruelle [90], was based on the construction of a Markov partition: a finite set of cross-sections that are sent to each other in a hyperbolic Markov fashion by Poincaré maps. In the singular hyperbolic case one has to allow for singular Poincaré maps of the type displayed by the geometric Lorenz models.

Problem 9.27. Construct quasi-Markov partitions for singular hyperbolic attractors, consisting of finitely many cross-sections that are mapped to each other by Poincaré maps, either in a Markov fashion or in a Lorenz fashion.

Another important issue is stochastic stability (see Appendix D):

Problem 9.28. Prove that singular hyperbolic attractors are stochastically stable under diffusion type random noise.

Kifer [242] proved that the geometric Lorenz models are stochastically stable, and that was extended by Metzger [292] to Lorenz-Rovella attractors. Metzger also reports progress in proving stochastic stability for the multidimensional singular attractors in [82].

9.4.3 From singular hyperbolicity back to robustness

As in the discrete time case (Chapters 7 and 10) we would like to have some sort of decomposition of the global dynamics of the flow into elementary pieces. In that direction, it is important to investigate the converse to Theorem 9.19:

Problem 9.29. When is a singular basic (transitive isolated) set Λ robustly transitive?

In other words, if U is an isolating block for Λ, which additional conditions ensure that $\Lambda_Y = \cap_{t \in \mathbb{R}} Y^t(U)$ is a transitive set for all vector fields Y in a C^1 neighborhood of X? Observe that every invariant subset of Λ_Y is singular hyperbolic. That is because partially hyperbolic splitting persist under perturbation (see Section B.1), and every vector field close to X has only Lorenz-like singularities inside U.

If Λ contains no singularities then, by Corollary B.13, it is uniformly hyperbolic. It is a classical fact that hyperbolic basic sets are robustly transitive, for instance, because they are structurally stable. So, the answer to Problem 9.29 includes every singular basic set without singularities. On the other hand, the following example from [309] shows that not all singular basic sets are robustly transitive:

Example 9.30. Let X be a Cherry [128] vector field on $\mathbb{T}^2$, with two hyperbolic singularities, a saddle s and an attractor a, and a cross-section Σ meeting every regular orbit and intersecting the local basin of a along a segment $[p, q]$. See Figure 9.7. Outside this segment there is a well-defined first return map to Σ, which extends to a continuous non-decreasing circle map ϕ of degree 1, with a plateau over $[p, q]$. Then ϕ has a well defined rotation number $\rho(\phi)$. Consider a one-parameter family X_μ of Cherry flows such that Σ is

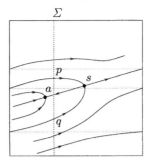

Fig. 9.7. A Cherry flow on the 2-dimensional torus

a cross-section for each X_μ, the rotation number $\rho(\phi_\mu)$ of the first return map ϕ_μ is irrational at $\mu = 0$, and is non-constant on any neighborhood of $\mu = 0$. Thus, there are parameters μ_n converging to 0 such that $\rho(\mu_n)$ is rational. For each μ, let $\Lambda(\mu)$ be the complement of the stable manifold of the attractor a_μ of X_μ. Then $\Lambda(0)$ is a transitive set without regular periodic orbits, whereas $\Lambda(\mu_n)$ may be taken to be the union of the stable manifold of the saddle s with a finite number of regular hyperbolic periodic orbits. Embed this family of vector fields in a manifold of dimension 3, in such a way that the initial torus $\mathbb{T}^2$ corresponds to a normally contracting invariant submanifold for the resulting one-parameter family Y_μ of vector fields. We may see each $\Lambda(\mu)$ as an invariant subset of Y_μ. For $\mu = 0$ this set contains dense orbits, since the rotation number is irrational, and so it is a singular basic set. However, it fails to be robustly transitive: each $\Lambda(\mu_n)$ contains periodic attractors and also intersects the stable manifold of the saddle, and so it can not be transitive.

Note that in this example, periodic orbits are not dense in the transitive set $\Lambda(0)$. Compare Problem 9.15. Related to this, Morales, Pacifico [303] prove that a singular hyperbolic attractor Λ with a unique singularity is C^r robustly transitive if periodic orbits are dense in the continuation Λ_Y for all vector fields Y in a C^r neighborhood.

The following result implies that, in general, the number of attracting pieces one may get from the breakup of a singular hyperbolic attractor is bounded by the number of singularities:

Theorem 9.31 (Morales [298]). *Let Λ be a singular hyperbolic attractor of a C^r vector field, $r \geq 1$, defined on a closed 3-dimensional manifold. Then there are neighborhoods U of the attractor Λ in the ambient manifold and $\mathcal{U}$ of X in the space of C^r vector fields such that every attractor in U of any vector field $Y \in \mathcal{U}$ contains at least one singularity.*

The proof of Theorem 9.31 is by contradiction. Recall that any invariant set of a nearby diffeomorphism inside U is singular hyperbolic. Suppose first that there are vector fields $X_n \to X$ exhibiting attractors $\Lambda_n \subset U$ that avoid a neighborhood of the singular set of X. Then, by Corollary B.13, the Λ_n are hyperbolic sets for X_n and their accumulation set $\tilde{\Lambda} \subset \Lambda$ is hyperbolic for X. Given any sequence of points $x_n \in \Lambda_n$ converging to $x \in \tilde{\Lambda}$, the unstable manifolds of the x_n converge to the unstable manifold of x. Observe also that all these unstable manifolds are contained in the corresponding hyperbolic attractors. Consequently, the orbit of $W^u(x)$ must avoid a neighborhood of the singular set of X. However, this is incompatible with the fact that Λ is transitive. To see this, consider a small segment I around x inside the unstable manifold, a rectangle $R = X^{[-\varepsilon,\varepsilon]}(I)$, consisting of orbit segments through the points of I, and a cube $C = W_\varepsilon^s(R)$ formed by local stable manifolds through the points in R. On the one hand, by construction, the orbit of C avoids a neighborhood of the singularities. On the other hand, C has nonempty interior in Λ, and so it is met by some dense orbit which, of course, accumulates on the singularities. This contradiction proves that the Λ_n must accumulate the singular set of X.

The final step of the proof is to see that Λ_n contains some singularity of X_n for all large n. This involves a local analysis based on the fact that the singularities are Lorenz-like. Let σ be a singularity accumulated by the Λ_n and Σ^1 and Σ^2 be a pair of small cross-sections close to it, such that every orbit passing through a neighborhood of σ intersects one of the Σ^i. We consider a continuous, not necessarily invariant, extension of the central bundle to a neighborhood of Λ, that we keep denoting as E^c. It is understood that the Σ^i are contained in that neighborhood. Then E^c induces a line field $E^{c,i} = E^c \cap T\Sigma^i$ on each of the cross-sections. Let Γ^i be the intersection of Σ^i with the local stable manifold of σ. Note that Γ^i is tangent to the stable subbundle, because σ is Lorenz-like, and so it is transverse to $E^{c,i}$. Let z be any point in $\Lambda \cap \Sigma^i$ and $W^{cu}(z)$ denote the unstable manifold of the orbit [5] of z. Note that $W^{cu}(z) \subset \Lambda$, because Λ is an attractor. Moreover, $W^{cu}(z)$ is tangent to E^c at z, and so $W^{cu}(z) \cap \Sigma^i$ is tangent to $E^{c,i}$ at z. This shows that $\Lambda \cap \Sigma^i$ contains a curve γ_z around z, tangent to $E^{c,i}$.

So far we have been considering the vector field X and the corresponding attractor Λ. However, this construction is stable, because dominated splittings vary continuously with the dynamics (see Section B.1), and so it extends to X_n and Λ_n for all large n. We denote the corresponding objects as σ_n, E_n^c, $E_n^{c,i}$, Γ_n^i. The previous arguments show that $\Lambda_n \cap \Sigma^i$ contains a curve γ_z tangent to $E_n^{c,i}$ around every point $z \in \Lambda_n \cap \Sigma^i$. It follows that z does not realize the minimum of

$$\mathrm{dist}(x, \Gamma_n^i), \quad x \in \Lambda_n \cap \Sigma^i,$$

unless z belongs to Γ_n^i. Since $\Lambda_n \cap \Sigma^i$ is compact, the minimum must be attained. Consequently, there exists $z \in \Lambda_n \cap \Gamma_n^i$. Since Λ_n is invariant and

---

[5] $W^{cu}(z)$ is the 2-dimensional manifold formed by the orbits through the points in $W^u(z)$, where the latter is the (one-dimensional) unstable manifold of z.

closed, it follows that $\sigma_n \in \Lambda_n$. This completes our outline of the proof of Theorem 9.31.

In a similar direction, Morales, Pacifico [302] have been investigating the decomposition of singular sets into homoclinic classes. They show that a connected attracting singular hyperbolic set with periodic orbits dense and a unique singularity is either transitive or a non-disjoint union of two homoclinic classes. The second case may indeed occur [49]. In this second case there are only finitely many distinct homoclinic classes. It is not known whether the same is true in the transitive case. More generally,

Problem 9.32 (Morales, Pacifico). Do singular hyperbolic sets contain finitely many distinct homoclinic classes? Are there explicit bounds for the number of homoclinic classes in terms of the number of singularities?

9.5 Singular Axiom A flows

Aiming at describing the global dynamics of vector fields in dimension 3, Morales, Pacifico, Pujals [309] proposed the following notions:

Definition 9.33. A vector field X is *singular Axiom A* if the non-wandering set $\Omega(X)$ of X is a disjoint union of finitely many singular basic sets and coincides with the closure of the critical elements of X. The family of singular basic sets is called the *singular spectral decomposition* of $\Omega(X)$.

Definition 9.34. A *cycle* is a subset $\Lambda_0 = \Lambda_m, \Lambda_1, \ldots, \Lambda_{m-1}$ of basic sets of the singular spectral decomposition such that for each $i = 1, \ldots, m$ there is a regular orbit whose α-limit is contained in Λ_{i-1} and whose ω-limit is contained in Λ_i.

Define $\mathcal{G}^r(M)$, $r \geq 1$ to be the interior of the set of C^r vector fields X on a manifold M such that all singularities and all periodic orbits of X are hyperbolic. There is a corresponding definition for diffeomorphisms, usually denoted by $\mathcal{F}^r(M)$. It has been shown by Aoki [17] and Hayashi [207], based on the work of Mañé [283], that $\mathcal{F}^1(M)$ coincides with the set of uniformly hyperbolic diffeomorphisms with no cycles. The next result provides a partial counterpart for vector fields:

Theorem 9.35 (Morales, Pacifico, Pujals [309]). *Let M be a closed manifold of dimension 3. Then every C^r singular Axiom A vector field with no cycles belongs to $\mathcal{G}^r(M)$ for all $r \geq 1$.*

To prove Theorem 9.35 one writes $\Omega(X) = \Lambda_1 \cup \cdots \cup \Lambda_m$ where each Λ_i is a singular basic set. For each set Λ_i there is a neighborhood U_i where the maximal invariant set $\Lambda_{i,Y}$ of any vector field Y close to X is singular hyperbolic. That is because dominated decompositions are persistent, see Section B.1, and all singularities remain hyperbolic. Then all periodic orbits in the union $\tilde{\Omega}(Y)$ of the $\Lambda_{i,Y}$ are hyperbolic, by Corollary B.13. To complete the proof it suffices to show that $\Omega(Y)$ coincides with $\tilde{\Omega}(Y)$, and so there are no other critical elements. This is analogous to the classical argument in the proof of the Ω-stability theorem (see Section 1.2): using the no-cycle condition one constructs a filtration such that each U_i is contained in the ith level and, in fact, $\Lambda_{i,Y}$ coincides with the maximal invariant set in the ith level of the filtration. The claim follows.

The converse to Theorem 9.35 is not true, in fact vector fields in $\mathcal{G}^1(M)$ need not be singular Axiom A. For instance, the vector fields with expanding singular cycles in [44] belong to $\mathcal{G}^1(M)$ because, as we seen in Section 9.2, the unfolding of the cycle does not give rise to non-hyperbolic periodic orbits. Yet, the critical elements are not dense in the non-wandering set. In this connection, Morales, Pacifico, Pujals [309] proposed the following

Conjecture 9.36. The subset of singular Axiom vector fields is open and dense in $\mathcal{G}^1(M)$.

There has been considerable progress in the direction of this conjecture:

Theorem 9.37 (Morales, Pacifico [305]). *Generic vector fields in $\mathcal{G}^1(M)$ are singular Axiom A and have no cycles.*

In fact, [305] uses a slightly stronger definition of singular Axiom A, where one requires every non-hyperbolic piece of the singular spectral decomposition $\Omega(X) = \Lambda_1 \cup \cdots \cup \Lambda_n$ to be either an attractor or a repeller. So, this theorem contains the statement that, generically in $\mathcal{G}^1(M)$, every singularity attached to (that is, non-isolated inside) the non-wandering set belongs to a non-trivial singular hyperbolic attractor or repeller.

The same authors prove in [304] that for generic vector fields in $\mathcal{G}^1(M)$ all non-trivial attractors and repellers are topologically mixing. This has been much improved by Abdenur, Avila, Bochi [2]: topological mixing holds for all non-trivial homoclinic classes of a residual subset of all C^1 vector fields (this subset is *not* open and dense). This implies that every robustly transitive attractor can be made mixing by an arbitrarily small perturbation. These developments extend a classical result of Bowen [88], who proved that hyperbolic basic sets Λ of generic Axiom A vector fields are topologically mixing. Recall that an invariant set Λ is *topologically mixing* if, given any pair of nonempty open subsets U and V, there exists $s > 0$ such that $X^t(U) \cap V \neq \emptyset$ for all $t > s$. Every mixing set is transitive, but the constant time suspension of a transitive diffeomorphism is an example of a flow for which the ambient space is transitive but not mixing. This construction also shows that not all Anosov flows are mixing: just use an Anosov diffeomorphism.

Let us say a few words about the proof of Theorem 9.37. One important step is to prove that for generic $X \in \mathcal{G}^1(M)$ every singularity σ accumulated by periodic orbits belongs to some singular hyperbolic attractor or repeller. The idea is to consider the closure of the one-dimensional invariant (unstable or stable, depending on whether the unstable index is 1 or 2) manifold $W^*(\sigma_Y)$ of the continuation of the singularity for nearby vector fields Y. This is a lower-semicontinuous function of the vector field, relative to the Hausdorff topology in the space of compact subsets of the manifold, because invariant manifolds vary continuously with the dynamics. Consequently, there exists a residual subset of vector fields which are continuity points for this function. For those vector fields the closure of $W^*(\sigma_Y)$ is a Lyapunov stable set (either for Y or for $-Y$, depending on whether $* = u$ or $* = s$): otherwise, one could use the connecting lemma (Theorem A.5) to explode the closure (see the proof of Lemma 9.22 for a similar argument), contradicting the assumption that we are at a continuity point. It follows that the closure of $W^*(\sigma_Y)$ is an attractor, either for Y or for $-Y$.

Recall also that, as a consequence of the closing lemma [364] (see also Appendix A), for generic vector fields periodic orbits are dense in the non-wandering set. So, using the previous paragraph, for the generic $X \in \mathcal{G}^1(M)$ we may find singular attractors and repellers $\Lambda_1, \ldots \Lambda_k$ containing all the singularities attached to the non-wandering set $\Omega(X)$. Let $\sigma_1, \ldots, \sigma_l$ be the non-attached singularities. Since Λ_i and $\{\sigma_1, \ldots, \sigma_l\}$ are open subsets of $\Omega(X)$, the complement H of $\Lambda_1 \cup \cdots \cup \Lambda_k \cup \{\sigma_1, \ldots, \sigma_l\}$ inside $\Omega(X)$ is a closed invariant subset containing no singularities. Consequently (Corollary B.13), H is a hyperbolic set. Hence, by the classical spectral decomposition theorem (Theorem 1.3), H splits into a finite disjoint union of hyperbolic basic sets. This proves that the vector field is singular Axiom A. To see that there are no cycles, notice first that a cycle must involve only hyperbolic pieces of the singular spectral decomposition, for the non-hyperbolic pieces are either attractors and repellers. The unfolding of such a cycle would lead to the appearance of non-hyperbolic periodic orbits, which is impossible for vectors in $\mathcal{G}^1(M)$. This gives the conclusion of the theorem.

As a consequence of Theorem 9.37 one also obtains the following C^1 generic dichotomy, that extends to vector fields in dimension 3 a classical result of Mañé for surface diffeomorphisms (Theorem 10.23):

Theorem 9.38 (Morales, Pacifico [305]). *Let M be a closed manifold of dimension 3. There is a residual subset $\mathcal{R}$ of $\mathcal{X}^1(M)$ consisting of vector fields X such that*

1. *either X has infinitely many regular attracting or repelling orbits*
2. *or X is a singular Axiom A vector field with no cycles.*

To deduce this result from Theorem 9.37, consider the (open and dense) subset $\mathcal{S}$ of vector fields whose singularities are all hyperbolic. In particular, for $X \in \mathcal{S}$ there are only finitely many singularities. Let $\Lambda(X)$ be the closure of the set of singularities and hyperbolic periodic orbits that are either attracting or repelling. The function $X \mapsto \Lambda(X)$ is lower-semicontinuous, relative to the Hausdorff topology in the space of compact subsets of the manifold, since every hyperbolic critical element persists in a neighborhood of the vector field.

Therefore, there exists a residual subset $\mathcal{R}$ of $\mathcal{S}$ formed by continuity points of this function. For any vector field in $\mathcal{R}$, either $\Lambda(X)$ is infinite, in which case the first alternative in the theorem holds, or there are only finitely many periodic orbits. Moreover, in the latter case all periodic orbits are hyperbolic, for any vector field in a neighborhood. That is because a non-hyperbolic periodic orbit may be unfolded to give rise to two or more hyperbolic singularities, which would contradict upper semi-continuity. In other words, in the second case the vector field is in $\mathcal{G}^1(M)$. Then, by Theorem 9.37, a residual subset consists of singular Axiom A vector fields with no cycles. This corresponds to the second alternative in the theorem.

9.6 Persistent singular attractors

We call persistent a dynamical phenomenon that occurs with positive probability, in the sense of Lebesgue measure in parameter space, on generic parametrized families through the initial flow. *Persistent singular attractors* include robust singular attractors, of course. However, a number of recent results shows that it is, in fact, a much broader class.

The first examples of persistent non-robust singular attractors were constructed by Rovella [388]. He started from a variation of the classical geometric Lorenz models, assuming that the eigenvalues $\lambda_1 < \lambda_2 < 0 < \lambda_3$ at the singularity satisfy $\lambda_2 + \lambda_3 < 0$ instead. For parametrized families of flows through that configuration, he proves that an open and dense subset of the parameter space consists of flows with uniformly hyperbolic attractors, but *flows with singular attractors occupy a positive Lebesgue measure subset of the complement*. This corresponds to a kind of Hénon-like dynamics for flows, interacting with the presence of singularities.

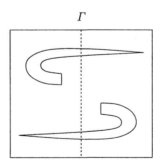

Fig. 9.8. Poincaré return map with folds

Another class of persistent singular attractors has been studied in recent works of Luzzatto, Viana [270, 269], and Luzzatto, Tucker [268]. These papers deal with Lorenz-like attractors with folds, an extension of the classical geometric Lorenz models where one allows for the image of the Poincaré map to fold, as in Figure 9.8. This kind of behavior seems to occur in Lorenz equations (9.1) for slightly larger values of the Rayleigh number $\rho \approx 30$. See Glendinning, Sparrow [191] for an interpretation. In fact, it was the observation of this phenomenon that lead Hénon [210] to introduce the map that is named after him, as a model for the dynamics of the Poincaré map near the fold. The main novelty in these geometric Lorenz models with folds is that they combine fold-type behavior interacting with the dynamics associated to the presence of singularities. At an early stage of the unfolding, robust Lorenz-like attractors are formed. The main result is that *these attractors persist, in the probabilistic sense above, even after the folds are formed.*

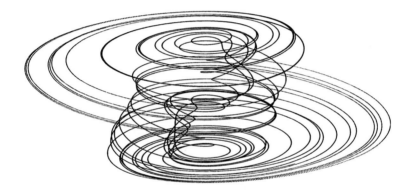

Fig. 9.9. A spiraling strange attractor

Another type of persistent attractor, with a much more complex geometry, occurs in connection with saddle-focus connections: a saddle-focus singularity θ having eigenvalues $\sigma > 0$ and $\rho \pm \imath\omega$ with $\rho < 0$, such that some branch of the unstable manifold of θ is contained in the stable manifold of θ. This dynamical configuration has been studied by many authors, starting from the pioneer work of Shil'nikov [408] showing that, in the expanding case $\sigma + \rho > 0$, every neighborhood of the homoclinic orbit contains infinitely many periodic orbits. In fact, see Tresser [427], there are infinitely many linked horseshoes in any neighborhood of the homoclinic orbit. Pumariño, Rodriguez [373] proved that unfolding of a saddle-focus homoclinic connection with $\sigma + \rho = 0$ leads to the appearance of persistent (non-singular) Hénon-like strange attractors. In fact, there may be infinitely many such attractors coexisting for the same parameter [374]. On the other hand, it had been conjectured by Arnéodo, Coullet, Tresser [29] that *global* attractors containing the unstable manifold of the saddle-focus singularity may occur in this context. A proof was given by Pacifico, Rovella, Viana [338, 337], who showed that such *spiraling attractors* do occur with positive Lebesgue probability in parameter space, for an open class of one-parameter families of flows.

10

Generic Diffeomorphisms

In this chapter we present a collection of recent results which provide a rather satisfactory picture of the global dynamics of generic C^1 diffeomorphisms on a closed manifold M. By *generic* we mean that we consider dynamical properties satisfied by a *residual* subset the space $\text{Diff}^1(M)$, that is, one containing a countable intersection of C^1 open and dense subsets. Sometimes we take a more local point of view: a property is *locally generic* if it holds on the intersection of a residual subset with some non-empty open subset of $\text{Diff}^1(M)$.

10.1 A quick overview

Recurrence:

Interesting dynamics is always associated with some sort of orbit recurrence. For instance, on a neighborhood of a wandering orbit the diffeomorphism is always smoothly conjugate to a simple translation in Euclidean space. So, in trying to understand the dynamical behavior one focusses on the recurrent part of the ambient space. Various notions of recurrence have been considered in Dynamics. In Section 10.2 we review the main ones: (in increasing order of generality) periodic points, recurrent points, limit points, non-wandering points, pseudo-periodic points, and chain recurrent points.

In principle, each of these notions corresponds to a different definition of the recurrent set. However, we shall see in Theorems 10.2 and 10.11 that *for generic diffeomorphisms all these definitions coincide.* For most of them this fact is a classical consequence of Pugh's closing lemma; a recent connecting lemma for pseudo-orbits extends it to the chain recurrent set. In particular, for generic diffeomorphisms the chain recurrent set coincides with the closure of the (hyperbolic) periodic points. The chain recurrent set varies upper semi-continuously with the dynamics, whereas the closure of the set of hyperbolic periodic points is lower semi-continuous. Since the former always contains the latter, every diffeomorphism for which the two sets coincide, is a continuity point for both of them.

This sets the stage for making precise what we mean by a global description of the dynamics:

Dynamical decomposition:

Motivated by the spectral decomposition theorem in uniformly hyperbolic theory (Theorem 1.3), itself inspired by the classical picture of gradient-like dynamics, we would like to break the recurrent set into different invariant pieces carrying independent dynamics, with an adapted filtration (Definition 1.6) guiding every forward and backward orbit to one of these pieces.

A precise solution was given in the late seventies by Conley: *two invariant sets may be separated by a filtration if and only if it is not possible to go from one to the other by pseudo-orbits with arbitrarily small jumps.* This showed that if one wants to split the dynamics into pieces separated by filtrations, then chain recurrence classes are the appropriate notion of elementary pieces: for any filtration, the maximal invariant set inside any filtration level contains the chain recurrence classes of all its points, and there exists a filtration (possibly infinite) that separates all chain recurrence classes. See Section 10.3.1.

This result holds for any homeomorphism on a compact metric space, in particular it needs no genericity assumption. On the other hand, little is known about the dynamics on chain recurrence classes. Moreover, the decomposition into chain recurrence classes may be very coarse, if these classes are far from being transitive. For generic diffeomorphisms, a number of recent results show that chain recurrence classes are intimately related with periodic orbits, and shed some light into their dynamical behavior:

Chain recurrence classes and periodic orbits:

A long series of works, using the connecting lemma and its variations, have shown that *for generic diffeomorphisms, the homoclinic class of any periodic point is a chain recurrence class*, or equivalently, any chain recurrence class containing a periodic point coincides with the corresponding homoclinic class. In particular, these classes are transitive sets. Since, for generic diffeomorphisms periodic points are dense in the chain recurrent set, it follows *any isolated chain recurrence class is a homoclinic class*. As we shall see, there exist locally generic diffeomorphisms with chain recurrence classes without periodic points. However, a subtle use of the connecting lemma shows that, for generic diffeomorphisms, chain recurrence classes are Hausdorff limits of periodic orbits. See Section 10.4.

Robustness and hyperbolicity of isolated classes:

Any isolated chain recurrence class is the maximal invariant set in a neighborhood given by the corresponding filtration level. For generic diffeomorphisms, this property is robust: for any nearby diffeomorphism the maximal invariant set in that neighborhood is a chain recurrence class. In other words, *the isolated chain recurrence classes of generic diffeomorphisms are robustly chain recurrent*. Then, results in [73], generalized in [1], imply that the isolated chain recurrence classes of generic diffeomorphisms are volume hyperbolic. See Section 10.4.3.

Finiteness of elementary pieces:

The issue of whether there are finitely many elementary dynamical pieces was, initially, approached in terms of homoclinic classes, and there is no harm in keeping that point of view here: *generic diffeomorphisms have finitely many homoclinic classes if and only if they have finitely many chain recurrence classes* (see Corollary 10.17). On the other hand, homoclinic classes do have some advantages, in that they are defined more constructively and they depend semi-continuously on the dynamics (because transverse intersections between the invariant manifolds are a robust phenomenon). See Section 10.4.

Inside a residual subset of diffeomorphisms, the number of disjoint homoclinic classes varies continuously with the dynamics (Theorem 10.24). Thus, generic diffeomorphisms naturally split into *tame* diffeomorphisms, having finitely many homoclinic classes, and *wild* diffeomorphisms, having an infinite number of homoclinic classes.

Tame dynamics:

We shall see in Section 10.5 that *for tame diffeomorphisms the homoclinic classes are the elementary dynamical pieces*, that is, there are no other chain recurrence classes. These elementary pieces are transitive, maximal, isolated, and they also have some degree of robustness and hyperbolicity (volume hyperbolicity). The dynamics is nicely organized by a finite adapted filtration. Moreover, there are only finitely many topological attractors (and repellers), and their basins cover an open dense subset of the ambient space. Thus, tame diffeomorphisms share many main properties of uniformly hyperbolic systems; in fact, on surfaces, tame diffeomorphisms are uniformly hyperbolic.

These results gives hope that a comprehensive theory of tame systems may be within reach, containing also a topological and statistical description of the dynamics on the elementary pieces and their neighborhoods. Building such a description is an important challenge.

Wild dynamics:

The first examples of wild dynamics were given by the locally C^2-generic surface diffeomorphisms with infinitely many periodic attractors, discovered by Newhouse (see Section 3.2). Actually, it is not yet known whether a corresponding C^1 statement holds for surface diffeomorphisms (recall Problem 3.8). However, coexistence of infinitely many sinks or sources is indeed known to be locally generic on any manifold of dimension ≥ 3 (Section 10.6.2). Moreover, by modifying this construction one obtains several other (locally generic) examples of coexistence of infinitely many, and even uncountably many, nontrivial chain recurrence classes without periodic points (Theorem 10.32 in Section 10.6).

Yet, more than three decades after the first examples appeared, very little is known about wild dynamics. That is another major obstacle on the way towards a global understanding of dynamical systems. For instance,

Problem 10.1. Does every wild diffeomorphism have some topological attractor (repeller)? Assuming attractors and repellers do exist, do their basins cover an open and dense subset of the ambient manifold?

In fact, we do not even know whether the (non-isolated) elementary dynamical pieces are actually transitive, not just weakly transitive. That is the case in the known examples.

10.2 Notions of recurrence

For studying the dynamics of diffeomorphisms one naturally focusses on the region of the manifold concentrating accumulation and recurrence of orbits. The simplest and strongest form of recurrence corresponds to periodic points, and that suffices for describing the dynamics of simple systems, such as the Morse-Smale diffeomorphisms. For more complicated chaotic dynamics, weaker notions of recurrence have been successively introduced. In this section we are going to see that, for C^1-generic diffeomorphisms, all these notions define the same (compact) subset of the manifold. Let us begin by listing the most common ones:

- *recurrent points,* whose forward or backward orbits return arbitrarily close to themselves;
- *limit points,* which are accumulated by the forward or backward orbit of some point;
- *non-wandering points,* such that some point in every neighborhood returns to that neighborhood;
- *pseudo-periodic points,* which can be made periodic by an arbitrarily small C^1-perturbation of the dynamics [1];
- *chain recurrent points,* whose orbit could be closed if one allowed arbitrarily small jumps at every iterate along the orbit.

More precisely, a point is chain recurrent if for every $\varepsilon > 0$ there is an ε-*pseudo-orbit* starting and ending at x, that is, a finite sequence $x = x_0, x_1, \ldots, x_n = x$ such that $d(f(x_i), x_{i+1}) < \varepsilon$ for all i. Every periodic point is recurrent,

---

[1]This notion, with a different name, appeared in the work of Liao [259] and played an important role in his study of the C^1-stability conjecture.

every recurrent point is a limit point, every limit point is non-wandering, every non-wandering point is pseudo-periodic (this is Pugh's closing lemma, Theorem A.1), and every pseudo-periodic point is chain recurrent.

Let $\mathrm{Per}(f)$, $\mathrm{Rec}(f)$, $\Omega(f)$, $\mathrm{Per}_*(f)$, and $\mathcal{R}(f)$ be the sets of periodic, recurrent, non-wandering, pseudo-periodic, and chain recurrent points, respectively. All these sets are f-invariant and generally different. For instance, for an Axiom A diffeomorphism with a cycle between its hyperbolic basic pieces, heteroclinic points associated to two basic sets in a cycle are wandering but chain recurrent. The non-wandering, pseudo-periodic, and chain recurrent sets are always closed. This is not the case for the sets of periodic, recurrent, and limit points, so that one usually considers their closure. The closure $L(f)$ of the set of limit points is called limit set.

Combining the Kupka-Smale theorem with Pugh's closing lemma one obtains the following genericity theorem:

Theorem 10.2. *For a residual subset of diffeomorphisms in* $\mathrm{Diff}^1(M)$,

1. *all periodic points are hyperbolic and all their invariant manifolds are transverse;*
2. *the non-wandering set coincides with the closure of the periodic points.*

The first part is just the Kupka-Smale theorem (see [342]). The closing lemma (Theorem A.1) means that $\Omega(f) \subset \mathrm{Per}_*(f)$. So, to prove the second part it suffices to show that periodic points are dense in $\mathrm{Per}_*(f)$ for generic diffeomorphisms. Let $U \subset M$ be a non-empty open set and f be a limit of maps f_n such that $\mathrm{Per}_*(f_n)$ intersects U. By definition, arbitrarily close to every f_n we may find a diffeomorphism g_n having some periodic point in U. By part 1 we may take the periodic point to be hyperbolic. Then it persists, inside U, for every diffeomorphism in a whole neighborhood. Since the g_n are also arbitrarily close to f, this proves that the closure of the set of diffeomorphisms g such that U intersects $\mathrm{Per}_*(g)$ but not $\mathrm{Per}(g)$ is a nowhere dense (closed) set. Taking the union over a countable basis of open sets U we obtain a meager set such that, for any diffeomorphism in the complement, periodic points are dense in the pseudo-periodic set.

This shows that $\overline{\mathrm{Per}}(f) = \mathrm{Per}_*(f) \subset \mathcal{R}(f)$ for generic diffeomorphisms. Furthermore, the connecting lemma for pseudo-orbits (Theorem A.7) states that any pseudo-periodic point can be made periodic by a small perturbation of the diffeomorphism. In other words, we always have $\mathrm{Per}_*(f) = \mathcal{R}(f)$. Hence, for generic diffeomorphisms we have $\overline{\mathrm{Per}}(f) = \mathcal{R}(f)$ and so all the previous recurrence regions coincide.

10.3 Decomposing the dynamics to elementary pieces

Our next goal is to present a dynamical decomposition theorem extending to generic diffeomorphisms the classical spectral decomposition Theorem 1.3. In order to explain this more precisely, let us recall that this theorem states that for uniformly hyperbolic systems (Axiom A with no cycles) the non-wandering set breaks into independent *hyperbolic basic sets*, which have a number of nice properties: they are

(a) *indecomposable*, in the sense that they contain dense forward orbits (*transitivity*);
(b) *maximal* indecomposable sets, meaning that they are not contained in a strictly larger transitive set;
(c) *isolated*, that is, they coincide with the maximal invariant set in a neighborhood;
(d) *robust*, in the sense that their continuations (as isolated sets) for any nearby system are still transitive;
(e) separated by a *filtration*: the hyperbolic basic sets are the maximal invariant sets inside each level;
(f) pairwise disjoint and they cover the whole $\Omega(f)$, so that they constitute a *partition* of the non-wandering set;
(g) *homoclinic classes*, that is, they coincide with the closure of the homoclinic points associated to any of their periodic points;
(h) finally, there are *finitely many* hyperbolic basic sets.

So, a central problem is to find a suitable notion of *elementary dynamical piece* for generic dynamical systems, sharing as much as possible of these features of hyperbolic basic sets. The theory developed by Conley [135] shows that, from the point of view of property (e), the optimal solution corresponds to chain recurrence classes. However, in this theory the notion of indecomposability is very weak, and these classes are also not robust, in general. Moreover, the relation between the chain recurrent set and the periodic orbits was poorly understood at that time, while the closing lemma (Theorem A.1) provided a clear link between periodic orbits and the non-wandering set. This motivated the search for other candidates:

• *robustly transitive sets* (property (d) and Chapter 7)
• *maximally transitive sets* (property (b))
• *homoclinic classes* of periodic points (property (e)).

The second one has been generalized by Arnaud [25] and Gan, Wen [190]:

• *maximal weakly transitive sets*: an invariant set Λ is weakly transitive if, given any pair of points in it there exist orbit segments, not necessarily in Λ, that start arbitrarily close to each of the points and end arbitrarily close to the other.

We shall see in Section 10.3.2 that, generically, maximal weakly transitive sets define a partition of the non-wandering set (property (f)). On the other hand, a priori they have no adapted filtration. However, Theorem 10.11 states that *for generic diffeomorphisms, the maximal weakly transitive sets coincide with the chain recurrence classes*, and so they do benefit from an adapted filtration.

10.3.1 Chain recurrence classes and filtrations

Let f be a homeomorphism on a compact metric space (X,d). Recall that an ε-*pseudo-orbit* of f is a sequence $x_i \in X$ such that all the *jumps* $d(f(x_i), x_{i+1})$ are smaller than ε. The chain recurrent set $\mathcal{R}(f)$ of f is the set of points x admitting closed pseudo-orbits starting and ending at x and having arbitrarily small jumps.

Using pseudo-orbits we define a natural binary relation between the points of X: we write $x \dashv y$ if there are finite pseudo-orbits with arbitrarily small jumps joining x to y. Note that a point x is chain recurrent if and only if $x \dashv x$. The relation $\dashv$ is clearly transitive and, in general, non-symmetric. Its symmetrization $\vdash\!\dashv$ (that is, $x \vdash\!\dashv y$ if and only if $x \dashv y$ and $y \dashv x$) induces an equivalence relation on $\mathcal{R}(f)$ whose equivalence classes are called the *chain recurrence classes*.

Notice that two points in the same chain recurrence class cannot be separated by a filtration (recall Definition 1.6), that is, they always belong to the same filtration level. Conley proved that the converse is also true: points in distinct chain recurrence classes can be separated by a filtration. More precisely:

Theorem 10.3 (Conley [135]). *Let f be a homeomorphism on a compact metric space (X,d). Then there is a continuous map $\psi\colon X \to \mathbb{R}$ satisfying*

1. *ψ is decreasing along the orbits of f, that is, $\psi(f(x)) \leq \psi(x)$ for all $x \in X$, and $\psi(f(x)) = \psi(x)$ if and only if $x \in \mathcal{R}(f)$;*
2. *for any pair (x,y) of chain recurrent points, $\psi(x)$ is equal to $\psi(y)$ if and only if x and y belong to the same chain recurrence class;*
3. *the set $\psi(\mathcal{R}(f))$ has empty interior in $\mathbb{R}$, and so it is a totally disconnected compact subset of the real line.*

Given real numbers $\alpha_1 > \cdots > \alpha_n$ in $\mathbb{R} \setminus \psi(\mathcal{R}(f))$ one gets a filtration associated to f by considering the sets $X_i = \psi^{-1}((-\infty, \alpha_i])$. Thus we may always separate any two given chain recurrence classes by a filtration. More than that, by considering a possibly infinite family of real numbers α_i such that each connected component of $\mathbb{R} \setminus \psi(\mathcal{R}(f))$ contains exactly one α_i, one obtains a filtration that separates all chain recurrence classes of f. Notice that a homeomorphism may have infinitely many, and even uncountably many, chain recurrence classes.

The previous theorem shows that, if we want to consider dynamical pieces which are separable by a filtration, then chain recurrence classes are the finest possible choice. However, in many cases this decomposition is much too coarse, as shown by the following simple example:

Example 10.4. Let f be an irrational rotation on the 2-dimensional sphere. Then $\Omega(f) = \mathcal{R}(f) = S^2$. Moreover, the whole sphere is the unique chain recurrence class. On the other hand, there is a natural finer decomposition of the sphere into pairwise disjoint f-invariant compact sets: the closures of the orbits, which are the parallel circles.

In the next section we introduce a different decomposition of $\Omega(f)$ whose classes, in this particular example, are exactly the parallel circles.

10.3.2 Maximal weakly transitive sets

Consider a diffeomorphism f on a compact manifold M. Motivated by the definition of non-wandering points, Arnaud [19] introduced the following binary relation [2] between points of M: $x \prec y$ if for any neighborhoods U of x and V of y there is $n > 0$ with $f^n(U) \cap V \neq \emptyset$. Notice that $x \prec x$ if and only if $x \in \Omega(f)$. Also, $x \prec y$ implies that $x \dashv y$.

In Example 10.4 above, $\prec$ is an equivalence relation and its equivalence classes are exactly the parallel circles. In general, this relation $\prec$ is neither transitive nor symmetric. However,

Theorem 10.5 (Arnaud [25]). *There is a residual subset of* $\mathrm{Diff}^1(M)$ *consisting of diffeomorphisms for which* $\prec$ *is a transitive relation.*

Then, for that residual subset, the symmetrization $\diamond$ of $\prec$ induces an equivalence relation on the non-wandering set. Its equivalence classes are the *maximal weakly transitive sets*. A compact f-invariant set Λ is *weakly transitive* if $x \diamond y$ for any two points $x, y \in \Lambda$. In other words, the forward orbit of any neighborhood of one point meets any neighborhood of the other point. Important examples of weakly transitive sets include: transitive sets; the ω-limit set $\omega(x)$ of any point $x \in M$; the Hausdorff limit of any sequence of periodic orbits.

An important difference between the notions of transitivity and weakly transitivity is that the second one is not intrinsic: the definition involves orbits in ambient space, which are not necessarily inside the invariant set itself. This has the consequence that the restriction of a diffeomorphism to a weakly transitive set may fail to be weakly transitive, that is, the set may not be weakly transitive for the restriction. Here is a simple illustration of this possibility, which is also an example of a weakly transitive set which is not transitive:

Example 10.6. Consider a diffeomorphism f of the annulus $S^1 \times [0, 1]$ such that every circle $S^1 \times \{a\}$ is f-invariant and f has a unique fixed point p, a saddle-node, in the circle $C = S^1 \times \{1/2\}$ say. Then both the backward orbit and the forward of any point of C converge to p. It is now immediate to check that the entire circle C is a weakly transitive set for f. Clearly, this set is not transitive. Moreover, C is not a weakly transitive set for the restriction f to the C, although it is a chain recurrent set for the restriction.

The definition implies that the closure of the union of an increasing sequence of weakly transitive sets is a weakly transitive set. So, by Zorn's lemma,

---

[2]This notion was also considered by Gan, Wen [190], under the name of *attainability relation*. Its symmetrization is called bi-attainability.

every weakly transitive set is contained in a maximal one. In general, maximal weakly transitive sets need not be disjoint. Theorem 10.5 asserts that, for generic diffeomorphisms, maximal weakly transitive sets are pairwise disjoint, and they induce a partition of the non-wandering set. This partition of $\Omega(f)$ refines the one obtained by intersecting the chain recurrence classes with the non-wandering set (because $x \prec y$ implies that $x \dashv y$, as we observed before).

Our next goal is to understand in more depth the relation between these two partitions in the generic setting.

10.3.3 A generic dynamical decomposition theorem

Inspired by the conclusions of the previous two sections, we now propose a general definition of elementary dynamical piece, and state precisely what we mean by a diffeomorphism having a decomposition into elementary pieces.

As a first step, we define a *filtrating neighborhood* of an f-invariant compact set Λ to be any compact neighborhood U of the form $U = M_1 \setminus \mathrm{int}(M_2)$, where M_1 and M_2 are compact submanifolds with boundary, of the same dimension as M, such that $f(M_i)$ is contained in the interior of M_i, $i = 1, 2$. Notice that this is a generalization of the notion of filtration level (recall Definition 1.6).

Remark 10.7. The following observations are easy consequences of the definitions:

1. If U is a filtrating neighborhood of some compact invariant set then the chain recurrence class of any point of $\mathcal{R}(f) \cap U$ is contained in U.
2. If Λ is the maximal invariant set in some filtrating neighborhood U, then it admits a basis of filtrating neighborhoods: it is enough to consider the intersections $\bigcap_{-n}^{n} f^i(U)$, $n \in \mathbb{N}$.
3. Theorem 10.3 implies that any chain recurrence class admits a basis of filtrating neighborhoods, and every isolated chain recurrence class is the maximal invariant set in some filtrating neighborhood of it.
4. If some chain recurrence class is accumulated by others (this must occur if there is an infinite number of classes), then it is not the maximal invariant set in any of its neighborhoods.
5. A weakly transitive set admits a basis of filtrating neighborhoods if and only if it is a chain recurrence class, in which case it is a maximal weakly transitive set.

Let us only say a few words about the proof of 5. The *if* half is contained in 3. To prove the converse, recall that a weakly transitive set Λ_0 is always contained in some chain recurrence class Λ. Assuming the existence of a basis of filtrating neighborhoods, it follows from part 1 in this remark that $\Lambda_0 = \Lambda$. Moreover, Λ is a maximal weakly transitive set.

The previous remark is the main motivation for the following definition:

Definition 10.8. An f-invariant compact set Λ is an *elementary dynamical piece* of the diffeomorphism f if it is, simultaneously, weakly transitive and a chain recurrence class.

In view of Remark 10.7, any elementary dynamical piece has a basis of filtrating neighborhoods. Moreover, any two different elementary dynamical pieces Λ_1 and Λ_2 are disjoint and separated by some filtration. To see the last assertion, fix some filtrating neighborhood $M_1 \setminus \mathrm{int}(M_2)$ of Λ_1, disjoint from Λ_2. For each $i = 1, 2$, the class Λ_2 is either contained in M_i or disjoint from M_i, because $f(M_i) \subset \mathrm{int}(M_i)$. Thus, Λ_2 is either contained in M_2 or disjoint from M_1. Therefore, (M, M_2), in the first case, or (M, M_1), in the second one, is the announced filtration.

Definition 10.9. A diffeomorphism f admits a *decomposition into elementary dynamical pieces* if its non-wandering set $\Omega(f)$ is the union of elementary dynamical pieces.

This includes, in particular, the uniformly hyperbolic diffeomorphisms (Axiom A with no cycles), and their elementary dynamical pieces are exactly the hyperbolic basic sets in Theorem 1.3.

Lemma 10.10. *If the global dynamics of a diffeomorphism f admits a decomposition into elementary dynamical pieces then $\Omega(f) = \mathcal{R}(f)$. Moreover, in such a case the induced equivalence relations $\vdash\!\dashv$ and $\diamond\!\!\!-\!\!\!\diamond$ coincide.*

Proof. It is enough to show that $\mathcal{R}(f) \subset \Omega(f)$. Let y be a point which does not belong to $\Omega(f)$. We want to prove that $y \notin \mathcal{R}(f)$. This forward limit set $\omega(y)$ is contained in $\Omega(f)$ and weakly transitive. So $\omega(y)$ is contained in some elementary dynamical piece Λ of f. By hypothesis, $y \notin \Lambda$. Consider a small filtrating neighborhood $U = M_1 \setminus \mathrm{int}(M_2)$ of Λ such that $y \notin U$. If $y \notin M_1$ then any pseudo-orbit starting at y and having small jumps will eventually enter in M_1 and remain in M_1 thereafter. So, this pseudo-orbit can never return to y, and this proves that $y \notin \mathcal{R}(f)$. If $y \in M_2$ then a similar argument proves that $y \notin \mathcal{R}(f)$. $\square$

We are now ready to state the main result, the generic dynamical decomposition theorem: C^1-generic diffeomorphisms have a decomposition into elementary pieces (not necessarily finite). More precisely,

Theorem 10.11 (Bonatti, Crovisier [70, 69]). *There is a residual subset of diffeomorphisms $f \in \mathrm{Diff}^1(M)$ for which the relations $\prec$ and $\dashv$ coincide. In particular, $\Omega(f) = \mathcal{R}(f)$, the chain recurrence classes are the maximal weakly transitive sets, and f has a decomposition into elementary dynamical pieces.*

The first claim follows from the connecting lemma for pseudo-orbits (Theorem A.7) through the Baire argument that we explain next. For any pair of open sets U and V in the manifold, the subset $\mathcal{O}(U, V)$ of diffeomorphisms such that, robustly, the forward orbit of U meets V or, robustly, the forward orbit of U does not meet V is open and dense. Consider the residual subset $\mathcal{R}$ of diffeomorphisms obtained by intersecting all these $\mathcal{O}(U,V)$ over a countable basis of open sets. The first claim in the theorem holds for every $f \in \mathcal{R}$. Indeed, let x, y be any points such that $x \dashv y$, and let $U \ni x$ and $V \ni y$ be any neighborhoods of the two points. Theorem A.7 states that there are arbitrarily close diffeomorphisms for which y is in the forward orbit of x, and so the forward orbit of U does meet V. By the definition of $\mathcal{R}$, it follows that the same is true for f (robustly). This means that $x \prec y$. All the other statements in the theorem are direct consequences of the first one.

10.4 Homoclinic classes and elementary pieces

Next we are going to analyze the relations between homoclinic classes of periodic orbits and the elementary dynamical pieces defined in the previous section. The *homoclinic class* of a hyperbolic periodic orbit is the closure of the transverse intersections of its invariant manifolds. Another way to define it is using the homoclinic relation introduced by Smale. Two hyperbolic periodic orbits are homoclinically related if the stable manifold of each orbit meets transversely the unstable manifold of the other. This implies that both periodic points have the same Morse index, that is, the same dimension of the unstable manifold. This is an equivalence relation in the set of periodic orbits: transitivity follows from the λ-lemma in [342]. Then the homoclinic class of a hyperbolic saddle is the closure of all periodic orbits in the equivalence class of the orbit of the saddle.

For Axiom A diffeomorphisms with no cycles, the homoclinic classes are exactly the hyperbolic basic pieces in the spectral decomposition theorem. In the general case, distinct homoclinic classes may intersect each other (see for instance [160]) and they may also fail to cover the entire closure of the set of periodic points. Nevertheless, as we are going to see, homoclinic classes do bear some relation to our elementary dynamical pieces. For generic diffeomorphisms the homoclinic class of a periodic orbit is the elementary dynamical piece that contains it. Moreover, if there are only finitely many homoclinic classes then the two notions just coincide. This case is treated in Section 10.5. On the other hand, we shall see in Section 10.6 that there are locally generic diffeomorphisms exhibiting infinitely many elementary dynamical pieces which contain no periodic orbit and, in particular, are not homoclinic classes.

10.4.1 Homoclinic classes and maximal transitive sets

A *maximal transitive set* of a diffeomorphism f is a transitive set which is a maximal element of the family of transitive sets of f ordered by inclusion. The next results sets the foundations of this concept: it allows us to use Zorn's lemma to deduce that any transitive set is contained in a maximal one.

Lemma 10.12. *The closure of an increasing family of transitive sets is transitive.*

Proof. Let $(K_i)_i$ be an increasing family of transitive sets and K its closure. To verify that K is transitive it is enough to see that for any pair U, V of non-empty open sets of K there is $n > 0$ with $f^n(U) \cap V \neq \emptyset$. By hypothesis, there is j such that $U \cap K_j \neq \emptyset$ and $V \cap K_j \neq \emptyset$. Thus, since K_j is transitive, $f^n(U) \cap V \neq \emptyset$ for some $n > 0$.   □

There are examples of distinct maximal transitive sets with non-empty intersection; see for instance [160]. These examples motivate the notion of saturated transitive set (this is called *maximal transitivity* in [118]): a transitive set is *saturated* if it contains any transitive set intersecting it. In other words, a saturated transitive set is an equivalence class for the equivalence relation generated by the binary relation *to be in the same transitive set*. By definition, saturated transitive sets are maximal transitive sets, and they are pairwise disjoint. However, there are maximal transitive sets which are not contained in any saturated one (see for instance [160]).

It was conjectured in [72] that, generically in $\mathrm{Diff}^1(M)$, the notions of homoclinic class and maximal transitive set coincide. We now know that this is partially false: there are locally generic diffeomorphisms having saturated transitive sets (which are also chain recurrence classes) that contain no periodic points and, hence, can not be homoclinic classes; this will be analyzed in Section 10.6. On the other hand, it is true that, generically, every homoclinic class is indeed a maximal transitive set. In what follows we comment on this and related results.

First, Bonatti, Díaz [72] used Hayashi's connecting lemma (Theorem A.4) to prove that, generically, two periodic points belong to the same transitive set if and only if their homoclinic classes coincide. Then

Theorem 10.13 (Arnaud [25]). *Generically in the space $\mathrm{Diff}^1(M)$, every transitive set containing a periodic orbit p is contained in the homoclinic class of p. In particular, generically, every homoclinic classes is a maximal transitive set.*

This result was then generalized by considering saturated transitive sets. The conclusion is very interesting for our purposes, for it shows that homoclinic classes of generic diffeomorphisms define a partition of a part of the non-wandering set:

Theorem 10.14 (Carballo, Morales, Pacifico [119]). *Generically in the space $\mathrm{Diff}^1(M)$, every homoclinic class is a saturated transitive set. In particular, two homoclinic classes either coincide or are disjoint.*

Later, [70, Remarque 1.10] gave a slightly stronger version, which is also very convenient here:

Theorem 10.15 (Bonatti, Crovisier [70, 69]). *Generically in* $\mathrm{Diff}^1(M)$, *every homoclinic class is a chain recurrence class. Equivalently, every chain recurrence class containing a periodic point p is the homoclinic class of p.*

Let us highlight some main ingredients in the proofs of these results. The first step was to prove that, *generically, the homoclinic class $H(p, f)$ of every saddle p coincides with the intersection of the closures of the stable manifold and the unstable manifold of p:*

$$\overline{W^s(p)} \cap \overline{W^u(p)} = H(p, f). \tag{10.1}$$

Recall that $H(p, f)$ is defined as the closure of the transverse intersections between $W^s(p)$ and $W^u(q)$, and so it is always contained in the intersection of the closures of the two invariant manifolds. The proof that they coincide relies on the connecting lemma. First, we observe that both the closures of the invariant manifolds and the homoclinic class $H(p, f)$ depend lower semi-continuously on f. Thus, for diffeomorphisms f in a residual set $\mathcal{R}$ of $\mathrm{Diff}^1(M)$, homoclinic classes depend continuously on f. Consider now a point x in the intersection of the closures of $W^s(p)$ and $W^u(p)$. Using Theorem A.5 one creates a transverse intersection between the invariant manifolds of p at x (this is a direct application if x is not periodic, but is rather more subtle in the periodic case). As the homoclinic class depends continuously on $f \in \mathcal{R}$, the point x belongs to $H(p, f)$. This shows that (10.1) does hold generically.

The remainder of the proof involves the notion of Lyapunov stability. An invariant set K is *Lyapunov stable* if for every neighborhood U of K there is another neighborhood V of K such that $f^n(V) \subset U$ for every $n \geq 0$. One shows that *generically, for every saddle p, the closure of the unstable manifold of p is Lyapunov stable and the closure of the stable manifold of p is Lyapunov stable for f^{-1}.* The proof goes as follows. Consider a diffeomorphism f in the residual set $\mathcal{R}$ above. Arguing by contradiction, assume there is a saddle p such that the closure of $W^u(p)$ is not Lyapunov stable. Then there are a neighborhood U of the closure of $W^u(p)$ and a sequence (x_n) of points converging to some $x \in \overline{W^u(p)}$ whose forward orbits escape from U. Considering the orbits of the points x_n one gets a new sequence converging to some point $q \notin U$. Theorem A.5 now implies (once more, the case when x is not periodic requires extra work) that, after a small perturbation of f, the point q is in the closure of $W^u(p)$. But this contradicts the fact that the closure depends continuously on $f \in \mathcal{R}$.

Finally, for $f \in \mathcal{R}$, consider any weakly transitive set K intersecting a homoclinic class $H(p, f)$. Lyapunov stability implies that K is contained in every neighborhood of the closure of the stable manifold and of the unstable manifold of p, and so $K \subset \overline{W^s(p)}$ and $K \subset \overline{W^u(p)}$. Using (10.1), we get that $K \subset H(p, f)$. This ends our discussion of Theorems 10.13 and 10.14. These arguments also give that, generically, homoclinic classes are maximal weakly transitive sets. Since, by Theorem 10.11, maximal weakly transitive sets are chain recurrence classes, generically, Theorem 10.15 also follows.

10.4.2 Homoclinic classes and chain recurrence classes

We have seen in Section 10.2 that, generically, hyperbolic periodic orbits are dense in the chain recurrent set. While it is not true that every chain recurrence class contains some periodic point (see Section 10.6), Crovisier has recently shown that periodic orbits accumulate every chain recurrence class with some topological regularity. This extends a corresponding result of Arnaud [27] for ω-limit sets.

Theorem 10.16 (Crovisier [142]). *For C^1-generic diffeomorphisms, any chain recurrence class is the Hausdorff limit of a sequence of periodic orbits.*

Together with the results in the previous sections, this has a number of interesting consequences:

Corollary 10.17. *For generic diffeomorphisms in* $\mathrm{Diff}^1(M)$,

1. *every isolated chain recurrence class is a homoclinic class;*
2. *every chain recurrence class with non-empty interior is a homoclinic class;*
3. *when they are finitely many, the chain recurrence classes coincide with the homoclinic classes;*
4. *the set of chain recurrence classes is infinite if and only if the set of homoclinic classes also is infinite.*

Notice that there are at most countably many homoclinic classes (the set of hyperbolic periodic points is countable), whereas we shall see in Section 10.6 that the chain recurrence classes may be uncountably many. Thus, the cardinals of the two sets in the last part of the corollary may not be the same.

10.4.3 Isolated homoclinic classes

We have seen in Remark 10.7 that any isolated chain recurrence class of a diffeomorphism is the maximal invariant set inside a filtrating neighborhood. Moreover, by Corollary 10.17, generically it is a homoclinic class and, hence, a transitive set. In this section we discuss some robustness properties for isolated homoclinic classes. A first natural question is

Problem 10.18. For generic diffeomorphisms, is every isolated homoclinic class a robustly transitive set?

The answer is not known, but there is some partial progress. An invariant compact set Λ is *robustly chain recurrent* if there are neighborhoods U of Λ in the ambient manifold M and $\mathcal{V}$ of f in $\mathrm{Diff}^1(M)$ such that, for every $g \in \mathcal{V}$, the maximal invariant set Λ_g of g in the closure of U is a chain recurrent set contained in U and $\Lambda_f = \Lambda$. As a consequence of Theorem A.7 one obtains

Corollary 10.19. *For C^1-generic diffeomorphisms, every isolated homoclinic class is robustly chain recurrent.*

One proves that, for generic diffeomorphisms, given any filtrating neighborhood U of an isolated chain recurrence class, every nearby diffeomorphism has a unique chain recurrence class intersecting U and it is contained in U.

For proving this, one considers a countable basis U_n of the topology of the manifold. For each U_n let $\mathcal{A}_n$ be the subset of diffeomorphism for which U_n intersects at most one homoclinic class and $\mathcal{B}_n$ be the subset of diffeomorphisms for which U_n intersects at least two homoclinic classes. Using that, generically, periodic points are dense in the chain recurrent, and having a homoclinic class intersecting an open set is a robust property, one proves that $\mathrm{int}(\mathcal{A}_n) \cup \mathrm{int}(\mathcal{B}_n)$ is dense. Consider the residual set $\mathcal{R}$ obtained by intersecting all these open dense sets with the residual set of diffeomorphisms for which isolated chain recurrence classes are homoclinic classes. Let c be an isolated chain recurrence class of a diffeomorphism $f \in \mathcal{R}$, and U a neighborhood of c and intersecting no other chain recurrence class. Now consider a covering of c by a finite family of open sets $U_{n_i} \subset U$. Then f belongs to the interior of the $\mathcal{A}_{n_i}$ and each U_{n_i} meets the homoclinic class c robustly. For any diffeomorphism in the union of the $\mathrm{int}(\mathcal{A}_{n_i})$, the recurrent set intersects the union $\cup_i U_{n_i}$ exactly on the chain recurrent class that contains the continuation of c.

Other results in the direction of Problem 10.18 involve a weaker form of robustness. Theorem 7.6 states that every robustly transitive set has a dominated splitting and is volume hyperbolic. The definition of robustly transitive set involves all diffeomorphisms in a neighborhood. Since here we are adopting a generic (or locally generic) point of view, it is more suitable to consider only diffeomorphisms in a relative neighborhood inside a residual subset. We say that compact isolated set Λ of a diffeomorphism f is a *generically transitive set* if there is an isolating block U for Λ, a neighborhood $\mathcal{U}$ of f in $\mathrm{Diff}^1(M)$, and a residual subset $\mathcal{R}$ of $\mathcal{U}$ containing f such that, for every diffeomorphism $g \in \mathcal{R}$, the maximal invariant set of g in the closure of U is a transitive set.

As isolated chain recurrent classes of generic diffeomorphisms are transitive, Corollary 10.19 implies that isolated chain recurrence classes of generic diffeomorphisms are generically transitive sets. Theorem 7.6 has also been extended to this set-up in [1]:

Proposition 10.20 (Abdenur [1]). *Consider a diffeomorphism f and a saddle p of f. Suppose that $H(p,f)$ is a generically transitive set. Then $H(p,f)$ has a dominated splitting and, in fact, is volume hyperbolic.*

Corollary 10.21. *For C^1-generic diffeomorphisms, every isolated homoclinic class is volume hyperbolic.*

As a matter of fact, at this point all known examples of generically transitive sets are also robustly transitive, so that the following question generalizing Problem 10.18 is open:

Problem 10.22. Is any generically transitive set a robustly transitive one?

10.5 Wild behavior *vs.* tame behavior

Having established the generic decomposition Theorem 10.11, the next stage in our program is to analyze the dynamics on the elementary dynamical pieces, for generic maps. This is very far from completion, but there has been some significant progress in the case when there are only finitely many elementary pieces. As we have seen in Corollary 10.17, that happens if and only if the number of homoclinic classes is finite. Moreover, in that case the two notions coincide.

The main result in this section, Theorem 10.28, provides an extension to arbitrary dimension of the following generic dichotomy between finitely many and infinitely many homoclinic classes, proved by Mañé for surface diffeomorphisms:

Theorem 10.23 (Mañé [278]). *For any closed surface M, there is a residual subset $\mathcal{R}$ of $\mathrm{Diff}^1(M)$ such that for every $f \in \mathcal{R}$*

- *either f is an Axiom A diffeomorphism with no cycles,*
- *or f simultaneously has infinitely many sinks or sources.*

As a consequence of Corollaries 10.19 and 10.21 we obtain the following local version of this result: *for generic diffeomorphisms on a closed surface, any isolated chain recurrence class is a basic set isolated by a filtrating neighborhood.*

10.5.1 Finiteness of homoclinic classes

Let us consider a residual subset $\mathcal{R}$ of $\mathrm{Diff}^1(M)$ such that every homoclinic class depends continuously on $f \in \mathcal{R}$ and any two homoclinic classes either are disjoint or coincide. Thus, the number $\#(f) \in \mathbb{N} \cup \{\infty\}$ of homoclinic classes of f is well defined. Since, locally in $\mathcal{R}$, the number of homoclinic classes only can increase, the function $\#(f)$ is lower semi-continuous. It follows that there exists a new residual subset of $\mathrm{Diff}^1(M)$ (also denoted by $\mathcal{R}$ for simplicity) where $\#(f)$ is continuous and even locally constant:

Theorem 10.24 (Abdenur [1]). *There is a residual subset $\mathcal{R}$ of $\mathrm{Diff}^1(M)$ such that the function $\#(\cdot)$ is well defined and locally constant on $\mathcal{R}$.*

This result means that the property of having $k \in \mathbb{N} \cup \{\infty\}$ homoclinic classes is stable in $\mathcal{R}$. This motivates that we consider the following two types of generic dynamics:

- *Tame diffeomorphisms*, that is, diffeomorphisms $f \in \mathcal{R}$ with finitely many homoclinic classes and such that every nearby diffeomorphism has exactly the same number of homoclinic classes. We denote this subset of $\mathcal{R}$ by $\mathcal{T}$.

- *Wild diffeomorphisms*, that is, diffeomorphisms $f \in \mathcal{R}$ having infinitely many homoclinic classes and such that the same holds for every $g \in \mathcal{R}$ close to f. We denote this subset of $\mathcal{R}$ by $\mathcal{W}$.

Theorem 10.24 means that $\mathcal{R} = \mathcal{T} \cup \mathcal{W}$. Examples of tame diffeomorphisms include, for instance, the Axiom A diffeomorphisms with no cycles and the robustly transitive diffeomorphisms in Chapter 7. One can also easily construct non-hyperbolic tame diffeomorphisms admitting a filtration such that the maximal invariant set in each level is a robustly transitive set: for instance, [152, 151, 158] show that such examples can be obtained from Morse-Smale diffeomorphisms via a heterodimensional cycle. The first examples of wild diffeomorphisms corresponded to Newhouse's phenomenon of coexistence of infinitely many sinks. See Section 10.6 for these and other examples.

It is not known if there is a open set of tame diffeomorphisms which is dense in the complement of the closure of $\mathcal{W}$, in which case one could choose $\mathcal{T}$ to be open. However, as a consequence of Corollary 10.19 and Proposition 10.20, we can prove

Corollary 10.25. *There is an open set $\mathcal{O}_{\mathcal{T}} \subset \mathrm{Diff}^1(M)$ containing $\mathcal{T}$ such that every $f \in \mathcal{O}_{\mathcal{T}}$ has finitely many chain recurrence classes. Moreover, the number of classes is locally constant, each class is volume hyperbolic and the maximal invariant set in a filtration level, and it contains some homoclinic class.*

Let us point out that we do not know whether the chain recurrence classes remain weakly transitive, and so we do not know whether they remain elementary dynamical pieces in the whole open set $\mathcal{O}_{\mathcal{T}}$. Notice that this is true generically, by Theorem 10.11. A positive answer to Problem 10.18 would imply that each recurrence class of a diffeomorphism $f \in \mathcal{O}_{\mathcal{T}}$ is a homoclinic class.

10.5.2 Dynamics of tame diffeomorphisms

There is a fair amount of understanding of the dynamics when $f \in \mathcal{T}$. Recall that, generically, homoclinic classes are chain recurrence classes. Thus, by Conley's Theorem 10.3, they are separated by some filtration. Moreover, it is clear that in the tame case all homoclinic classes are isolated. So, by Remark 10.7, they coincide with the maximal invariant sets in the corresponding filtration levels. This leads to the next result, which had been obtained previously by Abdenur through different arguments.

Theorem 10.26 (Abdenur [1]). *Consider a tame diffeomorphism $f \in \mathcal{T}$. Then there are a filtration $\mathcal{M} = (M_i)_{i=1}^k$, finitely many hyperbolic periodic points $P_f^1, \ldots, P_f^k$, and a neighborhood $\mathcal{U}$ of f in $\mathrm{Diff}^1(M)$ such that, for every $g \in \mathcal{U} \cap \mathcal{T}$,*

1. *the continuations $P_g^1, \ldots, P_g^k$ of those periodic points of f are defined,*

2. the maximal invariant set of g inside each filtration level $L_i = (M_i \setminus M_{i+1})$ is the homoclinic class of P_g^i,
3. and every homoclinic class of g is volume hyperbolic.

Let $\Lambda_1, \ldots, \Lambda_m$ be a finite family of different homoclinic classes of a diffeomorphism f. These classes form a *cycle* if $W^u(\Lambda_i) \cap W^s(\Lambda_{i+1}) \neq \emptyset$ for all $i = 1, \ldots, (m-1)$, and $W^u(\Lambda_m) \cap W^s(\Lambda_1) \neq \emptyset$. From Theorem 10.26 we immediately get

Corollary 10.27. *Given any tame diffeomorphism $f \in \mathcal{T}$ there are no cycles between homoclinic classes of f.*

Putting the previous results together one gets the following generic dichotomy, which generalizes Theorem 10.23 to arbitrary dimension.

Theorem 10.28. *There is a residual subset $\mathcal{R}$ of $\mathrm{Diff}^1(M)$ and a partition of $\mathcal{R}$ into disjoint open subsets $\mathcal{T}$ and $\mathcal{W}$, such that,*

(a) the number of homoclinic classes is locally constant in $\mathcal{R}$,
(b) if $f \in \mathcal{W}$ then f has infinitely many homoclinic classes,
(c) if $f \in \mathcal{T}$ then the chain recurrent set is the union of finitely many pairwise disjoint homoclinic classes, and there exists a filtration such that
  - *every homoclinic class is a generically transitive set in a filtration level,*
  - *every homoclinic class is volume hyperbolic,*
  - *there are no-cycles between the homoclinic classes.*

For tame diffeomorphisms, it is possible to go further in the description of the topological dynamics, namely in what concerns attractors and repellers. We say that a homoclinic class is a *topological attractor (repeller)* if its stable (respectively unstable) set is a neighborhood of the homoclinic class. Existence of a finite filtration separating all homoclinic classes, immediately, implies that there exists some topological attractor: the last level of the filtration is a neighborhood of the corresponding homoclinic class contained in its basin of attraction, and so that homoclinic class must be an attractor. Analogously, every tame diffeomorphism has some topological repeller. Then, it is natural to ask whether the union of the basins of the attractors (repellers) is dense in the ambient manifold. The answer is positive:

Theorem 10.29 (Carballo, Morales [117]). *For any $f \in \mathcal{T}$, the union of the basins of the attractors is a dense open subset of the whole manifold.*

The corresponding question for wild diffeomorphisms is wide open:

Problem 10.30. Does every generic diffeomorphism have an elementary dynamical piece that is a topological attractor or repeller?

Hurley [222] conjectured that a weaker form of the conclusion of Theorem 10.29 is true C^r-generically, for every r: *the union of all quasi-attractors is a residual subset of the ambient manifold.* A *quasi-attractor* is an intersection of topological (not necessarily transitive) attractors. Hurley [222] proved this conjecture in the case $r = 0$ (for other results on attractors of generic homeomorphisms see [223, 224]), and the case $r = 1$ was settled only very recently by Bonatti, Crovisier [70]. Another question naturally raised by Theorem 10.29 is

Problem 10.31. For tame diffeomorphisms is the union of the basins of the attractors a full measure subset of the ambient manifold?

10.6 A sample of wild dynamics

In contrast, little is known about wild diffeomorphisms, apart from the fact that they exist and are locally generic. Building on the known constructions, we are currently trying to uncover the most basic properties of these systems. Let us begin by recalling the original construction of C^r locally generic dynamics for $r \geq 2$:

10.6.1 Coexistence of infinitely many periodic attractors

The first examples of wild dynamics were discovered by Newhouse, [320, 321] in the setting of C^2 surface diffeomorphisms. He proved (recall Theorem 3.6) that there exist an open subset $\mathcal{U}$ of $\mathrm{Diff}^r(M)$, $r \geq 2$, and a residual subset $\mathcal{R}$ of $\mathcal{U}$ consisting of diffeomorphisms with infinitely many sinks or sources.

As explained in Section 3.2, this construction is intimately associated to homoclinic phenomena, in fact the underlying mechanism are persistent homoclinic tangencies. To draw a parallel with the forthcoming constructions, we also note that, up to reducing the neighborhood $\mathcal{U}$ and the residual subset $\mathcal{R}$ if necessary, for every $f \in \mathcal{R}$ the homoclinic class of the saddle involved in the tangency is contained in the closure of the set of sinks or sources.

In higher dimensions, a trivial way to construct C^2-locally generic wild diffeomorphisms is to consider a normally hyperbolic invariant surface S such that the restricted dynamics has infinitely many sinks or sources. Depending on the type of normal hyperbolicity of the surface S (attracting, repelling or of saddle type), this configuration may lead to infinitely many sinks, sources or saddles with trivial homoclinic classes.

A much deeper fact is that, in any dimension, persistence of tangencies does occur close to any diffeomorphism with a homoclinic tangency, [347, 387], and, under dissipativeness assumptions [347] this leads to coexistence of infinitely many sinks or sources for a C^r-locally residual subset, see Section 3.5.1. In fact, the argument in [387] is based on finding a normally hyperbolic surface restricted to which the tangency is unfolded.

10.6.2 C^1 coexistence phenomenon in higher dimensions

Next we discuss a construction of infinitely many sinks or sources coexisting for locally C^1-generic diffeomorphisms in dimension 3 (or higher), first proposed by Bonatti, Díaz [72]. This phenomenon is not explicitly related to the persistence of tangencies, but rather to persistent lack of domination, although homoclinic tangencies do occur in this context.

For any closed 3-manifold M, consider an open subset $\mathcal{U}$ of $\mathrm{Diff}^1(M)$ such that every diffeomorphism $f \in \mathcal{U}$ has a periodic saddle Q_f, depending continuously on the diffeomorphism, such that its homoclinic class $H(Q_f, f)$ does not admit any dominated splitting for any $f \in \mathcal{U}$. A simple way to get this property is the following.

Consider the model heterodimensional cycle in Section 6.1.2. Perturbing this cycle, one finds an open set $\mathcal{V}$ of C^1-diffeomorphisms such that the homoclinic class of the fixed point Q of Morse (unstable) index 2 in the cycle contains the fixed point P of index 1. This construction only involves semi-local properties of the bifurcating diffeomorphisms, and so it is compatible with other global assumptions on the dynamics. For instance, the point P may be heteroclinically related to some fixed point P' of index 1 having a pair of non-real contracting eigenvalues and, similarly, the point Q may be heteroclinically related to some fixed point Q' of index 2 with a pair of non-real expanding eigenvalues. Hence, for every diffeomorphism $g \in \mathcal{V}$,

$$H(P', g) = H(P, g) \subset H(Q, g) = H(Q', g) = \Lambda_g.$$

This implies that the homoclinic class Λ_g does not admit any dominated splitting $T_{\Lambda_g} M = E \oplus F$: the presence P' prevents the existence of a splitting such that F has dimension 2, and the presence of Q' means that E cannot have dimension 2 either.

It is proved in [72] that this dynamical configuration implies that, for a residual subset $\mathcal{R}$ of diffeomorphisms $g \in \mathcal{V}$, the homoclinic class $H(Q_g, g)$ is contained in the closure of the set of sources or sinks of g. Nowadays, this is a direct consequence of Theorem 7.6. Since the homoclinic class of Q is non-trivial, it follows that every $g \in \mathcal{R}$ has infinitely many sinks or sources. More precisely, if the Jacobian of f at some of the points P, P', Q, Q' is less than 1 then the homoclinic class is contained in the closure of the set of sinks. Analogously, if the Jacobian of f at some of these fixed points is larger than 1 then $H(Q, g)$ is contained in the closure of the set of sources. In particular, $H(Q, g)$ can be contained in the closures of both sinks and sources!

10.6.3 Generic coexistence of aperiodic pieces

Robustly non-dominated homoclinic classes are also at the heart of the following, more sophisticated, construction of diffeomorphisms exhibiting a wide range of wild phenomena.

Consider diffeomorphisms $f \in \mathrm{Diff}^1(M)$ on some 3-dimensional manifold, exhibiting a homoclinic class whose continuation to some C^1 neighborhood $\mathcal{U}$ of f never admits a dominated splitting. This was called a *wild homoclinic class* in [68]. Here, to simplify the statements that follow, we add to this notion the requirement that for every $g \in \mathcal{U}$ there are periodic saddles heteroclinically related to Q_g (in particular, with the same unstable index as Q_g) with Jacobian both larger than 1 and smaller than 1.

Theorem 10.32 (Bonatti, Díaz [68]). *Let $\mathcal{U}$ be a C^1 open set of diffeomorphisms on a closed 3-dimensional manifold M, such that there is a periodic saddle Q_f defined for every $f \in \mathcal{U}$ (depending continuously on the map) and the corresponding homoclinic class $H(Q_f, f)$ is wild for every $f \in \mathcal{U}$. Then there is a C^1 residual set $\mathcal{R} \subset \mathcal{U}$ such that for every $f \in \mathcal{R}$ the homoclinic class $H(Q_f, f)$ is contained in the closure of uncountably many pairwise disjoint compact invariant sets $\mathcal{M}_i$ satisfying*

- *Every $\mathcal{M}_i$ is a maximal transitive set without periodic points. Moreover, the dynamics of g in $\mathcal{M}_i$ is minimal and uniquely ergodic.*
- *Every $\mathcal{M}_i$ is Lyapunov stable both for f and for f^{-1}. Consequently, $\mathcal{M}_i$ is a saturated transitive set.*

Let us recall that an invariant set Σ is *minimal* if the orbit of every point of Σ is dense in the whole Σ. An invariant set Σ is *uniquely ergodic* if it supports a unique invariant measure (which is necessarily ergodic). Lyapunov stability of the sets $\mathcal{M}_i$ for f and f^{-1} implies that they are weakly transitive sets, and chain recurrence classes since these two notions coincide generically. Hence, the $\mathcal{M}_i$ are elementary dynamical pieces in the sense of Section 10.3.3. We shall see in the construction that they are also, simultaneously, quasi-attractors and quasi-repellers. Recall, from Section 3.2.1, that uniquely ergodic quasi-attractors also occur in the unfolding of homoclinic tangencies on surfaces. Finally, the previous result implies that there are locally generic diffeomorphisms with chain recurrence classes, and even saturated transitive sets, which are not homoclinic classes.

Universal dynamics:

The main step for proving Theorem 10.32 is to construct a residual subset of diffeomorphisms $f \in \mathcal{U}$ with *universal dynamics* at the wild homoclinic class $H(Q_f, f)$. In order to define this notion, let B^3 be the compact unitary ball in 3-dimensional Euclidean space, and $\mathrm{Diff}^+(B^3)$ be the set of orientation preserving C^1-diffeomorphisms $\phi\colon B^3 \to \mathrm{int}(B^3)$. A diffeomorphism f has *universal dynamics* at some invariant set Λ if, for any open set $\mathcal{O} \subset \mathrm{Diff}^+(B^3)$ and for both $g = f$ and $g = f^{-1}$, there are periodic disks $D \subset M$ (this means that there is $n \geq 1$ such that $g^n(D) \subset \mathrm{int}(D)$ and the disks $g^i(D)$, $0 \leq i < n$ are pairwise disjoint) inside any neighborhood of Λ such that $g^n\colon D \to D$ is differentiably conjugate to some element of $\mathcal{O}$.

To prove that such a residual subset of $\mathcal{U}$ does exist, and given that $\mathrm{Diff}^+(B^3)$ has a countable basis of open sets, we only have to show that for every $\mathcal{O}$ there exists an open dense subset of diffeomorphisms in $\mathcal{U}$ such that some element of $\mathcal{O}$ may be realized as a return map to some periodic disk. The argument relies on methods from [73]. Let f be any element of $\mathcal{U}$. Since the homoclinic class $H(Q_f, f)$ is robustly non-dominated, the arguments in Section 7.2.2 ensure that, up to a C^1-small perturbation, f has some periodic point p with period $\kappa \geq 1$ such that $Df^\kappa(p)$ is a homothetic transformation.

In fact, we may choose this periodic point such that $Df^\kappa(p)$ is close to the identity. That is because there are saddle points in the homoclinic class with both Jacobian bigger than 1 and Jacobian smaller than 1: one selects orbits with an appropriate number of iterates close to the saddle points which are volume expanding and volume contracting, so that multiplying the derivatives along the orbit one obtains a homothetic transformation with Jacobian close to 1 and, hence, close to the identity. Then, up to another small perturbation, there is a small disk D containing p such that $D, f(D), \ldots, f^{\kappa-1}(D)$ are pairwise disjoint disks and the restriction of f^κ to D is the identity.

Now we use the fact that every element of $\mathrm{Diff}^+(B^3)$ may be written as a composition of (several) diffeomorphisms in any fixed neighborhood of the identity. So, perturbing the identity on D, we may find a small periodic disk $D' \subset D$, with arbitrarily large period, such that the first return map of f to D' is any prescribed diffeomorphism in $\mathcal{O}$. Since $\mathcal{O}$ is open, the return map to D' of any diffeomorphism close to f is still an element of $\mathcal{O}$. This completes the argument.

Wild homoclinic classes:

Now, to prove Theorem 10.32, we only have to show that universal dynamics leads to the conclusion of the theorem. For this, fix any open set $\mathcal{O} \subset \mathrm{Diff}^+(B^3)$ of diffeomorphisms exhibiting wild homoclinic classes (for instance, the one defined in Section 10.6.2). For any diffeomorphism f in the residual set $\mathcal{R}$ of $\mathcal{U}$ constructed in the previous paragraph there exists a periodic disk D_0 such that the return map of f to D_0 is conjugate to some diffeomorphism in $\mathcal{O}$. In particular, the orbit of D_0 contains a wild homoclinic class.

Applying to this new homoclinic class the argument in the previous paragraph, we conclude that for a residual subset of a neighborhood of f this return map has universal dynamics. Thus, reducing the residual set $\mathcal{R}$ if necessary, we may find a periodic disk $D_1 \subset D_0$ such that the return map to D_1 is conjugate to some element of $\mathcal{O}$. Iterating this procedure, we end up with a slightly smaller residual subset of $\mathcal{U}$, that we still denote $\mathcal{R}$, such that for each $f \in \mathcal{R}$ there is a nested sequence of disks D_k with diameters going to zero, and a sequence of natural numbers $n_k \to \infty$, satisfying

- for every k, the disks $D_k, \ldots, f^{n_k-1}(D_k)$ are pairwise disjoint, and $f^{n_k}(D_k)$ is contained in the interior of D_k,

- the diameters of the disks of the forward orbit of D_k are bounded by some d_k, where $d_k \to 0$ as $k \to \infty$,
- the restriction of f^{n_k} to D_k is conjugate to some element of $\mathcal{O}$.

By construction, the set

$$\mathcal{M}_f = \bigcap_k \bigcup_{i=0}^{n_k-1} f^i(D_k).$$

is an f-invariant Cantor set without periodic orbits. Moreover, $\mathcal{M}_f$ is Lyapunov stable for f and every forward orbit in $\mathcal{M}_f$ is dense in it. To ensure $\mathcal{M}_f$ is also Lyapunov stable for f^{-1} one needs to consider disks D_k such that $f^{n_k}(D_k) \subset \operatorname{int}(D_k)$ and $f^{-n_k}(D_k) \subset \operatorname{int}(D_k)$, for alternate values of k. Incidentally, this is the reason why the definition of universal dynamics involves both f and f^{-1}.

Remark 10.33. Notice also that these $\mathcal{M}_f$ are infinitely renormalizable sets, more precisely, $f \mid \mathcal{M}_f$ is conjugate to an adding machine.

Extensions and further comments:

The following important principle emerges from the definition of universal dynamics and the procedure outlined above: *Generic diffeomorphisms with a wild homoclinic class display infinitely many times any given robust feature of diffeomorphisms of the disk B^3.* This gives the following strong form of the coexistence phenomenon for diffeomorphisms on 3-dimensional manifolds:

Theorem 10.34. *Let M be a 3-dimensional closed manifold and $\mathcal{W}_h$ be the open subset of $\operatorname{Diff}^1(M)$ formed by the diffeomorphisms with some wild homoclinic class. Then there is a residual subset $\mathcal{R}$ of $\mathcal{W}_h$ such that every $f \in \mathcal{R}$ simultaneously has infinitely many*

1. *sinks and sources,*
2. *saddles with trivial homoclinic classes (independent saddles),*
3. *non-trivial uniformly hyperbolic attractors and repellers,*
4. *non-trivial partially hyperbolic attractors and repellers,*
5. *wild homoclinic classes*
6. *maximal saturated transitive Cantor sets which are minimal sets for the dynamics and, in particular, contain no periodic points.*

In fact, this construction extends to higher dimensions in two different ways. The first one is, simply, by multiplying the 3-dimensional example by a transverse uniformly hyperbolic dynamics. Another, non-trivial, extension is by considering robustly non-dominated homoclinic classes with convenient additional hypotheses on the Jacobians.

We finish this section with some questions concerning wild dynamics, especially the basins of their attractors, the geometrical distribution of independent saddles, and Smale's ordering of saddle points:

Problem 10.35. 1. Let f be a locally generic diffeomorphism having infinitely many independent saddles but only finitely many non-trivial homoclinic classes. Are there finitely many (local?) hyper-surfaces containing all the independent saddles?
2. Is it possible to have (generically) an infinite sequence of independent saddles $(s_i)_i$ with $s_1 \prec s_2 \prec \cdots s_i \prec \cdots$, where $r \prec s$ means that the unstable manifold of r meets the stable manifold of s.
3. Generically, is the union of the basins of the topological attractors a dense and open subset of the ambient manifold?
4. Do locally generic diffeomorphisms having non-transitive (aperiodic) elementary dynamical pieces exist?

•

11

SRB Measures and Gibbs States

For conservative systems the notion of typical trajectory may be, naturally, defined to correspond to full Lebesgue measure subsets in the ambient manifold. For instance, if the system is ergodic then, by the ergodic theorem,

$$\lim_{n \to +\infty} \frac{1}{n} \sum_{j=0}^{n-1} \phi(f^j(x)) = \int \phi \, d\mu \qquad \text{for any continuous function } \phi \quad (11.1)$$

and μ-almost every point x, where μ is normalized Lebesgue measure. In the general dissipative case where, a priori, there is no distinguished invariant probability measure, it is much more subtle what one should mean by *describing the behavior of almost all orbits*. The point of view in this chapter, and in most of the literature in this field, going back to Sinai and Ruelle, is to emphasize invariant measures which are "physically observable", in the following sense.

A *Sinai, Ruelle, Bowen (SRB) measure* for a diffeomorphism $f : M \to M$ is an invariant probability measure μ such that (11.1) holds for a positive Lebesgue measure set of points x. This set is called the *basin* of μ and denoted by $B(\mu)$. Every uniformly hyperbolic C^2 flow or diffeomorphism admits a finite number of SRB measures, and the union of their basins contains Lebesgue almost every point in the ambient manifold [90, 390, 416]. We are going to discuss some recent extensions of this fundamental result beyond the hyperbolic world, especially for dominated, or even partially hyperbolic, diffeomorphisms.

The classical construction of SRB measures for uniformly hyperbolic systems involves, as an intermediate step, invariant measures which are absolutely continuous with respect to Lebesgue measure along the unstable direction, and which we call *Gibbs u-states*. In fact, in that context, the two notions coincide. In Section 11.2 we consider partially hyperbolic systems with an invariant splitting of type $E^e \oplus E^{cs}$. The main conclusions are that they admit Gibbs u-states, this class of measures includes any existing SRB measure, and we also discuss when a Gibbs u-state is an SRB-measure.

The situation becomes much more delicate in the absence of a uniformly expanding direction. Section 11.3 treats systems with a dominated splitting involving a center unstable direction E^{cu} in which the vectors are non-uniformly expanded by forward iterates. Using a notion of *hyperbolic times* (iterates at which E^{cu} "seems" uniformly expanding) one is able to recover most of the previous theory. In particular, one constructs invariant measures absolutely continuous with respect to Lebesgue measure along the center unstable direction, that we call *Gibbs cu-states*.

These results give a fair picture of the statistical behavior of a large class of systems with a dominated splitting separating positive and negative Lyapunov exponents. Situations where the finest dominated splitting involves sub-bundles mixing Lyapunov exponents with different signs are not covered by the present methods, and constitute a great challenge.

On the other hand, an outstanding recent result of Tsujii for surface maps, that we present in Section 11.4, extends the conclusions of Sections 11.2 and 11.3 to Gibbs u-state admitting Lyapunov exponents equal to zero, thus proving that, *existence and finiteness of SRB measures whose basins cover Lebesgue almost every point is generic among partially hyperbolic surface endomorphisms*. His methods suggest that an ergodic theory of generic partially hyperbolic diffeomorphisms with 1-dimensional central direction maybe within reach.

For uniformly hyperbolic systems, Gibbs states, including u-states as a particular case, are part of the broader theory of thermodynamical formalism, which has several other applications. See the book of Bowen [86]. The previous results on SRB measures raise hopes that it may be possible to extend a good part of this formalism to very general non-hyperbolic systems. Some progress in this direction is presented in Section 11.5.2, concluding this chapter.

11.1 SRB measures for certain non-hyperbolic maps

Right now, let us start by describing some of the examples that have been motivating the development of this theory.

11.1.1 Intermingled basins of attraction

In the uniformly hyperbolic context, each attractor Λ supports exactly one SRB measure μ, and the basin $B(\mu)$ is a full measure subset of the (open) stable set $B(\Lambda)$ of the attractor. The construction we describe here, due to Kan [229], shows that the topology of the basins of SRB measures may be much more complicated, in general: in this example both basins are dense in the ambient space.

For simplicity, we describe the example as a map of the 2-dimensional cylinder preserving both boundary components. An interesting point is that in this setting the construction is C^2 robust. At the end we mention a version for diffeomorphisms of the 3-torus.

Consider $F : S^1 \times [0,1] \to S^1 \times [0,1]$ of the form $F(x,t) = (kx \bmod \mathbb{Z}, f_x(t))$ where k is some integer number with $|k| \geq 3$. Then $x \mapsto kx \bmod \mathbb{Z}$ has at least two fixed points, denoted p and q. Assume F is C^2 and

(1) F preserves the boundary components: $f_x(0) = 0$ and $f_x(1) = 1$;
(2) f_p and f_q have exactly two fixed points each, a source at $t = 1$ (respectively $t = 0$) and a sink at $t = 0$ (respectively $t = 1$);
(3) $|f'_x(t)| < k$ at every point $(x,t) \in S^1 \times [0,1]$;
(4) $\int \log f'_x(0)dx < 0$ and $\int \log f'_x(1)dx < 0$.

Proposition 11.1. *Every map satisfying (1)–(4) admits exactly two SRB measures. Both basins are dense, and their union has total Lebesgue measure in the cylinder $S^1 \times [0,1]$.*

Proof. Let μ_0 and μ_1 be the normalized arc-length measures along each of the two boundary components of the circle. These are ergodic F-invariant probabilities, and hypothesis (4) means that their Lyapunov exponents along the t-direction are negative. Consequently, both μ_0 and μ_1 are SRB measures for F: Pesin local stable manifolds of typical points in $t = 0$ form a positive area subset B_0 of $B(\mu_0)$, and similarly for $t = 1$.

To prove the remaining statements, consider any smooth segment γ in the open cylinder, transverse to the t-direction. We are going to show both basins intersect γ on positive Lebesgue measure subsets whose union has full measure in γ. Up to replacing γ by some forward iterate, we may suppose that it intersects $W^s(p,0) = \{p\} \times [0,1)$ and $W^s(q,0) = \{q\} \times (0,1]$. Then the orbit of γ accumulates on both boundary components. It follows (see [84, Section 2]) that some forward iterate $f^n(\gamma)$ intersects both $B_0 \subset B(\mu_0)$ and $B_1 \subset B(\mu_1)$ on positive measure subsets. Then $\gamma \cap f^{-n}(B_i)$ is contained in $B(\mu_i)$ and, in fact, its points have Pesin local stable manifolds (short segments in the t-direction) whose union has positive area and is contained in the basin of μ_i. As a consequence, every segment sufficiently C^1 close to γ intersects $B(\mu_i)$ on a subset whose measure is at least half that of $\gamma \cap f^{-n}(B_i)$.

One can easily deduce that γ is essentially contained in $B(\mu_0) \cup B(\mu_1)$. Indeed, suppose there was a positive measure subset in the complement of the union. By considering a density point and using the fact that forward iterates of F do not distort Lebesgue measure much, we find a sequence $\gamma_n \subset F^n(\gamma)$ of segments with constant length and such that the relative measure of the union of the basins inside γ_n goes to zero when n goes to infinity. Restricting to a subsequence if necessary, we may suppose that the γ_n converge to some segment β contained in the closed cylinder. This contradicts the conclusion of the previous paragraph, for β in the place of γ.   □

Along the same lines we may construct diffeomorphisms of the 3-torus with intermingled basins. Roughly, one replaces the expanding map of S^1 by an Anosov diffeomorphism on T^2, and then glues along the boundary two diffeomorphisms of $T^2 \times [0,1]$ obtained in this way. These examples are still some robust, but only restricted to maps that preserve the two invariant tori obtained from the boundary components.

11.1.2 A transitive map with two SRB measures

Here we refine the previous construction, to exhibit a dynamical system where the topological and the ergodic notions of indecomposability are strikingly different: the map is transitive, yet it admits two different SRB measures. These examples are robust as endomorphisms of the compact cylinder preserving the boundary.

Consider a map $F : S^1 \times [0,1] \to S^1 \times [0,1]$ satisfying, in addition to properties (1)–(4), that there exist two compact sectors $I_j \times [0,1]$, $j = 1, 2$, such that

(5) F maps each $I_j \times [0,1]$ onto the whole cylinder, the image of the boundary being disjoint from $(I_1 \cup I_2) \times \{0,1\}$.
(6) There exists a vector field X which vanishes exactly at $t = 0$ and $t = 1$, and a negative irrational number α such that the transformation f_x is given by the time-1 map of X on the first sector, and by the time-α map of X on the second sector.

Proposition 11.2. *Every transformation F satisfying (1)–(6) is transitive.*

Proof. Any open subset U has an iterate which crosses the sector $I_1 \times [0,1]$. Consider a small square inside this iterate and intersecting the stable manifold of a fixed point in the sector. Using that F is given by a product on $I_1 \times [0,1]$, we get that some further iterate $F^m(U)$ contains a rectangle $I_1 \times [t_1, t_2]$. Now, using that α is irrational, we may find sub-sectors $J_\pm \times [0,1]$ of $I_1 \times [0,1]$, and iterates $F^{n_\pm}$ which map the corresponding sub-sector bijectively onto $I_1 \times [0,1]$ and whose vertical components are given by the time-$s_\pm$ maps of X with $s_- < 0 < s_+$ arbitrarily close to zero. For this, notice that for a point that spends k iterates in $I_1 \times [0,1]$ and l iterates in $I_2 \times [0,1]$ the vertical component of F^{k+l} is given by the time-$(k + l\alpha)$ map of X; since α is negative and irrational, we may always choose k and l so that $k + l\alpha$ is close to zero. Then the orbit of $F^m(U)$ under $F^{n_\pm}$ contains the sector $I_1 \times (0,1)$. It follows that the complete forward orbit of U contains the whole open cylinder. Clearly, this implies transitivity. □

As in the previous section, this construction extends to diffeomorphisms of the 3-dimensional torus. To ensure transitivity, just ask that the halves of the 3-torus bounded by the invariant 2-tori be mapped to one another. Such example *do not* fill an open subset of the space of all diffeomorphisms on T^3. In fact, no robust examples are known of transitive diffeomorphisms on boundariless manifolds with more than one SRB measure.

11.1.3 Robust multidimensional attractors

We describe a construction, by Viana and Alves, of open sets of non-uniformly hyperbolic maps with multidimensional expanding direction. It applies also to diffeomorphisms, but we restrict ourselves to a non-invertible situation.

Let $F : S^1 \times [-2, 2] \to S^1 \times [-2, 2]$ be a C^3 map of the form

$$F(\theta, x) = (k\theta \bmod \mathbb{Z}, x^2 - a_0 + a(\theta)),$$

where $a : S^1 \to (1, 2)$ is a Morse function, for instance $a(\theta) = \alpha \sin(2\pi\theta)$, and a_0 is a Misiurewicz parameter (the critical point $x = 0$ is non-recurrent and there is no periodic attractor) for the quadratic transformation $x \mapsto x^2 - a_0$.

Theorem 11.3. *Assume k is large enough.*

1. (Viana [440]) *For every $\alpha > 0$ sufficiently small there exists $c_0 > 0$ and a C^3 neighborhood $\mathcal{U}$ of F in the space of endomorphisms of the cylinder, such that every $G \in \mathcal{U}$ has only positive Lyapunov exponents*

$$\liminf_{n \to +\infty} \frac{1}{n} \log \|DG^n(\theta, x)^{-1}\| \le -c_0 < 0 \qquad (11.2)$$

   at Lebesgue almost every point $(\theta, x) \in S^1 \times [-2, 2]$.
2. (Alves [7]) *Every $G \in \mathcal{U}$ admits some ergodic invariant probability measure μ_G absolutely continuous with respect to Lebesgue measure.*
3. (Alves, Viana [14]) *The measure μ_G is unique and depends continuously on the map $G \in \mathcal{U}$.*

In the sequel we are going to outline the proof of these statements. Notice that [440] considered $k \ge 16$, but Buzzi, Sester, Tsujii [114] show that $k \ge 2$ suffices.

Beforehand, let us mention that the ergodic properties of these systems have been studied by several authors very recently. Especially, Alves, Luzzatto, Pinheiro [13] prove that the correlations of Hölder continuous observables decay super-polynomially i.e. faster than $Cn^{-\gamma}$ for any $\gamma > 0$. See Appendix E for definitions and background information. Their result was then improved by Baladi, Gouezel [42, 196], who prove that the decay of correlations is at least stretched-exponential, that is, faster than $Ce^{-c\sqrt{n}}$. No lower bounds are known and, in particular, the following question remains open:

Problem 11.4. Does (G, μ_G) have exponential decay of correlations in the space of Hölder continuous functions? Same question for $G = F$.

Let us explain how part 1 of the theorem is obtained for $G = F$. One writes the Lyapunov exponent along the vertical direction as

$$\liminf_{n \to +\infty} \frac{1}{n} \sum_{j=0}^{n-1} \log |\partial_x F(x_j)| = \liminf_{n \to +\infty} \frac{1}{n} \sum_{j=0}^{n-1} \log(2|x_j|)$$

where $(\theta_j, x_j) = F^j(\theta, x)$ for $j \geq 0$. It suffices to prove (11.2) for almost every point on any admissible (nearly horizontal) curve. The assumption on k ensures that the family of such curves is invariant under forward iteration by F. The assumption that $a(\cdot)$ is a Morse function, together with a delicate analysis of iterates of admissible curves, shows that at each iterate j the set of points (θ, x) in any admissible curve for which

$$\log|x_j| \in [-r-1, -r]$$

decays exponentially with r. Combining this with the fact that the horizontal dynamics $x \mapsto kx$ is rapidly mixing, via a large deviations argument, one finds $c_0 > 0$ such that

$$\frac{1}{n} \sum_{j=0}^{n-1} \log(2|x_j|) > c_0 n,$$

but for a subset of the admissible curve whose Lebesgue measure decreases as $e^{-c\sqrt{n}}$ with n. Using a Borel-Cantelli argument, one obtains the conclusion of part 1 for $F = G$.

To obtain the same conclusion for general G in a neighborhood, one uses the fact that the vertical fibration is normally expanding for F. Consequently, by a version of the stability Theorem B.10 for non-invertible maps, it gives rise to an invariant central foliation, for every nearby G. Then the argument is carried out as before, with the tangent to this central foliation in the role of vertical direction. The condition of C^3-closeness ensures that the critical set of G, where the derivative along the central direction vanishes, is a circle C^2 close to the original $\{x = 0\}$.

The proof of part 2 is based on the idea of *hyperbolic time*. The precise definition will appear later, here we just mention a consequence: if n is a hyperbolic time for a point $z = (\theta, x)$ then $DG^k(G^{n-k}(z))$ is an expansion

$$\|DG^k(G^{n-k}(z))^{-1}\| \leq e^{-ck}, \quad c > 0$$

for every $1 \leq k \leq n$. From the proof of (11.2) one gets that Lebesgue almost every point has infinitely many hyperbolic times, in fact, the set of hyperbolic times has positive density at infinity. This is based on the following remarkably useful (see e.g. [12] for a proof)

Lemma 11.5 (Pliss [359]). *Given $A \geq c_0 > c$ there exists $\theta = (c_0 - c)/(A - c)$ such that, given any real numbers $a_1, \ldots, a_N$ with*

$$\sum_{i=1}^{N} a_i \geq c_0 N \quad \text{and} \quad a_i \leq A \text{ for every } i,$$

there exist $l > \theta N$ and $1 \leq n_1 < \cdots < n_l \leq N$ such that

$$\sum_{i=n+1}^{n_j} a_i \geq c(n_j - n) \quad \text{for all } 0 \leq n < n_j \text{ and } j = 1, \ldots, l.$$

From the construction of hyperbolic times, one shows that every map $G \in \mathcal{U}$ induces a map $\tilde{G}$, of the form $\tilde{G}(z) = G^{i(z)}(z)$ which is uniformly expanding (though not a Markov map). One proves that the Perron-Frobenius operator of $\tilde{G}$ has a non-negative fixed point h_G in the space of bounded variation functions. Then h_G is the density of a $\tilde{G}$-invariant probability measure $\tilde{\mu}_G$. Now

$$\mu = \sum_{j=0}^{\infty} G_*^j(\tilde{\mu}_G | \{z : i(z) > j\})$$

defines an absolutely continuous G-invariant measure μ_G. The fact that μ is finite follows from $e^{-c\sqrt{n}}$ estimates as above. By normalization and ergodic decomposition of μ one easily gets an ergodic G-invariant probability, absolutely continuous with respect to Lebesgue measure.

Concerning part 3 of Theorem 11.3, we just mention that continuity follows from analyzing the variation of the fixed point h_G of the Perron-Frobenius operator as a function of the dynamics.

11.1.4 Open sets of non-uniformly hyperbolic maps

We describe a construction of robustly non-hyperbolic maps, proposed in [12, Appendix], which includes the examples in Sections 7.1.2 and 7.1.4. We start with the version for endomorphisms.

Let $f : M \to M$ be a C^1 map on a manifold with finite volume, such that there exist $p, q \in \mathbb{N}$, $\delta_1 > 0$, $\sigma_1 > q$, and a covering of M by measurable subsets $B_1, \ldots, B_p, B_{p+1}, \ldots, B_{p+q}$ satisfying

(a) f is injective on B_j for every $1 \leq j \leq p + q$;
(b) $|\det(Df)| \geq \delta_1$ at every point, and $|\det(Df)| \geq \sigma_1$ on $B_{p+1}, \ldots, B_{p+q}$.

Assume, in addition, that there exist $\sigma_2 < 1$ and $\delta_2 > 0$ such that

(c) $\|Df^{-1}\| \leq e^{\delta_2}$ at every point, and $\|Df^{-1}\| \leq \sigma_2$ on $B_1, \ldots, B_p$.

Proposition 11.6. *Given $p, q, \delta_1, \sigma_1, \sigma_2$ there exists $\bar{\delta}_2 > 0$ and $c_0 > 0$ such that any map satisfying conditions (a), (b), (c) for some $\delta_2 \leq \bar{\delta}_2$ has*

$$\limsup_{n \to +\infty} \frac{1}{n} \sum_{j=0}^{n-1} \log \|Df(f^j(x))^{-1}\| \leq -c_0 < 0$$

at Lebesgue almost every point $x \in M$.

First, one shows that under hypotheses (a) and (b) almost every point spends at least a fraction $\varepsilon_0 = \varepsilon_0(p, q, \delta_1, \sigma_1) > 0$ of time inside $B_1 \cup \cdots \cup B_p$: for every $n \geq 1$ the measure of the set of points with less than $\varepsilon_0 n$ visits to $B_1 \cup \cdots \cup B_p$ during the first n iterates is bounded by

$$\mathrm{Leb}(M)\sigma_1^{-(1-\varepsilon_0)n}\delta_1^{-\varepsilon_0 n}e^{\gamma_0 n}p^{\varepsilon_0 n}q^n$$

where $\gamma_0 = \gamma_0(\varepsilon_0)$ goes to zero when ε_0 goes to zero. Since $q < \sigma_1$, this expression decreases exponentially with n, if ε_0 is chosen small enough. For all other points, using (c) and taking δ_2 small (depending on ε_0),

$$\sum_{j=0}^{n-1} \log \|Df(f^j(x))^{-1}\| \leq (\varepsilon_0 n) \log \sigma_2 + (1 - \varepsilon_0) n \delta_2 \leq -c_0 n$$

with $c_0 = (\varepsilon_0/2) |\log \sigma_2|$. The proposition follows, using the Borel-Cantelli lemma.

Here is a typical application of Proposition 11.6. Let $f_0 : M \to M$ be a uniformly expanding map on the d-dimensional torus M:

$$\|Df_0^{-1}\| \leq \sigma_2 < 1.$$

Let V be some small domain in M, and f_1 be any volume expanding map in M coinciding with f_0 outside V and only mildly contracting inside V: $\|Df_1^{-1}\| \leq e^{\delta_2}$ for some small δ_2. Then *every map f in a C^1 neighborhood of f_1 is non-uniformly expanding*, in the sense that it satisfies the conclusion of Proposition 11.6. Just take $q = 1$ and $B_{p+1} = V$.

Along the same lines, one may give robust examples of C^1 diffeomorphisms admitting a dominated splitting $TM = E^{cu} \oplus E^{cs}$ with E^{cu} non-uniformly expanding and E^{cs} non-uniformly contracting:

$$\limsup_{n \to +\infty} \frac{1}{n} \sum_{j=0}^{n-1} \log \left\| (Df \mid E_{f^j(x)}^{cu})^{-1} \right\| \leq -c_0 < 0$$

and

$$\limsup_{n \to +\infty} \frac{1}{n} \sum_{j=0}^{n-1} \log \left\| Df \mid E_{f^j(x)}^{cs} \right\| \leq -c_0 < 0 \qquad (11.3)$$

at Lebesgue almost every point x. In fact, one may construct these examples so that they have simultaneous hyperbolic times, in the sense of Section 11.3.3. Thus the methods in there prove existence and finiteness of SRB measures for these diffeomorphisms (assuming C^2 regularity).

Start with an Anosov diffeomorphism $f_0 : M \to M$ on the d-dimensional torus $M = T^d$, with hyperbolic splitting $TM = E^u \oplus E^s$. Let $V \subset M$ be a small domain: both V and $f_0(V)$ are contained in the projections of unit cubes of $\mathbb{R}^d$. Let f be a diffeomorphism such that

1. f admits thin invariant cone fields $\mathcal{C}^{cu}$ and $\mathcal{C}^{cs}$ containing E^u and E^s, respectively;
2. f contracts volume along any disk D^{cs} tangent to $\mathcal{C}^{cs}$ and expands volume along any disk D^{cu} tangent to $\mathcal{C}^{cu}$:

$$|\det(Df \mid TD^{cs})| < \sigma_1^{-1} < 1 \quad \text{and} \quad |\det(Df \mid TD^{cu})| > \sigma_1 > 1;$$

3. f is C^1 close to f_0 outside V, so that $f(V)$ is still contained in the projection of a unit cube, and

$$\|Df \mid TD^{cs}\| < \sigma_2 < 1 \quad \text{and} \quad \|(Df \mid TD^{cu})^{-1}\| < \sigma_2 < 1$$

for any disks D^{cu} and D^{cs} as before;

4. inside V the behavior of the two sub-bundles may reversed, but only mildly:

$$\|Df \mid TD^{cs}\| < e^{\delta_2} \quad \text{and} \quad \|(Df \mid TD^{cu})^{-1}\| < e^{\delta_2}$$

for any disks D^{cu} and D^{cs} as before.

Notice that 1–4 are C^1 open conditions. Arguing as in the proof of Proposition 11.6 one shows that they imply (11.3), if δ_2 is small enough.

11.2 Gibbs u-states for $E^u \oplus E^{cs}$ systems

Let Λ be a compact invariant subset for a C^r diffeomorphism $f : M \to M$ with a dominated splitting $T_\Lambda M = E^u \oplus E^{cs}$ where the sub-bundle E^u is uniformly expanding and has positive dimension. Through every point $x \in \Lambda$ there exists a unique C^r injectively immersed strong unstable manifold $\mathcal{F}^u(x)$ tangent to E^u_x and which is uniformly contracted by the negative iterates of f. The strong unstable foliation $\mathcal{F}^u = \{\mathcal{F}^u(x) : x \in \Lambda\}$ is invariant under the diffeomorphism. In all that follows we assume that the set Λ consists of entire strong unstable leaves. That is the case, for instance, if Λ is a topological attractor, that is, the maximal invariant set in a positively invariant neighborhood.

11.2.1 Existence of Gibbs u-states

We are going to see that if f is of class C^2 then it admits a special class of invariant probability measures characterized by having conditional measures along (local) strong unstable leaves which are absolutely continuous with respect to the corresponding Lebesgue measure. We shall refer to these measures as Gibbs u-states, by analogy with the uniformly hyperbolic case, where they correspond to the Gibbs states associated to a potential defined by the Jacobian along the unstable direction. There is a dual notion of Gibbs s-state, when Λ admits a uniformly contracting direction E^s and consists of entire strong stable leaves.

We begin by introducing the notion of *foliated box* for the strong unstable foliation $\mathcal{F}^u$. Let X be any small unstable disk, that is, contained in some strong unstable leaf, and let Σ be a small cross section to the foliation through the center p of X. Denote by Y the intersection of Σ with the invariant set Λ. Then there exists $\phi : X \times Y \to \Lambda$ which is a homeomorphism onto its image, such that ϕ maps each horizontal $X \times \{y\}$ diffeomorphically to an unstable domain through y. We may choose ϕ such that $\phi(p, y) = y$ for all $y \in Y$ and $\phi(x, p) = x$ for all $x \in X$. In what follows we identify $X \times Y$, and each $X \times \{y\}$, with their images under such a chart ϕ.

Definition 11.7. An invariant probability measure μ is a *Gibbs u-state* if, for every foliated box $X \times Y$ such that $\mu(X \times Y) > 0$, the conditional measures of $\mu \mid X \times Y$ with respect to the partition into unstable plaques $\{X \times \{y\} : y \in Y\}$ are absolutely continuous with respect to Lebesgue measure along the corresponding plaque.

See Appendix C for a definition of conditional measures. For any domain D inside a strong unstable leaf, we shall denote by m_D the Lebesgue measure induced on D by some Riemannian structure on the manifold.

Theorem 11.8 (Pesin, Sinai [354]). *For any unstable domain $D \subset \Lambda$, every accumulation point of the sequence of probability measures*

$$\mu_n = \frac{1}{n} \sum_{j=0}^{n-1} f_*^j \left(\frac{m_D}{m_D(D)} \right)$$

is a Gibbs u-state with densities $\rho(\cdot)$ with respect to Lebesgue measure along unstable plaques satisfying

$$\frac{\rho(z_1)}{\rho(z_2)} = \prod_{j=0}^{\infty} \frac{\det(Df^{-1} \mid E_{f^{-j}(z_1)}^u)}{\det(Df^{-1} \mid E_{f^{-j}(z_2)}^u)} \tag{11.4}$$

for any points z_1 and z_2 in the same unstable plaque.

The proof of Theorem 11.8 is based on the fact that the distortion of Lebesgue measure along strong unstable leaves by the iterates of f is bounded. More precisely,

Lemma 11.9. *Given any $L > 0$ there exists $C_0 > 0$ such that if $x, y \in D$, and $N \geq 1$ are such that $\mathrm{dist}(f^N(x), f^N(y)) \leq L$ then*

$$\frac{\det(Df^N | E_x^u)}{\det(Df^N | E_y^u)} \leq C_0 .$$

Here $\mathrm{dist}(\cdot, \cdot)$ means distance along the corresponding strong unstable leaf. The proof is a classical argument (see e.g. [84, Lemma 3.3]). One ingredient is the uniform contraction of unstable leaves by backward iterates of f. Another is uniform Lipschitz (or Hölder) variation of the unstable Jacobian $\det(Df \mid E^u)$ along strong unstable leaves (with bounded diameter). The latter is a consequence of the fact that strong unstable manifolds have uniformly bounded curvature, when f is C^2. See [97, 216].

Using the distortion Lemma 11.9 one proves a uniform bound for the measure induced by every iterate $f_*^n(m_D/m_D(D))$ on each foliated box $X \times Y$:

$$f_*^n\Big(\frac{m_D}{m_D(D)}\Big)(A \times Y) \le C\frac{m_X(A)}{m_X(X)} + \varepsilon_n \qquad (11.5)$$

for every measurable subset A of X, where $C > 0$ and $\varepsilon_n > 0$ depend only on f, D, X, and ε_n goes to zero when n goes to infinity. The last term bounds the total weight of connected components of $f^n(D) \cap (X \times Y)$ that do not cross the foliated box completely; convergence to zero is exponential. Property (11.5) passes to the limit: every accumulation measure μ satisfies

$$\mu(A \times Y) \le C\frac{m_X(A)}{m_X(X)} \qquad (11.6)$$

for every measurable $A \subset X$. This suffices to prove that μ is a Gibbs u-state: the densities of the conditional measures with respect to Lebesgue measure along the plaques are bounded by $C/\mu(X \times Y)$, on every foliated box which has positive weight for μ. This ends our outline of the proof of Theorem 11.8.

Remark 11.10. If Λ is a topological attractor, the center stable sub-bundle E^{cs} admits a unique continuous invariant extension to the basin $B(\Lambda)$. This extension, which we also denote E^{cs} may be found as the limit of backward iterates of any continuous extension of the center stable sub-bundle to a small neighborhood of Λ: the limit exists by a contraction argument. The subspaces E_y^{cs} are characterized by the fact that their forward iterates grow slower than any vector in the complement.

Remark 11.11. When Λ is an attractor, including the case $\Lambda = M$, the conclusion of Theorem 11.8 extends to any C^2 disk contained in the basin of attraction of Λ and transverse to the center stable direction E^{cs}.

11.2.2 Structure of Gibbs u-states

Our notion of Gibbs u-state seems more general than the one of Pesin, Sinai [354], who require (11.4) as part of the definition. Observe that this implies that the densities are uniformly bounded from zero and infinity, with bounds independent of the Gibbs u-state. In particular, the support consists of entire strong stable manifolds. In this section we show that the two definitions do coincide. It follows that the set of Gibbs u-states is a convex compact space (relative to the weak* topology).

The first step is to show that Theorem 11.8 remains valid if we replace D by a positive Lebesgue measure subset E:

Lemma 11.12. *Given any unstable domain D and any positive Lebesgue measure set $E \subset D$, every accumulation point of*

$$\mu_{n,E} = \frac{1}{n}\sum_{j=0}^{n-1} f_*^j\Big(\frac{m_E}{m_D(E)}\Big)$$

is a Gibbs u-state with densities satisfying (11.4).

Proof. Given any $\delta > 0$ we may find pairwise disjoint domains $D_1, \ldots, D_s$ in D such that the relative Lebesgue measure of E inside each D_i is larger than $1 - \delta$ and the total measure of E outside the union of the D_i is less than $\delta m_D(E)$. This follows from a Lebesgue density argument: just decompose D into small "cubes", and pick as D_i those which are entirely contained in D (no intersection with the boundary) and have a large fraction occupied by E. Then, for any $j \geq 1$, we may write

$$f_*^j\left(\frac{m_E}{m_D(E)}\right) = \sum_{i=1}^{s} \frac{m_D(D_i)}{m_D(E)} f_*^j\left(\frac{m_{D_i}}{m_D(D_i)}\right)$$
$$+ \frac{1}{m_D(E)} f_*^j m_{(E\setminus\cup_{i=1}^s D_i)} - \frac{1}{m_D(E)} \sum_{i=1}^{s} f_*^j m_{D_i\setminus E}.$$

The total masses of both the second and the third term do not depend on j, and are less than δ. Therefore, every accumulation point of $\mu_{n,E}$ differs from an accumulation point of

$$\sum_{i=1}^{s} \frac{m_D(D_i)}{m_D(E)} \frac{1}{n} \sum_{j=0}^{n-1} f_*^j\left(\frac{m_{D_i}}{m_D(D_i)}\right)$$

by a measure whose total mass is less than δ. By Theorem 11.8 applied to each domain D_i, every accumulation point of this last sequence is a Gibbs u-state, with densities satisfying (11.4); in particular, they are uniformly bounded. Since this is a closed property, making δ go to zero we get that every accumulation point of $\mu_{n,E}$ is a Gibbs u-state with densities satisfying (11.4). $\square$

Lemma 11.13. *Ergodic components of any Gibbs u-state μ are Gibbs u-states whose densities satisfy* (11.4).

Proof. Let $B(f)$ be the full μ-measure set of points where forward and backward time averages converge and coincide. Since μ is absolutely continuous along strong unstable manifolds, μ-almost every point x is contained in an unstable disk which intersects $B(f)$ on a positive Lebesgue measure subset D_x containing x. The limit time averages of all points $y \in D_x$ coincide with μ_x. Thus,

$$\lim_{n\to\infty} \frac{1}{n} \sum_{j=0}^{n-1} f_*^j\left(\frac{m_{D_x}}{m_D(D_x)}\right) = \lim_{n\to\infty} \frac{1}{m_D(D_x)} \int_{D_x} \frac{1}{n} \sum_{j=0}^{n-1} \delta_{f^j(y)} \, dm_D(y) = \mu_x.$$

On the other hand, by Lemma 11.12 the limit on the left is a Gibbs u-state with densities satisfying (11.4). $\square$

Now we can prove that our Gibbs u-state satisfy all the conditions in [354]:

Corollary 11.14. *If μ is any Gibbs u-state then*

1. It has densities with respect to Lebesgue measure along unstable plaques satisfying (11.4).
2. The densities are uniformly bounded from zero and infinity, and the support of μ consists of entire strong unstable leaves.

Proof. For part 1 just note that each strong unstable leaf intersects at most one ergodic component, so that the density is given by that of the ergodic component. The first statement in 2 is a direct consequence of part 1. The densities being bounded from zero implies that the support consists of entire leaves. $\qquad\square$

Remark 11.15. It follows that the set of Gibbs u-states is closed, hence compact, relative to the weak* topology in the space $\mathcal{M}$ of probabilities. If Λ is a topological attractor (maximal invariant set in a forward invariant neighborhood), we even have that the set of pairs (μ, g) such that μ is a Gibbs u-stated supported in the continuation of Λ for a nearby diffeomorphism g is closed in the product of $\mathcal{M}$ by a neighborhood of f.

11.2.3 Every SRB measure is a Gibbs u-state

When Λ is a topological attractor with partially hyperbolic splitting $T_\Lambda M = E^u \oplus E^{cs}$, it is easy to see that every SRB measure μ supported in Λ must be a Gibbs u-state. Indeed, consider any disk D inside $B(\Lambda)$ transverse to the center stable sub-bundle and intersecting the basin of μ on a positive Lebesgue measure subset D_0. On the one hand,

$$\frac{1}{n}\sum_{j=0}^{n-1} f_*^j\left(\frac{m_{D_0}}{m_D(D_0)}\right) = \frac{1}{m_D(D_0)}\int_{D_0} \frac{1}{n}\sum_{j=0}^{n-1} \delta_{f^j(x)}\, dm_D(x)$$

converges to μ when $n \to \infty$. On the other hand, combining Remark 11.11 and Lemma 11.12, the limit must be a Gibbs u-state.

Here we are going to prove a stronger fact: *for Lebesgue almost every point on any unstable disk, all accumulation points of time averages are Gibbs u-states.* We only assume that Λ admits a partially hyperbolic splitting $T_\Lambda M = E^u \oplus E^{cs}$ and consists of entire strong unstable leaves.

Theorem 11.16. *There exists $E \subset \Lambda$ intersecting every unstable disk on a full Lebesgue measure subset, such that for any $x \in E$ every accumulation point ν of*

$$\nu_{n,x} = \frac{1}{n}\sum_{j=0}^{n-1} \delta_{f^j(x)}$$

is a Gibbs u-state.

In the remainder of this section we prove this theorem. Let D be any unstable domain, and $X \times Y$ be a foliated box. Our goal is to obtain a bound like (11.6) for every measurable subset A of X. It is no restriction to take X with radius less than $1/10$. and to suppose that A itself is a disk, contained in the unstable disk with the same center as X and radius 10 times smaller. We say that A is far from the boundary of X.

Fix $C_* > 0$ such that the Lebesgue measure of any unstable disk is at most C_* times larger than the Lebesgue measure of any unstable disk with half the radius. C_* may be chosen depending only on the dimension of E^u and a bound on the curvature of strong unstable leaves. Define $K_0 = C_0 C_*^2$ where C_0 is the distortion bound given by Lemma 11.9.

Proposition 11.17. *Under the previous hypotheses,there exists $c > 0$ such that, given any sub-disk A far from the boundary of X, there exists $n_0 \geq 1$ such that, for all $n \geq n_0$,*

$$m_D \left(\{ x \in D : \nu_{n,x}(A \times Y) \geq 10 K_0 \frac{m_X(A)}{m_X(X)} \} \right) \leq e^{-cn}.$$

The first step in the proof of this proposition is a Markov type construction. Replacing f by some iterate f^ℓ, if necessary, we may assume that the least expansion $\sigma = \|Df^{-1} \mid E^u\|^{-1}$ along the unstable direction satisfies

$$\sigma \operatorname{diam}(A) > 1.$$

Observe that we have the right to do this, since the limits of time averages $\nu_{n,x}$ of f are given by the limits of time averages $\nu_{n,f^i(x),\ell}$ of f^ℓ, averaged over $0 \leq i < \ell$.

Lemma 11.18. *There exist families $\{\hat{X}_y : y \in Y\}$ and $\{\hat{A}_y : y \in Y\}$ of unstable domains such that*

1. *$X \times \{y\} \subset \hat{X}_y$ and $\hat{X}_y$ is contained in the unstable disk with the same center as $X \times \{y\}$ and radius twice as large;*
2. *if $f^n(X \times \{y\})$, $n \geq 1$ intersects some $X \times \{z\}$ then $f^n(\hat{X}_y)$ contains $\hat{X}_z$*
3. *$A \times \{y\} \subset \hat{A}_y \subset X \times \{y\}$ and $\hat{A}_y$ is contained in the unstable disk with the same center as $A \times \{y\}$ and radius twice as large;*
4. *if $f^n(A \times \{y\})$, $n \geq 1$ intersects some $X \times \{z\}$ then $f^n(\hat{A}_y)$ contains $\hat{X}_z$.*

Proof. First we take $\hat{X}_y$ to be the (increasing) union of the sequence $(\hat{X}_y^n)_n$ defined by $\hat{X}_y^0 = X \times \{y\}$ and $\hat{X}_y^n =$ the union of $\hat{X}_y^{n-1}$ with all pre-images $f^{-n}(\hat{X}_z^0)$, $z \in Y$ that intersect $\hat{X}_y^{n-1}$. The diameter of each $f^{-n}(\hat{X}_z^0)$ is bounded by $\sigma^{-n} \operatorname{diam}(X)$. As $\sigma > (\operatorname{diam} A)^{-1}$ is much larger than 1, it follows that $\hat{X}_y^n$ is contained in the neighborhood of radius

$$\sum_{i=1}^{n} \sigma^{-i} \operatorname{diam}(X) < \frac{1}{2} \operatorname{diam}(X)$$

around $X \times \{y\}$. Claim 1 follows immediately.

By definition, if $f^{-n}(\hat{X}_z^0)$ intersects $\hat{X}_y^0$ then it is contained in $\hat{X}_y^n$. Now let $i \geq 1$ and assume we have shown that $f^{-n}(\hat{X}_z^{i-1})$ is contained in $\hat{X}_y^{n-1+i}$. Consider any w such that $f^{-i}(\hat{X}_w^0)$ intersects X_z^{i-1} and, thus, is part of $\hat{X}_z^i$. Then $f^{-n-i}(\hat{X}_w^0)$ intersects the set $f^{-n}(\hat{X}_z^{i-1})$ and, consequently, also $\hat{X}_y^{n+i-1}$. So, by definition, $f^{-n-i}(\hat{X}_w^0)$ is part of $\hat{X}_y^{n+i}$. This proves that $f^{-n}(\hat{X}_z^i)$ is contained in $\hat{X}_y^{n+i}$. By induction, we have established this fact for every $i \geq 1$. This implies that $f^{-n}(\hat{X}_z) \subset \hat{X}_y$, and so claim 2 is proved.

Next, define $\hat{A}_y$ as the union of $A \times \{y\}$ with all the pre-images $f^{-n}(\hat{X}_z)$, $n \geq 1$ that intersect $A \times \{y\}$. The diameter of any of these pre-images is bounded by $\sigma^{-1} \operatorname{diam}(\hat{X}_z) < \operatorname{diam}(A) \operatorname{diam}(\hat{X}_z) \leq \frac{1}{2} \operatorname{diam}(A)$. Hence, $\hat{A}_y$ is contained in a disk with the same center as A and radius twice as large. In particular, it is contained in $X \times \{y\}$, as claimed in part 3.

Finally, if $f^{-n}(\hat{X}_z^0) = f^{-n}(X \times \{z\})$ intersects $A \times \{y\}$, so does $f^{-n}(\hat{X}_z)$. Then, by definition, $f^{-n}(\hat{X}_z)$ is contained in $\hat{A}_y$. This proves claim 4. $\qquad\square$

Lemma 11.19. *Let $1 \leq k \leq n$ and $0 \leq n_1 < \cdots < n_k < n$ be fixed. Then the m_D-measure of the set of points $x \in D$ which hit $A \times Y$ at times $n_1, \ldots, n_k$ is bounded by $(K_0 m_X(A)/m_X(X))^k$.*

Proof. Let Δ be the union of D with all the pre-images $f^{-n_1}(\hat{X}_y)$ that intersect it. The diameter of each pre-image is bounded by $\sigma^{-n_1} 2 \operatorname{diam}(X) \ll 1$, and so Δ is only slightly larger than D. Since $D \subset \Delta$, it suffices to prove that the m_Δ measure of the set of points $x \in \Delta$ which hit A at times $n_1, \ldots, n_k$ is bounded by $(K_0 m_X(A)/m_X(X))^k$.

By construction, if $f^{n_1}(\Delta)$ intersects some $X \times \{y_1\}$ then it contains it. Let $\Delta_X(y_1) = \Delta \cap f^{-n_1}(X \times \{y_1\})$ and $\Delta_A(y_1) = \Delta \cap f^{-n_1}(A \times \{y_1\})$. The union Δ_A^1 of the $\Delta_A(y_1)$ over all these y_1 contains the set of points $x \in \Delta$ which hit $A \times Y$ at time n_1. Given any $1 \leq i < k$, suppose we have subsets $\Delta_A(y_1, \ldots, y_i) \subset \Delta_X(y_1, \ldots, y_i)$ of Δ such that

(a) f^{n_i} maps $\Delta_X(y_1, \ldots, y_i)$ and $\Delta_A(y_1, \ldots, y_i)$ diffeomorphically onto $\hat{X}_{y_i}$ and $\hat{A}_{y_i}$, respectively;

(b) the union Δ_A^i of the $\Delta_A(y_1, \ldots, y_i)$ over all $y_1, \ldots, y_i$ for which it is defined contains the set of points $x \in \Delta$ which hit $A \times Y$ at times $n_1, \ldots, n_i$.

For each such subset, consider all $y_{i+1} \in Y$ such that $f^{n_{i+1}-n_i}(A \times \{y_i\})$ intersects $A \times \{y_{i+1}\}$. By part 4 of Lemma 11.18,

$$f^{n_{i+1}}(\Delta_A(y_1, \ldots, y_i)) = f^{n_{i+1}-n_i}(\hat{A}_{y_i}) \supset \hat{X}_{y_{i+1}}. \qquad (11.7)$$

Define $\Delta_X(y_1, \ldots, y_i, y_{i+1})$ and $\Delta_A(y_1, \ldots, y_i, y_{i+1})$ to be the pre-images of $\hat{X}_{y_{i+1}}$ and $\hat{A}_{y_{i+1}}$ under $f^{n_{i+1}}$. Observe that the union Δ_A^{i+1} of the sets $\Delta_A(y_1, \ldots, y_i, y_{i+1})$ over all choices of $y_1, \ldots, y_i, y_{i+1}$ for which it is defined contains the set of points $x \in \Delta$ which hit $A \times Y$ at times $n_1, \ldots, n_i, n_{i+1}$. This follows from the inductive hypotheses, together with the observation that if $f^{n_{i+1}-n_i}(A \times \{y_i\})$ is disjoint from $A \times \{y\}$ then there are no points in $\Delta_A(y_1, \ldots, y_i)$ hitting $A \times Y$ at times $n_1, \ldots, n_i$ and $A \times \{y\}$ at time n_{i+1}. Thus, we have recovered at time n_{i+1} the inductive properties (a) and (b) above.

After k steps we obtain a family of sets $\Delta_A(y_1, \ldots, y_k) \subset \Delta_X(y_1, \ldots, y_k)$ of Δ satisfying (a) and (b) for $i = k$. Using Lemma 11.9 and parts 1 and 3 of Lemma 11.18,

$$\frac{m_\Delta(\Delta_A(y_1, \ldots, y_i))}{m_\Delta(\Delta_X(y_1, \ldots, y_i))} \leq C_0 \frac{m_X(\hat{A}_{y_i})}{m_X(\hat{X}_{y_i})} \leq K_0 \frac{m_X(A)}{m_X(X)}$$

for every $1 \leq i \leq k$. Recall the definition of $K_0 = C C_*^2$ preceding the statement of Proposition 11.17. The relation (11.7) implies that $\Delta_A(y_1, \ldots, y_i)$ contains $\Delta_X(y_1, \ldots, y_i, y_{i+1})$. So, denoting the union of all $\Delta_X(y_1, \ldots, y_i, y_{i+1})$ as Δ_X^{i+1}, we deduce that

$$\frac{m_\Delta(\Delta_A^{i+1})}{m_\Delta(\Delta_A^i)} \leq \frac{m_\Delta(\Delta_A^{i+1})}{m_\Delta(\Delta_X^{i+1})} \leq K_0 \frac{m_X(A)}{m_X(X)}$$

for every $1 \leq i < k$. This is also true for $i = 0$, if one interprets $\Delta_A^0 = \Delta$. Multiplying these inequalities over all i, we find that

$$m_\Delta(\Delta_A^i) \leq \left(K_0 \frac{m_X(A)}{m_X(X)}\right)^k$$

and this implies the claim.   □

We are ready to give the proof of Proposition 11.17:

Proof. By definition, $n\nu_{n,x}(A \times Y)$ is the number of times a point x hits $A \times Y$ during the first n iterates. Thus, in view of Lemma 11.19, the m_D-measure of the set of points with $\nu_{n,x}(A \times Y) > 10 K_0 m_X(A)/m_X(X)$ is bounded by

$$\sum_{k > 10 n K_0 m_X(A)/m_X(X)} \binom{n}{k} \left(K_0 \frac{m_X(A)}{m_X(X)}\right)^k.$$

The first factor counts the sequences $0 \leq n_1 < \cdots < n_k < n$ of hitting times. To conclude the proof, use Lemma 11.20 below with $\alpha = K_0 m_X(A)/m_X(X)$ and $\beta = 10\alpha$.   □

Lemma 11.20. *Given $0 < 10\alpha \leq \beta \leq 1/2$ there exist constants $C > 1 > c > 0$ such that*

$$\sum_{k > \beta n} \binom{n}{k} \alpha^k \leq C e^{-cn} \quad \text{for all } n \geq 1.$$

Proof. Stirling's formula [1] implies, for every $0 \leq j \leq k$,

$$\binom{n}{k} \leq A\left(\frac{n}{k}\right)^k \left(\frac{n}{n-k}\right)^{n-k}$$

where A is a universal constant. We claim that

$$\left(\frac{n}{k}\right)^k \left(\frac{n}{n-k}\right)^{n-k} \alpha^k \leq e^{-\beta k} \quad \text{for all } k > \beta n. \tag{11.8}$$

Let us assume this for a while. Then,

$$\sum_{k>\beta n} \binom{n}{k} \alpha^k \leq \sum_{k>\beta n} A e^{-\beta k} \leq C e^{-cn}$$

where $c = \beta^2$ and $C > 1$ depends only on A and β. Thus we obtain the conclusion of the lemma. Hence, we are left to prove the claim (11.8). For this purpose, we write the inequality as

$$x \log \frac{1}{x} + (1-x) \log \frac{1}{1-x} - x \log \frac{1}{\alpha} \leq -\beta x \quad \text{for } x > \beta,$$

where $x = k/n$ and we have taken n:th roots and logarithms on both sides. The sum of the first two terms is a concave function of x, whereas the other two are linear. So, the statement holds if only if it holds at the initial point $x = \beta$:

$$\beta \log \frac{1}{\beta} + (1-\beta) \log \frac{1}{1-\beta} - \beta \log \frac{1}{\alpha} \leq -\beta^2 .$$

Since β is not far from zero, $\log 1/(1-\beta) \leq 2\beta$. Hence, for the previous inequality it suffices that

$$\beta \log \frac{1}{\beta} + (1-\beta)2\beta - \beta \log \frac{1}{\alpha} \leq -\beta^2 \quad \Leftrightarrow \quad \log \frac{\alpha}{\beta} + 2 \leq \beta .$$

The assumption $10\alpha \leq \beta$ implies that this last inequality is indeed true. Hence, (11.8) is also true, and that completes the proof of the lemma. $\qquad \square$

Finally, we give the proof of Theorem 11.16:

Proof. Let D be an unstable domain. Consider any foliated box $X \times Y$ as before. Proposition 11.17, together with a Borel-Cantelli argument, proves that, given any disk $A \subset X$ far from the boundary of X, there exists a full m_D-measure subset of points $x \in D$ such that

$$\nu_{n,x}(A \times Y) \leq 10 K_0 \frac{m_X(A)}{m_X(X)}$$

---

[1]That is, $n! \approx \sqrt{2\pi n}\,(n/e)^n$ where $\approx$ means the quotient tends to 1 as $n \to \infty$.

for all but, at most, finitely many $n \geq 1$. Consequently, the same upper bound holds for $\nu(A \times Y)$, for any accumulation measure ν of the sequence $\nu_{n,x}$. Applying this conclusion to a countable generating family of disks far from the boundary of X, and then taking the intersection of the corresponding full measure subsets of D, we get a full m_D-measure subset of points $x \in D$ for which every accumulation measure ν satisfies

$$\nu(A \times Y) \leq 10K_0 \frac{m_X(A)}{m_X(X)}.$$

for any measurable set A far from the boundary of X.

To complete the proof, cover the manifold by foliated boxes $X \times Y$ such that the radius of X is fairly small, and the foliated boxes obtained replacing each X by the disk $\bar{X}$ with the same center and half the radius still cover M. If ν does not give weight to $\bar{X} \times Y$, there is nothing to prove. Otherwise, by the previous paragraph,

$$\frac{\nu(A \times Y)}{\nu(\bar{X} \times Y)} \leq \frac{10K_0}{\nu(\bar{X} \times Y)} \frac{m_X(A)}{m_X(X)} \leq \frac{10K_0 C_*}{\nu(\bar{X} \times Y)} \frac{m_X(A)}{m_X(\bar{X})},$$

for every measurable set $A \subset \bar{X}$ (the constant C_* was defined prior to the statement of Proposition 11.17). This implies that ν is a Gibbs u-state, as claimed. $\Box$

Remark 11.21. When $\Lambda = M$ the set E has full Lebesgue measure in M. This follows immediately by absolute continuity of the strong unstable foliation.

Theorem 11.16 has the following immediate consequence for conservative diffeomorphisms. Simply note that, by the ergodic theorem, forward and backward time averages exist and coincide at almost every point.

Corollary 11.22. *Suppose $f : M \to M$ is a volume preserving C^2 diffeomorphism admitting a partially hyperbolic splitting $TM = E^u \oplus E^c \oplus E^s$ where E^u and E^s have positive dimensions. For Lebesgue almost every point $x \in M$, the time average*

$$\lim_{n \to \pm\infty} \frac{1}{n} \sum_{j=0}^{n-1} \delta_{f^j(x)}$$

is, simultaneously, a Gibbs u-state and a Gibbs s-state.

This motivates the following

Conjecture 11.23. If f is a volume preserving partially hyperbolic C^2 diffeomorphism having the accessibility property ,then it is ergodic with respect to Lebesgue measure.

11.2.4 Mostly contracting central direction

In general, Gibbs u-states are not SRB measures, even if they are ergodic. However,

Proposition 11.24. *Any ergodic Gibbs u-state μ whose Lyapunov exponents along the center stable sub-bundle E^{cs} are all negative is an SRB-measure.*

This is a simple consequence of Pesin theory (see Appendix C): by absolute continuity of the stable foliation and the definition of Gibbs u-states, the union of the local stable manifolds through the points $x \in \Lambda \cap B(\mu)$ form a positive Lebesgue measure subset of M contained in the basin of μ.

In this section we pursue this idea further on, leading to certain general criteria for existence and finiteness of SRB measures, and control of their basins. We consider Λ a topological attractor of a C^2 diffeomorphism $f : M \to M$, admitting a dominated splitting $T_\Lambda M = E^u \oplus E^{cs}$ such that E^u is uniformly expanding and has positive dimension.

Recall that the center stable sub-bundle admits a unique continuous invariant extension to the basin of Λ. For any $x \in B(\Lambda)$ we define

$$\lambda^c(x) = \limsup_{n \to +\infty} \frac{1}{n} \log \|Df^n \mid E_x^{cs}\|.$$

The main result is

Theorem 11.25 (Bonatti, Viana [84]). *Assume that $\lambda^c(x) < 0$ for a positive Lebesgue measure subset of every unstable disk D. Then there exist finitely many SRB measures $\mu_1, \dots, \mu_N$ supported in Λ, which are the ergodic Gibbs u-states of f. Moreover, the union of their basins is a full Lebesgue measure subset of the topological basin of attraction $B(\Lambda)$.*

The proof goes as follows. By Theorem 11.8 and Lemma 11.13, there exist ergodic Gibbs u states μ_α. The hypothesis $\lambda^c < 0$ implies that the central Lyapunov exponents of every μ_α are all negative. Hence, by Proposition 11.24, every ergodic Gibbs u-state is an SRB measure. If there were infinitely many μ_α then we could find a sequence μ_{α_n} accumulating some ergodic Gibbs state μ; recall Remark 11.15. Using that all the supports are saturated by strong unstable manifolds, by Corollary 11.14, we would find that the stable manifolds of points in $B(\mu)$ would capture a positive measure subset of points in $B(\mu_{\alpha_n})$ for all large n, which is a contradiction. Thus, there is only a finite number of ergodic Gibbs u-states $\mu_1, \dots, \mu_N$. Finally, suppose there was a positive Lebesgue measure subset $A \subset B(\Lambda)$ disjoint from the basins of all these SRB measures μ_i. Consider a disk transverse to the center stable sub-bundle and intersecting A on a positive Lebesgue measure subset. Using Lemmas 11.12 and 11.13, we would construct a new ergodic Gibbs u-state, contradicting the definition of N.

Kan's construction in Section 11.1.1 shows that one can not expect the SRB measure to be unique, even if the attractor is transitive. Indeed, his examples are partially hyperbolic with splitting $E^u \oplus E^{cs}$ and the SRB measures are ergodic Gibbs u-states with negative central Lyapunov exponents (in the endomorphism version one should consider the natural extension).

On the other hand, as a fairly direct consequence of Theorem 11.25, one gets

Theorem 11.26 (Bonatti, Viana [84]). *Assume the strong unstable foliation of Λ is minimal, and there exists some unstable disk D such that $\lambda^c(x) < 0$ for a positive Lebesgue measure subset of D. Then Λ supports a unique SRB measure μ, and $B(\mu) = B(\Lambda)$ up to zero Lebesgue measure.*

The minimality assumption gives that every unstable disk D' has an iterate close to D. This permits to recover the assumptions of Theorem 11.25 by considering the intersection of the iterate with stable manifolds of points in D. Thus, there exist finitely many SRB measures, which are ergodic Gibbs u-states with negative central exponents. Using minimality once more, one concludes that the SRB measure is unique.

For the conclusion of Theorem 11.25 it is not enough to suppose that $\lambda^c < 0$ Lebesgue almost everywhere in $B(\Lambda)$. For instance, [46, 168] give such an example with $\Lambda = M$ and an infinite number of SRB measures. The problem is that absolute continuity of the strong unstable foliation ensures $\lambda^c < 0$ at m_D-almost every point only inside *almost every* (not every) unstable disk D. On the other hand, the much stronger methods we are going to discuss in Section 11.3 permit to prove

Theorem 11.27. *Assume the center stable sub-bundle E^{cs} has a dominated splitting $E^{cs} = E^c \oplus E^s$ where E^c is one-dimensional and E^s is uniformly contracting. Let $c > 0$ and $H \subset B(\Lambda)$ be a positive Lebesgue measure set such that $\lambda^c(x) < -c$ for almost every $x \in H$. Then there are finitely many SRB measures, which are ergodic Gibbs u-states, such that H is contained $\mod 0$ in the union of their basins.*

The stable sub-bundle E^s may have dimension zero. The theorem will be part of a more general result, Theorem 11.35, which needs no assumption on the dimension of the central sub-bundle.

11.2.5 Differentiability of Gibbs u-states

Another important problem concerns the variation of SRB measures as a function of the dynamical system. In this direction, Dolgopyat [164] proves a differentiability result at certain partially hyperbolic maps f admitting a unique SRB measure, an ergodic Gibbs u-state, whose basin has full Lebesgue measure in the ambient. We refer the reader to [164] for the precise statement.

Here we just mention a striking application of Dolgopyat's main result to the time-1 map f of the geodesic flow in the unit tangent bundle $M = T^1 S$ of a surface S with negative curvature. It is well-known that f is volume preserving and partially hyperbolic, with splitting $TM = E^u \oplus E^c \oplus E^s$. The central bundle E^c corresponds to the direction of the flow. The Lebesgue measure is ergodic and, hence, the unique SRB measure for f. Furthermore, [200] prove that f is stably ergodic: the Lebesgue measure is ergodic for every volume preserving map in a C^2 neighborhood of f (Theorem 8.1).

Let us consider smooth one-parameter families of diffeomorphisms $(f_t)_t$ passing through $f_0 = f$, volume preserving or not. Of course, f_t is partially hyperbolic for every small t. For any f_t-invariant probability ν, define the integrated central Lyapunov exponent

$$\lambda^c(f_t, \nu) = \int \log |Df_t \mid E^c| \, d\nu \, .$$

Theorem 11.28 (Dolgopyat [164]). *There exists a quadratic form $c(\cdot)$ in the space of vector fields on M such that*

$$\lambda^c(f_t, \nu_t) = c\left(\frac{\partial f_t}{\partial t}\right) t^2 + o(t^2)$$

for every Gibbs u-state ν_t of f_t. Moreover, $c(\cdot)$ is not identically zero, even restricted to divergence free vector fields.

As usual, $o(t^2)$ represents an expression going to zero faster than t^2 when $t \to 0$. Here are some consequences of this theorem:

1. For generic families of volume preserving diffeomorphisms through f, and for all small t, either f_t or f_t^{-1} has negative central Lyapunov exponent.
2. For generic families of diffeomorphisms through f, volume preserving or not, and for all small t, the diffeomorphism f_t has a unique SRB measure.
3. Moreover, (f_t, m) has exponential decay of correlations and satisfies the central limit theorem, for any small t, and so does (g, m) for any volume preserving diffeomorphism g in a neighborhood of f_t.

Assertion (1) is a direct consequence of the previous theorem. See also Theorem 12.36 below for a related statement. Conclusion (2) is in line with the results in Section 11.2.4. Related to conclusion (3), Castro [122] and Dolgopyat [167, 163] had proven exponential decay of correlations and limit theorems for other classes of partially hyperbolic diffeomorphisms with splitting $E^u \oplus E^{cs}$, especially when the central sub-bundle is mostly contracting. See Appendix E for definitions and further information.

11.3 SRB measures for dominated dynamics

In this section we drop the requirement of existence of a uniformly expanding sub-bundle: the only standing assumptions are existence of a dominated splitting $T_K M = E^{cu} \oplus E^{cs}$ over a compact forward invariant set K, and C^2 regularity of the diffeomorphism.

This makes the situation much more subtle since, in the absence of uniform expansion, distortion control is not guaranteed. Recall that expansion was a key ingredient for Lemma 11.9. On the other hand, domination does provide the other key ingredient, namely Hölder control of the Jacobian:

Proposition 11.29. *There exists $\gamma > 0$ and, given $L > 0$ and any C^2 disk $D \subset K$ transverse to the center stable direction E^{cs}, there exists $C > 0$ such that*

$$x \mapsto \log |\det(Df \mid T_y f^n(D))|$$

is (C, γ)-Hölder on every domain of diameter less than L inside any $f^n(D)$, $n \geq 1$.

The crucial idea to overcome the lack of uniform expansion is the concept of *hyperbolic times*, introduced by Alves [7] and which we already mentioned in Section 11.1.3.

11.3.1 Non-uniformly expanding maps

To explain how the use of hyperbolic times permits to handle distortion, without having to deal with additional geometric difficulties, we begin by exploring this idea in a context of non-invertible dynamics.

A local diffeomorphism $g : M \to M$ on a compact manifold is *non-uniformly expanding* on a set H (not necessarily invariant), if there exists $c_0 > 0$ such that

$$\limsup_{n \to +\infty} \frac{1}{n} \log \sum_{j=0}^{n-1} \|Dg(g^j(z))^{-1}\| < -c_0 \qquad (11.9)$$

at Lebesgue almost every $z \in H$. Note that $Dg(\zeta)^{-1}$ represents the inverse of the derivative $Dg(\zeta)$ at a point $\zeta \in M$. If $\dim M = 1$, this condition just means that the (lower) Lyapunov exponent is bounded from zero on H:

$$\liminf_{n \to +\infty} \frac{1}{n} \log |Dg^n(z)| > c_0.$$

Theorem 11.30 (Alves, Bonatti, Viana [12]). *Suppose $g : M \to M$ is a C^2 local diffeomorphism non-uniformly expanding on some positive Lebesgue measure set H. Then H is contained $\mod 0$ in the union of the basins of finitely many SRB measures, which are ergodic measures absolutely continuous with respect to Lebesgue measure.*

In particular, *if the limit in (11.9) is negative Lebesgue almost everywhere in M then Lebesgue almost every point is in the basin of some SRB measure.*

Observe that this implies that time averages converge Lebesgue almost everywhere. In addition, if the limit is less than some $-c_0 < 0$, the SRB measures are finitely many.

Let us comment on the proof of Theorem 11.30. Fix some $c \in (0, c_0)$. An iterate n is a *hyperbolic time* for a point z if

$$\prod_{j=n-k}^{n-1} \|Dg(g^j(z))^{-1}\| \le e^{-ck} \quad \text{for every } 1 \le k \le n.$$

On the one hand, using Lemma 11.5 one proves that for every point z as in (11.9) the set of hyperbolic times has positive density at infinity: a lower bound on the density depends only on c_0, c, and a bound on the C^1 norm of f. On the other hand, at hyperbolic times we have uniform control of the distortion:

Lemma 11.31. *There exists $\rho > 0$ and $C_0 > 0$ such that, given any pair (z, n) such that n is a hyperbolic time for z, and given any $1 \le k \le n$, the inverse branch g_k of g^k mapping $g^n(z)$ to $g^{n-k}(z)$ satisfies*

1. *g_k is well defined and is a $e^{-ck/2}$-contraction on the ρ-ball around $g^n(z)$;*
2. *$\det(Dg_k(x)) \le C \det(Dg_k(y))$ for any x and y in the ρ-ball around $g^n(z)$.*

The first claim is proved by recurrence on k, using uniform continuity of the derivative and its inverse. The second one is a consequence, using also that the logarithm of the Jacobian is Lipschitz continuous (Hölder suffices).

Now, existence of absolutely continuous invariant probabilities can be proved as follows. Let m be Lebesgue measure on M, and

$$\mu_n = \frac{1}{n} \sum_{j=0}^{n-1} g_*^j m.$$

For each j, let H_j be the set of points $z \in H$ for which j is a hyperbolic time. Then we decompose $\mu_n = \xi_n + \eta_n$ where

$$\xi_n = \frac{1}{n} \sum_{j=0}^{n-1} g_*^j (m \mid H_j) \quad \text{and} \quad \eta_n = \frac{1}{n} \sum_{j=0}^{n-1} g_*^j (m \mid H_j^c).$$

Using part 2 of Lemma 11.31 one sees that every ξ_n is absolutely continuous with respect to Lebesgue measure, with uniformly bounded density. Using that the set of hyperbolic times has positive density at infinity, for almost every point in H, we find $\kappa > 0$ such that $\xi_n(M) \ge \kappa$ for all large n. Taking the limit along some convenient subsequence we find $\mu = \xi + \eta$ where μ is an g-invariant probability, and ξ is an absolutely continuous measure with total mass $\xi(M) \ge \kappa$. Let $\mu = \mu_a + \mu_s$ be the decomposition of μ as the sum of an absolutely continuous measure and a totally singular measure, relative to Lebesgue measure. Then both μ_a and μ_s are g-invariant and

$\mu_a(M) \geq \xi(M) \geq \kappa$. Thus, normalizing μ_a we obtain an absolutely continuous g-invariant probability.

The proof proceeds by ergodic decomposition. Using Lemma 11.31 and a density point argument, we see that any g-invariant set with positive ξ-measure must contain a full Lebesgue subset of some ρ-ball. Thus, up to zero measure subsets, there are only finitely many of these invariant sets. Taking the normalized restriction of μ_a to such an invariant set with minimal measure, we obtain an ergodic absolutely continuous probability measure μ_1, and we also check that the basin $B(\mu_1)$ intersects H on a positive Lebesgue measure subset. Repeating the whole argument with $H \setminus B(\mu_1)$ in the place of H, we exhaust H by basins of ergodic absolutely continuous probabilities, after a finite number of steps.

Remark 11.32. A version of Theorem 11.30 for piecewise smooth maps, possibly with critical points, is also given in [12]. One requires g to be a local diffeomorphism only on the complement of some singular set $\mathcal{S} \subset M$, with polynomial C^2 bounds close to $\mathcal{S}$: the norms of $Dg(z)$, $Dg^{-1}(z)$, $D^2g(z)$ are less than $C\operatorname{dist}(z,\mathcal{S})^{-c}$. In this setting, to control distortion it is necessary to keep a hold on visits to the neighborhood of $\mathcal{S}$. In addition to (11.9), one assumes that given $\varepsilon > 0$ there exists $\delta > 0$ such that

$$\liminf_{n\to+\infty} \frac{1}{n} \sum_j \log \operatorname{dist}(g^j(z), \mathcal{S}) \geq -\varepsilon \tag{11.10}$$

for Lebesgue almost every $z \in H$, where the sum is over all $0 \leq j \leq n-1$ for which the distance is less than δ. Then the full statement of Theorem 11.30 remains valid for these maps.

Let us also point out that Alves, Araújo [9] have recently proven stochastic stability for a large class of such transformations. An example are the cylinder maps treated in Section 11.1.3, where the singular set is a circle.

11.3.2 Existence of Gibbs cu-states

Now we apply these ideas to C^2 diffeomorphisms having a dominated splitting $E^{cu} \oplus E^{cs}$ over some positively invariant set K.

An invariant probability measure μ is called a *Gibbs cu-state* if the $\dim E^{cu}$ largest exponents are positive at μ-almost every point and μ is absolutely continuous along the corresponding Pesin strong unstable manifolds. See Appendix C. We say that E^{cu} is *non-uniformly expanding* on a set $H \subset K$ if there is $c_0 > 0$ such that

$$\limsup_{n\to+\infty} \frac{1}{n} \sum_{j=0}^{n-1} \log \left\| \left(Df \mid E^{cu}_{f^j(x)}\right)^{-1} \right\| \leq -c_0 < 0 \tag{11.11}$$

for Lebesgue almost every $z \in H$.

Theorem 11.33 (Alves, Bonatti, Viana [12]). *Let E^{cu} be non-uniform-ly expanding on a positive Lebesgue measure set $H \subset K$. Then f has some ergodic Gibbs cu-state supported inside K.*

For the proof one considers iterates $\mu_n = n^{-1} \sum_{j=0}^{n-1} f_*^j m_D$ of Lebesgue measure along some disk $D \subset K$ transverse to the center stable direction E^{cs}. Proposition 11.29 provides control of the geometry of the iterates of D. Otherwise, the argument goes along the same lines as that of Theorem 11.30. As before, the key idea is a notion of hyperbolic time: given $c > 0$ we say that n is a *cu-hyperbolic time* for a point $x \in K$ if for every $1 \le k \le n$

$$\prod_{j=n-k}^{n-1} \|(Df \mid E_{f^j(x)}^{cu})^{-1}\| \le e^{-ck}.$$

The hypothesis (11.11) implies that x has infinitely many *cu*-hyperbolic times, with positive density at infinity. Decomposing $\mu_n = \xi_n + \eta_n$ according to *cu*-hyperbolic times, and considering convenient accumulation points, one constructs some Gibbs *cu*-state for f. The last step in the proof is an ergodic decomposition argument.

Just as in Proposition 11.24, we have that every ergodic Gibbs *cu*-state such that the remaining dim E^{cs} Lyapunov exponents are negative is an SRB measure. In the next section we discuss some criteria that ensure this property.

11.3.3 Simultaneous hyperbolic times

We say that E^{cs} is *non-uniformly contracting* on a set H if the condition

$$\limsup_{n \to +\infty} \frac{1}{n} \sum_{j=0}^{n-1} \log \|(Df \mid E_{f^j(x)}^{cs})\| \le -c_0 < 0 \tag{11.12}$$

holds almost everywhere on H. Observe that, just as in (11.11), we consider $n \to +\infty$. As before, if x satisfies (11.12) then the set of *cs-hyperbolic times*, defined as the iterates n such that

$$\prod_{j=n-k}^{n-1} \|Df \mid E_{f^j(x)}^{cs}\| \le e^{-ck} \quad \text{for all } 1 \le k \le n,$$

has positive density at infinity. We say that n is a *simultaneous hyperbolic time* for x if it is both a *cu*- and a *cs*-hyperbolic time.

Theorem 11.34 (Alves, Bonatti, Viana [12]). *Let $H \subset K$ be a positive Lebesgue measure set such that almost every point in H has positive density at infinity of simultaneous hyperbolic times. Then H is covered* mod 0 *by the basins of SRB measures which are ergodic Gibbs cu-states with* dim E^{cs} *negative Lyapunov exponents.*

In the remainder of this section we highlight applications cases of this theorem, corresponding to the cases when either of the invariant sub-bundles is uniform.

Clearly, if E^{cu} is uniformly expanding then every $n \geq 1$ is a hyperbolic time. So, in this case, to ensure the hypotheses of Theorem 11.34 we only need E^{cs} to be non-uniformly contracting. Moreover, in this case there are only finitely many SRB measures.

Theorem 11.35. *Suppose the sub-bundle E^{cu} is uniformly expanding on K, and E^{cs} is non-uniformly contracting on a positive Lebesgue measure set $H \subset K$. Then H is covered* mod 0 *by the basins of finitely many SRB measures, which are ergodic Gibbs cu-states with* $\dim E^{cs}$ *negative Lyapunov exponents.*

Notice that this contains Theorem 11.27. Indeed, if E^{cs} splits as $E^c \oplus E^s$ with the $\dim E^c = 1$ and E^s uniformly contracting, then condition (11.12) reduces to

$$\limsup_{n \to \infty} \frac{1}{n} \log |Df^n \mid E^c_x| < -c_0 .$$

Moreover, in this context the notions of Gibbs cu-state and u-state coincide.

Similarly, if E^{cs} is uniformly contracting then any ergodic Gibbs cu-state is an SRB measure. Moreover, in this case there are only finitely many SRB measures.

Theorem 11.36. *Suppose the sub-bundle E^{cs} is uniformly contracting on K, and E^{cu} is non-uniformly expanding on a positive Lebesgue measure set $H \subset K$. Then H is covered* mod 0 *by the basins of finitely many SRB measures, which are ergodic Gibbs cu-states with* $\dim E^{cs}$ *negative Lyapunov exponents.*

Remark 11.37. When E^{cu} splits as $E^u \oplus E^c$ with E^u uniformly expanding, we may speak of both Gibbs cu-states and u-states. In fact, every cu-state is also a u-state, as a consequence of the absolute continuity of the strong unstable foliation. In particular, the support of a Gibbs cu-state always consists of complete strong unstable manifolds. In general, the support is not saturated by center unstable leaves, even assuming a center unstable foliation does exist.

Example 11.38. We describe a volume preserving diffeomorphism of the 3-torus with partially hyperbolic splitting $TM = E^u \oplus E^c \oplus E^s$, admitting several SRB measures which are ergodic Gibbs cu-states. Start with $f_0 = h \times \mathrm{id}$ where h is a linear Anosov diffeomorphism of the 2-dimensional torus and id is the identity map on the circle. Using the techniques in [45, 412] one may obtain a small C^1 perturbation f, coinciding with f_0 outside the union of pairwise disjoint horizontal strips $T^2 \times I_j$ and such that the average of the central Lyapunov exponent on each strip is positive. By the previous theorem, each $K = T^2 \times I_j$ contains the support of some ergodic Gibbs cu-state with 2 positive and 1 negative Lyapunov exponent.

11.3.4 Stability of cu-Gibbs states

In a very nice recent work, Vásquez [436] extends several results from Section 11.2 to Gibbs cu-states of topological attractors with a dominated splitting $E^{cu} \oplus E^{cs}$. He proves that the densities along Pesin strong unstable manifolds satisfy (11.4), at almost every point, and he deduces that the support consists of entire strong unstable manifolds. Assuming the center unstable bundle E^{cu} is non-uniformly expanding, in the sense of (11.11), Theorem 11.33 gives a constructive proof of the existence of Gibbs cu-states. Vásquez shows that, in fact, this construction produces *all* Gibbs cu-states. Most interesting, assuming the constant c_0 remains uniform in a neighborhood of the diffeomorphism, he proves continuity of cu-states with the dynamics in the C^2 topology: if $g_n \to g$ and μ_n is a Gibbs cu-state for g_n for each n, then any weak* limit μ is a Gibbs cu-state for g. He also extends the connection between Gibbs states and SRB measures in Section 11.2.3 to this generality: SRB measures must be Gibbs cu-states. These conclusions allow him to prove that the SRB measure varies continuously with the diffeomorphism, for an open class of systems with dominated decomposition containing the construction in Section 7.1.4. Finally, Vásquez proves that the conclusion of Theorem 11.34 remains valid if the simultaneous hyperbolic times condition is replaced by

$$\limsup_{n \to +\infty} \frac{1}{n} \log \|Df^n \mid E_x^{cs}\| < 0$$

on a full Lebesgue measure subset of every disk transverse to the center stable bundle E^{cs} (compare Theorem 11.25).

Alves, Araújo [9] have recently proven stochastic stability for a large class of transformations as in Remark 11.32. Moreover, combining their methods with those in the previously mentioned work of Vásquez, the three authors [11] were able to extend stochastic stability to certain diffeomorphisms with a dominated splitting. The precise statement is as follows.

Let $f : M \to M$ be a diffeomorphism for which there is a forward f-invariant open set $U \subset M$, in the sense that U contains the image of its closure. Assume there is a dominated Df-invariant splitting $E^s \oplus E^{cu}$ of $T_U M$, the tangent bundle over U. Assume E^s is uniformly contracting and E^{cu} is nonuniformly expanding: there is $c > 0$ such that for

$$\limsup_{n \to \infty} \frac{1}{n} \sum_{j=0}^{n-1} \log |Df^{-1} \mid E_{f^j(x)}^{cu}| \leq -c.$$

for Lebesgue almost all $x \in U$.

Theorem 11.39 (Araújo, Alves, Vásquez [11]). *Let f be as above. Then f is stochastically stable under random map type perturbations if, and only if, f is non-uniformly expanding along the center-unstable direction E^{cu} for random orbits: if there is $c > 0$ such that*

$$\limsup_{n\to\infty} \frac{1}{n} \sum_{j=0}^{n-1} \log \|Df^{-1} \mid E_{f_{z_j}}^{cu}\| le - c$$

for almost every random orbit and every small $\varepsilon > 0$.

11.4 Generic existence of SRB measures

In a remarkable recent work, Tsujii [428] greatly extends the arguments in the preceding sections to prove existence and finiteness of SRB measures under generic assumptions, within a class of partially hyperbolic systems. He states his main result for surface maps, but his methods are likely to extend to partially hyperbolic diffeomorphisms in any dimension, at least when the central sub-bundle is one-dimensional. The main novelty is that he replaces control of the central Lyapunov exponent by a geometric transversality condition, so that he is able to construct SRB measures also when the central exponent vanishes.

Let M be a C^∞ compact surface, possibly with boundary, and $f : M \to M$ be a C^r transformation, $r \geq 2$, possibly with critical points, i.e. points where the Jacobian vanishes. We say that f is *partially hyperbolic* if there exists a continuous invariant unstable (open) cone field $\{C(x) : x \in M\}$: there are $c > 0$ and $\sigma > 1$ such that, for all $x \in M$,

$$\overline{Df(x)C(x)} \subset C(f(x)) \cup \{0\} \quad \text{and} \quad \|Df^n(x)v\| \geq c\sigma^n \|v\| \text{ for all } v \in C(x).$$

See Appendix B. Then

$$E_x^c = \{w \in T_x M : Df^n(x)w \notin C(f^n(x)) \text{ for all } n \geq 0\}$$

is a Df-invariant line field, transverse to the cone field. Let $\mathcal{PH}^r$ be the set of C^r partially hyperbolic transformations, and $\mathcal{PH}_0^r$ be the subset of transformations without critical points.

Theorem 11.40 (Tsujii [428]). *For $r \geq 2$ there exists a residual subset $\mathcal{R}$ of $\mathcal{PH}_0^r$ such that every $f \in \mathcal{R}$ has a finite number of ergodic SRB measures, whose basins cover Lebesgue almost every point of M. The same result is true on $\mathcal{PH}^r$ for $r \geq 20$. In both settings, these physical measures are absolutely continuous with respect to Lebesgue measure if the sum of their Lyapunov exponents is positive.*

Problem 11.41. Is $\mathcal{R}$ open and dense? Are the SRB measures generically hyperbolic?

The next question is related to issues we discussed in Chapter 10:

Problem 11.42. For generic partially hyperbolic diffeomorphisms (with invariant splitting $TM = E^u \oplus E^c \oplus E^s$) with $\dim E^c = 1$ on a compact manifold are there

1. finitely many physical measures, whose basins contain Lebesgue almost every point?
2. finitely many attractors, whose basins cover an open full Lebesgue measure set?
3. finitely many pairwise disjoint homoclinic classes?

The strategy of the proof of Theorem 11.40 may be sketched as follows. Let H be a positive Lebesgue measure set of points for which the central Lyapunov exponent is well-defined:

$$\lambda^c(x) = \lim_{n \to \infty} \frac{1}{n} \log |Df^n \mid E_x^c| \,.$$

For every $\varepsilon > 0$ the set $H_\varepsilon^- = \{\lambda^c < -\varepsilon\}$ is covered, up to zero Lebesgue measure, by the basins of finitely many physical measures. This follows from a version of Theorem 11.27 for non-invertible maps. Similarly, using Theorem 11.30, for every $\varepsilon > 0$ the set $H_\varepsilon^+ = \{\lambda^c > \varepsilon\}$ is covered, up to zero Lebesgue measure, by the basins of finitely many absolutely continuous ergodic invariant measures.

The truly novel case corresponds to $\lambda^c \in [-\varepsilon, \varepsilon]$ on a positive Lebesgue measure subset. Tsujii proves that in this case, under a generic transversality condition, that subset is covered by the basins of finitely many SRB measures, which are still absolutely continuous ergodic invariant measures. We discuss some of the main issues by means of a simplified model taken from [430], which we study in the next sections.

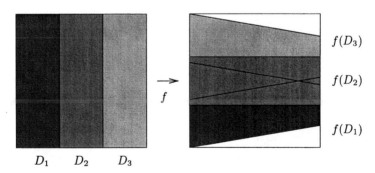

Fig. 11.1. Piecewise affine surface endomorphism

11.4.1 A piecewise affine model

Let $D_i = \{(i-1)/3 \le x \le i/3\}$ for $i = 1, 2, 3$. Then consider $f : [0,1]^2 \to [0,1]^2$ a piecewise affine map of the form

$$f(x, y) = \begin{cases} (3x, \alpha y + \beta x) & (x, y) \in D_1 \\ (3x - 1, \frac{1}{2} + \alpha(y - \frac{1}{2})) & (x, y) \in D_2 \\ (3x - 2, 1 - \alpha(1 - y) + \beta(x - \frac{2}{3})) & (x, y) \in D_3 \end{cases}$$

where $\alpha \in (1/3, 1)$ and $\beta > 0$ (the fact that f is multi-valued is harmless). Denote $f_i = f \mid D_i$ for $i = 1, 2, 3$.

The crucial properties are

(a) (volume expansion) $\det(Df) = 3\alpha > 1$ at every point.
(b) (partial hyperbolicity) There is a (constant) invariant unstable cone field:

$$C = \{(\dot{x}, \dot{y}) : |\dot{y}| < \theta|\dot{x}|\}, \qquad \theta = \frac{\beta}{3 - \alpha}.$$

This follows from the observation that

$$Df_1 : (\dot{x}, \dot{y}) \mapsto (3\dot{x}, \alpha\dot{y} + \beta\dot{x}) \qquad Df_3 : (\dot{x}, \dot{y}) \mapsto (3\dot{x}, \alpha\dot{y} - \beta\dot{x})$$

and

$$Df_2 : (\dot{x}, \dot{y}) \mapsto (3\dot{x}, \alpha\dot{y}).$$

(c) (transversality) $Df_i(C) \cap Df_j(C) = \emptyset$ for all $i \neq j$.

To check this, observe that $Df_i(C)$ is the cone bounded by the lines with slopes

$$\begin{cases} (\beta \pm \theta\alpha)/3 & \text{if } i = 1 \\ (\pm\theta\alpha)/3 & \text{if } i = 2 \\ (-\beta \pm \theta\alpha)/3 & \text{if } i = 3 \end{cases} \tag{11.13}$$

and we have (recall that $\alpha < 1$ and $\theta = \beta/(3 - \alpha)$)

$$\beta + \theta\alpha > \beta - \theta\alpha > \theta\alpha > -\theta\alpha > -\beta + \theta\alpha > -\beta - \theta\alpha.$$

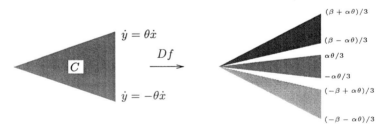

Fig. 11.2. Transverse invariant cone fields

To relate this model to outline of the proof of Theorem 11.40, notice that if a partially hyperbolic map has central Lyapunov exponent close to zero then it is asymptotically volume expanding, corresponding to property (a) of the model. Also, property (c) corresponds to the generic transversality condition mentioned in the previous section.

Theorem 11.43. *The map f admits an absolutely continuous invariant probability.*

11.4.2 Transfer operators

For proving Theorem 11.43 we consider the Perron-Frobenius operator

$$\mathcal{L}\psi(y) = \sum_{i=1}^{3} \mathcal{L}_i\psi(y) = \sum_{x: f(x)=y} \frac{1}{3\alpha}\psi(x),$$

where

$$\mathcal{L}_i\psi(y) = \begin{cases} \frac{1}{3\alpha}(\psi \circ f_i^{-1})(y) & \text{if } y \in f_i(D_i) \\ 0 & \text{otherwise} \end{cases}$$

for $i = 1, 2, 3$. Its dynamical relevance comes from

Lemma 11.44. *The push-forward of an absolutely continuous measure $\psi dxdy$ is the absolutely continuous measure $(\mathcal{L}\psi)dxdy$.*

Proof. Let E be any measurable set and ψ be an integrable function. Denote $E_i = f_i^{-1}(E)$.

$$f_*(\psi\,dxdy)(E) = (\psi\,dxdy)(f^{-1}(E)) = \sum_{i=1}^{3}\int_{E_i}\psi\,dxdy$$

$$= \sum_{i=1}^{3}\int_{f(E_i)}\frac{1}{3\alpha}(\psi \circ f_i^{-1})\,dxdy$$

$$= \sum_{i=1}^{3}\int_E \mathcal{L}_i\psi\,dxdy = \int_E \mathcal{L}\psi\,dxdy$$

This proves the lemma. □

Let $\mathcal{X}$ be the set of all functions $\psi : [0,1] \times \mathbb{R} \to \mathbb{R}$ of the form

$$\psi(x,y) = \sum_{\kappa \in K}\psi_\kappa(y - \kappa x)$$

where $\psi_\kappa \geq 0$ are L^2 functions, $|\kappa| \leq \theta$, and the sum is over any finite set K.

Lemma 11.45. $\mathcal{L}(\mathcal{X}) \subset \mathcal{X}$.

Proof. Let $\psi(x,y) = \psi_\kappa(y - \kappa x)$, that is, ψ is constant along lines of slope κ. The family of such lines is mapped by f_i to the family of lines of slope

$$\kappa_i = \begin{cases} (\alpha\kappa + \beta)/3 & \text{if } i = 1 \\ \alpha\kappa/3 & \text{if } i = 2 \\ (\alpha\kappa - \beta)/3 & \text{if } i = 3 \end{cases} \tag{11.14}$$

Therefore, $\mathcal{L}_i\psi(x,y) = (\psi \circ f_i^{-1})(x,y)$ is constant along lines of slope κ_i, that is, it may be written as $\psi_{\kappa_i}(y - \kappa_i x)$. This proves that $\mathcal{L}_i\psi_\kappa \in \mathcal{X}$ for any $i = 1, 2, 3$. The claim in the lemma follows immediately. □

Proposition 11.46. *There exists $C > 0$ such that for every $\psi \in \mathcal{X}$*

$$\|\mathcal{L}\psi\|_2^2 \leq \frac{1}{3\alpha}\|\psi\|_2^2 + C\|\psi\|_1^2 \,.$$

The crucial ingredient in the proof of this proposition is

Lemma 11.47. *Let $\psi_i : [0,1] \times \mathbb{R} \to \mathbb{R}$ be constant on the lines of slope κ_i, for $i = 1,2$, with $\kappa_1 \neq \kappa_2$. Then*

$$\psi_1 \cdot \psi_2 \leq \frac{1}{|\kappa_1 - \kappa_2|}\|\psi_1\|_1\|\psi_2\|_1$$

and the equality holds if the functions are non-negative.

Proof.

$$\psi_1 \cdot \psi_2 = \int \psi_1(x,y)\psi_2(x,y)\,dx\,dy$$

$$= \int \psi_1(0, y - \kappa_1 x)\psi_2(0, y - \kappa_2 x)\,dx\,dy$$

By changes of variables $\tilde{y} = y - \kappa_1 x$ and $\tilde{x} = \tilde{y} + (\kappa_1 - \kappa_2)x$ we find

$$\psi_1 \cdot \psi_2 = \int \psi_1(0, \tilde{y})\psi_2(0, \tilde{y} + (\kappa_1 - \kappa_2)x)\,dx\,d\tilde{y}$$

$$= \frac{1}{|\kappa_1 - \kappa_2|}\int \psi_1(0, \tilde{y})\psi_2(0, \tilde{x})\,d\tilde{x}\,d\tilde{y}.$$

Since ψ_i is non-negative and constant on lines of slope κ_i, we have

$$\int \psi_i(0, \tilde{y})\,d\tilde{y} = \int \psi_i(x, \tilde{y})\,dx\,d\tilde{y} \leq \|\psi_i\|_1$$

(the equality holds if $\psi_i \geq 0$). If follows that

$$\psi_1 \cdot \psi_2 \leq \frac{1}{|\kappa_1 - \kappa_2|}\|\psi_1\|_1\|\psi_2\|_1$$

and the equality holds if $\psi_1, \psi_2 \geq 0$, as claimed. □

Lemma 11.48.

$$\sum_{i=1}^{3} \|\mathcal{L}_i\psi\|_2^2 = \frac{1}{3\alpha}\|\psi\|_2$$

for every $\psi \in L^2$.

Proof. Changing variables $x = f_i^{-1}(y)$,

$$\|\mathcal{L}_i\psi\|_2^2 = \int_{f(D_i)} \frac{1}{(3\alpha)^2}(\psi \circ f_i^{-1})^2\,dxdy = \int_{D_i} \frac{1}{3\alpha}\psi^2\,dxdy.$$

Adding over $i = 1, 2, 3$, we obtain the conclusion. □

Now we are ready to give the proof of Proposition 11.46:

Proof.

$$\|\mathcal{L}\psi\|_2^2 = \sum_{i=1}^{3} \|\mathcal{L}_i\psi\|_2^2 + \sum_{i\neq j} \mathcal{L}_i\psi \cdot \mathcal{L}_j\psi\,.$$

By Lemma 11.48 the first term on the right is bounded by $(1/3\alpha)\|\psi\|_2^2$. Write $\psi(x,y) = \sum_{\kappa\in K} \psi_k(y - \kappa x)$. Then

$$\mathcal{L}_i\psi \cdot \mathcal{L}_j\psi = \sum_{\kappa,\kappa'} \mathcal{L}_i\psi_\kappa \cdot \mathcal{L}_j\psi_{\kappa'}$$

The functions $\mathcal{L}_i\psi_\kappa$ and $\mathcal{L}_j\psi_{\kappa'}$ are constant, respectively, along the lines of slopes κ_i and κ'_j given by (11.14). In the definition of $\mathcal{X}$ we took κ and κ' smaller than θ in norm. By (11.13), this gives that $\|\kappa_i - \kappa'_j\| \geq (\beta - 2\alpha\theta)/3$. Using Lemma 11.47,

$$\mathcal{L}_i\psi_\kappa \cdot \mathcal{L}_j\psi_{\kappa'} \leq \frac{3}{\beta - 2\alpha\theta} \int \psi_\kappa\, dxdy \int \psi_{\kappa'}\, dxdy.$$

Summing over κ and κ',

$$\mathcal{L}_i\psi \cdot \mathcal{L}_j\psi \leq \frac{3}{\beta - 2\alpha\theta} \int \psi\, dxdy \int \psi\, dxdy = \frac{3}{\beta - 2\alpha\theta}\|\psi\|_1^2\,.$$

Putting these bounds together,

$$\|\mathcal{L}\psi\|_2^2 \leq \frac{1}{3\alpha}\|\psi\|_2^2 + C\|\psi\|_1^2\,,$$

with $C = 18/(\beta - 2\alpha\theta)$. □

11.4.3 Absolutely continuous invariant measure

Lemma 11.44 has the following direct consequence:

Corollary 11.49. $\|\mathcal{L}\psi\|_1 \leq \|\psi\|_1$ *for any integrable function* ψ.

Proof. If $\psi \geq 0$ then $\mathcal{L}\psi \geq 0$, and we have

$$\|\mathcal{L}\psi\|_1 = \int_M \mathcal{L}\psi\, dxdy = f_*(\psi\, dxdy)(f^{-1}(M)) = (\psi\, dxdy)(M),$$

which is the same as

$$(\psi\, dxdy)(M) = \int_M \mathcal{L}\psi\, dxdy = \|\psi\|_1\,.$$

The general case is reduced to this one, noticing that $|\mathcal{L}\psi| \leq \mathcal{L}|\psi|$. □

We also need the following consequence of Proposition 11.46:

Corollary 11.50. *There exists $C' > 0$ such that*

$$\|\mathcal{L}^j\psi\|_2^2 \le (3\alpha)^{-j}\|\psi\|_2^2 + C'\|\psi\|_1^2$$

for all $\psi \in \mathcal{X}$ and $j \ge 1$.

Proof. By Proposition 11.46 and Corollary 11.49,

$$\|\mathcal{L}^j\psi\|_2^2 \le (3\alpha)^{-1}\|\mathcal{L}^{j-1}\psi\|_2^2 + C\|\mathcal{L}^{j-1}\psi\|_1^2 \le (3\alpha)^{-1}\|\mathcal{L}^{j-1}\psi\|_2^2 + C\|\psi\|_1^2$$

for all $j \ge 1$. By recurrence,

$$\|\mathcal{L}^j\psi\|_2^2 \le (3\alpha)^{-j}\|\psi\|_2^2 + C\sum_{j=0}^{n-1}(3\alpha)^{-j}\|\psi\|_1^2$$

and it suffices to take $C' = C\sum_{j=0}^{\infty}(3\alpha)^{-j}$. $\qquad\square$

Now we are ready to prove the existence of an absolutely continuous invariant probability. Consider the sequence $\mu_n = n^{-1}\sum_{j=0}^{n-1} f_*^j(dxdy)$. Each μ_n is an absolutely continuous probability with density

$$\rho_n = \frac{1}{n}\sum_{j=0}^{n-1}\mathcal{L}^j 1.$$

By Corollary 11.50 the square of the L^2-norm of $\mathcal{L}^j 1$ is bounded by $1 + C'$ for all j. This implies that the sequence ρ_n is bounded in L^2. Consequently (theorem of Alaoglu) it is relatively compact for the weak* topology: there exists $\rho \in L^2$ and a subsequence n_k such that

$$\phi \cdot \rho_{n_k} \to \phi \cdot \rho \qquad \text{for every } \phi \in L^2.$$

Let $\mu = \rho\, dxdy$. Then μ is the weak* limit of μ_{n_k}:

$$\int \phi\, d\mu = \int \phi\rho\, dxdy = \phi \cdot \rho = \lim_k \phi \cdot \rho_{n_k} = \lim_k \int \phi\, d\mu_{n_k}$$

for every $\phi \in L^2$. Consequently, μ is an invariant probability, absolutely continuous. This proves the existence part in Theorem 11.43.

Remark 11.51. The measure μ is ergodic. To see this is to consider the *natural extension* (or *inverse limit*) $\hat{f} : \hat{M} \to \hat{M}$ of f, and the lift $\hat{\mu}$ of μ to $\hat{M}$. Recall that $\hat{M}$ is the space of backward orbits, that is, sequences $(x_n)_{n \le 0}$ in $M = [0,1]^2$ with $f(x_n) = x_{n+1}$ for all n, and

$$\hat{f}(\dots, x_n, \dots, x_0) = \hat{f}(\dots, x_n, \dots, x_0, f(x_0)).$$

Notice that $\pi \circ \hat{f} = f \circ \pi$ where π is the projection to the 0th coordinate. The lift is the $\hat{f}$-invariant probability defined by $\hat{\mu} = \lim_n (\hat{f}^n \circ \iota)_* \mu$ where $\iota : M \to \hat{M}$ satisfies $\pi \circ \iota = \mathrm{id}$. Then it suffices to check that $\hat{\mu}$ is ergodic for $\hat{f}$. In the present situation this follows from observing that $\hat{\Lambda} = \cap_{n>0} \hat{f}^n(\hat{M})$ is a transitive uniformly hyperbolic attractor (homeomorphic to the product $\{1, 2, 3\}^{\mathbb{N}} \times [0, 1]$) and $\hat{\mu}$ is the corresponding SRB measure.

11.5 Extensions and related results

We close this chapter with a brief discussion of related results, about invariant measures obtained as zero-noise limits of random perturbations of the map, and about an extension to non-hyperbolic systems of the classical thermodynamical formalism ([86, 395, 445]).

11.5.1 Zero-noise limit and the entropy formula

The development of a theory of non-uniformly hyperbolic systems (stable manifold theorem, absolute continuity of foliations) allowed Pesin [356] to prove that if a $C^{1+\alpha}$ diffeomorphism $f : M \to M$ preserves a probability μ absolutely continuous with respect to Lebesgue measure, then the entropy of (f, μ) is given by

$$h_\mu(f) = \lambda_\mu^+(f), \qquad (11.15)$$

where $\lambda_\mu^+(f)$ is the integral, with respect to μ, of the sum of all positive Lyapunov exponents at each point (with multiplicity).

This remarkable formula holds, more generally, if μ has absolutely continuous conditional measures along unstable manifolds; see [256]. Furthermore, this sufficient condition is also necessary for the entropy formula to hold, as shown by Ledrappier, Young [253, 257]. Combining this with the absolute continuity Theorem C.4, one concludes that ergodic probabilities that satisfy the entropy formula are SRB measures, if the Lyapunov exponents are all different from zero: local stable manifolds through a typical local unstable manifold form a positive Lebesgue measure set contained in the basin.

Cowieson, Young [141] recently proposed to construct invariant probabilities satisfying the entropy formula as zero-noise levels of stationary measures of random perturbations of the system. They consider C^∞ diffeomorphisms with some attractor Λ, and perturbations of such diffeomorphisms by random maps, determined by probability distributions ν_ε supported in the ε-neighborhood of f (see Appendix D.2). Given a stationary probability μ_ε for the associated Markov chain, let h_{μ_ε} be the corresponding "random" entropy, and $\lambda_{\mu_\varepsilon}^+$ be the sum of positive "random" Lyapunov exponents (for definitions see Sections 2.1 and 3.1 in Kifer [241]).

Theorem 11.52 (Cowieson, Young [141]). *Assume* $f : M \to M$ *is a* C^∞ *diffeomorphism. Given a perturbation scheme by random maps as above, with transition probabilities* $p_\varepsilon(\cdot \mid x)$ *absolutely continuous with respect to Lebesgue measure, let* μ_ε *be stationary measures for the corresponding Markov chains and* μ *be an accumulation point of* μ_ε *as* $\varepsilon \to 0$. *If* $\lambda^+_{\mu_\varepsilon} \to \lambda^+_\mu$ *as* $\varepsilon \to 0$ *then* μ *satisfies the entropy formula:* $h_\mu = \lambda^+_\mu$.

The proof combines three main ingredients. The first one is a random version of the entropy formula stating that $h_{\mu_\varepsilon} = \lambda^+_{\mu_\varepsilon}$. This uses the hypothesis that the noise is absolutely continuous. A second ingredient is the fact that h_{μ_ε} depends lower semi-continuously on ε, and hence $h_\mu \geq \limsup_{\varepsilon \to 0} h_{\mu_\varepsilon}$. This uses a random version of arguments of Yomdin [455] and Newhouse [322] and depends on considering the C^∞ topology. The final main ingredient is Ruelle's inequality [391], which says that $h_\mu \leq \lambda^+_\mu$.

As an application, they consider the case when the attractor Λ admits a dominated splitting $T_\Lambda M = E^+ \oplus E^-$ such that for every point $x \in \Lambda$ and every vector $v \in E^+$ (respectively, $v \in E^-$) such that

$$\lim_{|n| \to \infty} \frac{1}{n} \log \| D f^n(x) v \|$$

exists and is non-negative (respectively, non-positive). This means that E^+ and E^- are precisely sums of the Oseledets subspaces corresponding to non-negative and non-positive Lyapunov exponents, respectively. Then, using the definition of the integrated Lyapunov exponent (recall also Remark C.2), we get

$$\lambda^+_\mu = \int \log |\det Df \mid E^+| \, d\mu \tag{11.16}$$

Let us point out that the inequality $\geq$ maybe obtained by a different argument that uses only domination. Indeed, let F_x be the sum of the Oseledets subspaces associated to positive Lyapunov exponents. The fact that the splitting $E^+ \oplus E^-$ is dominated implies that either $F_x \subset E^+$ or $F_x \supset E^+$. In either case, the sum of the Lyapunov exponents along E^+_x is less or equal than the sum of all positive Lyapunov exponents. Taking the integral over the ambient space (recall also Remark C.2) one gets the inequality $\geq$ in (11.16).

This observation is useful because, we need to consider perturbations of f and, while the condition on the Lyapunov exponents is not robust, dominated splittings do extend to a neighborhood of the map. Indeed, along these lines one gets an averaged version of the previous inequality

$$\lambda^+_{\mu_\varepsilon} \geq \int \log |\det(Dg \mid E^+_{g,x})| \, d\mu_\varepsilon(x) \, d\nu_\varepsilon(g)$$

where $E^+_g \oplus E^-_g$ denotes the continuation of the dominated splitting for g. Taking $\mu_\varepsilon \to \mu$ along some sequence of values of ε we get $\liminf \lambda^+_{\mu_\varepsilon} \geq \lambda^+_\mu$, and this suffices for the argument in Theorem 11.52.

A related paper, by Araújo, Tahzibi [20] appeared at about the same time. The results are similar in flavor, but they use different methods and assumptions: the maps are just $C^{1+\alpha}$ and the perturbation scheme is by random maps with additive absolutely continuous noise (they assume the manifold is parallelizable). Moreover, the assumption on convergence of Lyapunov exponents is replaced by a stronger assumption of expansion everywhere except at a finite number of points for the map f.

Theorem 11.53 (Araújo, Tahzibi [20]). *Let $f : M \to M$ be a $C^{1+\alpha}$ local diffeomorphism such that*

1. $\|Df(x)^{-1}\| \leq 1$ *for all* $x \in M$;
2. *the set* $K = \{x \in M : \|Df(x)^{-1}\| = 1\}$ *is finite and* $|\det Df(x)| > 1$ *for every* $x \in K$.

Then, given any non-degenerate additive random perturbation there exists a unique ergodic stationary probability measure μ_ε for each $\varepsilon > 0$. Moreover, as $\varepsilon \to 0$ these measures μ_ε converge, in the weak* topology, to an absolutely continuous f-invariant probability measure μ_0, whose basin has full Lebesgue measure in M.

This is a consequence of the following result, that relies on the entropy formula, both for deterministic and for random maps: under the assumptions of the theorem, every weak* accumulation point μ of μ^ε satisfies

$$h_\mu(f) = \int \log |\det Df(x)| \, d\mu(x).$$

Moreover, any measure that satisfies this entropy formular is a convex linear combination of an absolutely continuous invariant probability measure with finitely many Dirac measures concentrated on periodic orbits where the Jacobian equals 1.

11.5.2 Equilibrium states of non-hyperbolic maps

Let $f : M \to M$ be a continuous transformation on a compact space, and $\phi : M \to \mathbb{R}$ be a continuous function. An f-invariant probability measure is an *equilibrium state* of f for ϕ if it maximizes the functional

$$F_{f,\phi}(\nu) = h_\nu(f) + \int \phi \, d\nu$$

among all f-invariant probability measures. By classical results of Sinai, Ruelle, Bowen [86, 395, 416], equilibrium states always exist if f is a uniformly expanding transformation or a uniformly hyperbolic diffeomorphism. Moreover, the equilibrium state is unique if the *potential* ϕ is Hölder continuous. In this setting the SRB measure coincides with the equilibrium state for the potential $\phi = -\log |\det Df \mid E^u|$.

Problem 11.54. Investigate the existence and uniqueness of equilibrium states for non-uniformly expanding maps and non-uniformly hyperbolic diffeomorphisms and flows.

Important contributions have been given recently by several authors: Bruin, Keller [98], Denker, Urbanski [150, 433], Wang, Young [447] for interval maps, rational functions of the sphere, and Hénon-like maps; and Bressaud [92], Buzzi, Maume, Sarig [109, 112, 113, 404] and Yuri [463, 464, 465], for countable Markov shifts and for piecewise expanding maps, to mention just a few of the most recent works.

One difficulty is to formulate the assumption of non-uniform hyperbolicity in a suitable fashion. Assuming (11.9) at *Lebesgue* almost every point permits to prove existence of SRB measures, as we have seen in Section 11.3.1. But it is not clear how that could be useful for the construction of other equilibrium states, since most equilibrium states are likely to be singular with respect to Lebesgue measure.

It is tempting to require (11.9) to hold for almost every point, relative to any invariant probability, but that is much too strong: it implies that f is uniformly expanding, as observed by Alves, Araújo, Saussol [10]. This has been improved by Cao [115]: *if every f-invariant probability has only positive Lyapunov exponents then f is uniformly expanding*. His argument goes as follows.

Let ν be any f-invariant probability. By assumption, there exists $\delta > 0$ such that

$$\frac{1}{m} \int \log \|Df^m(x)^{-1}\|^{-1} \, d\nu(x) > \delta \tag{11.17}$$

for any large $m \geq 1$. A priori, m and δ depend on ν. However, it is clear that we may fix them such that the previous relation remains true for every probability in a neighborhood of ν, in the weak* topology. So, since the space of invariant probabilities is weak*-compact, we may indeed choose m and δ so that (11.17) is true for every f-invariant probability. Replacing f by f^m, we may suppose that $m = 1$. Now, if f is not uniformly expanding, given any $k \geq 1$ there exists $x_k \in M$ such that

$$\prod_{j=0}^{k-1} \|Df(f^j(x_k))^{-1}\|^{-1} \leq \|Df^k(x_k)^{-1}\|^{-1} \leq 2.$$

Let ν_k be the average of Dirac measures over the first k iterates of x_k and ν be any accumulation point of the sequence ν_k. Then ν is an f-invariant probability and

$$\int \log \|Df(x)^{-1}\|^{-1} \, d\nu = \lim_k \int \log \|Df(x)^{-1}\|^{-1} \, d\nu_k$$

$$= \lim_k \frac{1}{k} \sum_{j=0}^{k-1} \log \|Df(f^j(x_k))^{-1}\|^{-1} \leq \lim_k \frac{2}{k} = 0.$$

This contradicts (11.17), and so f must be uniformly expanding.

Nevertheless, a substantial contribution to the solution of Problem 11.54 was recently made by Oliveira:

Theorem 11.55 (Oliveira [332]). *There exists $\tau > 0$ and an open set of local C^2 diffeomorphisms f of the d-dimensional torus, $d \geq 2$ which are not uniformly expanding (they exhibit periodic saddles) and which have equilibrium states for every continuous potential ϕ satisfying $\max \phi - \min \phi \leq \tau h_{top}(f)$.*

Actually, Oliveira uses a weaker condition

$$P_f(\phi) \geq \max \phi + (1 - \tau)h_{top}(f)$$

where $P_f(\phi)$ is the topological pressure [2] of ϕ. The way he overcomes the difficulty discussed above is by identifying a suitable subset $\mathcal{K}$ of invariant probabilities whose elements are known *a priori* to be non-uniformly expanding. More precisely, using the hypothesis that the potential is not far from constant, he proves that every element in this set $\mathcal{K}$ gives small weight to the regions where the map is not expanding. Consequently, (11.9) holds almost everywhere for any measure in $\mathcal{K}$, which is what we mean here by non-uniform expansion. Building on this property, he concludes that

- f is ν-expansive (uniformly): there exists a partition $\mathcal{P}$ of M such that ν-almost every point $x \in M$ is determined by its itinerary relative to $\mathcal{P}$, for every $\nu \in \mathcal{K}$.

Expansivity implies a kind of semi-continuity, which ensures that the supremum of this functional on the compact set $\mathcal{K}$ is attained (compare [86] or [445]). In fact,

- the supremum of $F_{f,\phi}$ on $\mathcal{K}$ is attained and coincides with the supremum on the whole set of f-invariant probabilities.

Then, any measure $\nu \in \mathcal{K}$ realizing the supremum is an equilibrium state.

In a more recent work Oliveira, Viana [333] propose a different approach which also proves uniqueness of equilibrium states, under similar hypotheses (τ may be any number < 1). In both cases, the statements apply to systems of the type described in Section 11.1.4.

A very natural question in this context is whether equilibrium states do exist also for potentials with large variation. Currently, there is some evidence that the equilibrium states do exist but fail to be non-uniformly expanding, in general.

---

[2]The pressure $P_f(\phi)$ is the supremum of $h_\nu(f) + \int \phi \, d\nu$ over all invariant probabilities ν. Notice that $P_f(\phi) \geq h_{top}(f) + \min \phi$.

12

Lyapunov Exponents

As we have seen, Lyapunov exponents play a key role in understanding the ergodic behavior of a dynamical system, including the construction of SRB measures. An invariant probability μ is *hyperbolic* for a diffeomorphism $f : M \to M$ if the Lyapunov exponents of f are all different from zero at μ-almost every point: this means that the tangent space splits as the sum of two subspaces which are exponentially contracted or expanded by all *large enough* iterates of the derivative. The theory initiated by Pesin [355] provides detailed geometric information on the dynamics of these *non-uniformly hyperbolic* systems (f, μ), building on which several deep results have been proved, such as the entropy formula for smooth measures (Pesin [356]), the existence of uniformly hyperbolic sets carrying large entropy (Katok [231]), and the converse to the entropy formula (Ledrappier, Young [253, 257]). Another result in this line, Barreira, Pesin, Schmeling [48] is discussed in Section 12.6.

Thus, it is of primary importance to be able to control Lyapunov exponents, in particular, to understand when one can avoid vanishing exponents. We are going to discuss a number of recent results in the direction of the following two fundamental problems: *How frequently are Lyapunov exponents different from zero? How do Lyapunov exponents vary with the dynamics?* These results establish a surprising connection between measurable objects such as Lyapunov exponents and Oseledets subspaces, and geometric and topological features of the dynamics such as existence of (Hölder) continuous invariant subbundles, opening the way for the use of geometric and topological tools in this intrinsically ergodic setting.

A survey on recent results about Lyapunov exponents has appeared recently, by Bochi, Viana [65], and the discussion that follows has some intersection with their own. Most results we discuss in this chapter have been motivated by the following result of Mañé and Bochi [63, 279]:

Theorem 12.1. C^1-*generic volume preserving diffeomorphisms on a compact surface have only zero Lyapunov exponents at Lebesgue almost every point, unless they are uniformly hyperbolic (Anosov).*

Section 12.2 is devoted to an extension of this result to arbitrary dimension: *generically, the Oseledets splitting is either dominated or trivial on almost every orbit* (Theorem 12.9). Here, trivial means that there exists a unique Lyapunov exponent, which is necessarily zero. The heart of the proof is the observation in Section 12.1 (Theorem 12.3) that, if the Oseledets splitting is neither trivial nor dominated, Lyapunov exponents can be made to vary discontinuously under small perturbations of the diffeomorphism. On the other hand, a simple argument shows that Lyapunov exponents vary continuously on generic volume preserving diffeomorphisms.

The tangent map of a diffeomorphism is a particular case of a linear cocycle over a map. Before understanding the behavior of the Lyapunov exponents under C^1-perturbation of the diffeomorphism, it is often easier to analyze the perturbations of the linear cocycle, keeping the underlying dynamics unchanged. Results in this direction are presented in Sections 12.3 and 12.4. The first ones correspond to extensions to this broader setting of the results in the preceding sections: for C^0-generic linear cocycles, in the absence of a dominated splitting, all Lyapunov exponents vanish.

Very much in contrast, Section 12.4 gives results in the opposite direction for Hölder continuous cocycles over an ergodic hyperbolic measure μ: generic cocycles have non-zero Lyapunov exponents and, in some cases, one even has that every Lyapunov exponent has multiplicity one. In Section 11.5 we mention a number of techniques to remove zero Lyapunov exponents.

12.1 Continuity of Lyapunov exponents

Let M be a compact manifold with dimension $d \geq 1$ and $f : M \to M$ be a C^r transformation, $r \geq 1$. The *largest Lyapunov exponent* of f at a point $x \in M$ is the exponential rate of growth

$$\lambda(f, x) = \lim_{n \to \infty} \frac{1}{n} \log \|Df^n(x)\|$$

of the iterates of the derivative over the orbit of x. It follows from the subadditive ergodic theorem [243] that the limit exists almost everywhere, with respect to any f-invariant probability measure μ. It is clear that it does not depend on the choice of the Riemannian structure $\| \cdot \|$ on M. We denote

$$\lambda(f, \mu) = \int \lambda(f, x) \, d\mu.$$

A more detailed statement is provided by Oseledets theorem: Assuming f is a diffeomorphism, for μ-almost all $x \in M$ there exist $k = k(f, x) \in \mathbb{N}$, real numbers $\hat{\lambda}_1(f, x) > \cdots > \hat{\lambda}_k(f, x)$, and a splitting

$$T_x M = E_x^1 \oplus \cdots \oplus E_x^k \tag{12.1}$$

of the tangent space, such that the splitting is invariant under Df, and

$$\lim_{n\to\pm\infty} \frac{1}{n} \log \|Df^n(x)v_i\| = \hat{\lambda}_i(f,x) \quad \text{for every non-zero } v_i \in E_x^i.$$

When f is non-invertible, instead of a splitting one gets a filtration into vector subspaces $T_x M = E_x^1 > \cdots > E_x^k > 0 = E_x^{k+1}$ such that

$$\lim_{n\to+\infty} \frac{1}{n} \log \|Df^n(x)v_i\| = \hat{\lambda}_i(f,x) \quad \text{for every } v_i \in E_x^i \setminus E_x^{i+1}.$$

In either case, the *Lyapunov exponents* $\hat{\lambda}_i(f,x)$ and the *Oseledets subspaces* E_x^i are unique μ-almost everywhere, and they vary measurably with the point x. Moreover, the Lyapunov exponents are constant on orbits, and so they are constant μ-almost everywhere if μ is ergodic. We also have $\lambda(f,x) = \hat{\lambda}_1(f,x)$ wherever they are defined. See Appendix C.

We begin by introducing a proper set-up for discussing dependence of Lyapunov exponents on the dynamical system. Let

$$\lambda_1(f,x) \geq \lambda_2(f,x) \geq \cdots \geq \lambda_d(f,x), \quad d = \dim M,$$

be the Lyapunov exponents of f at x *counted with multiplicity* (the dimension of the corresponding Oseledets subspace), and define

$$\lambda_i(f,\mu) = \int \lambda_i(f,x)\, d\mu(x) \quad \text{for } 1 \leq i \leq d.$$

We think of the $\lambda_i(f,\mu)$ as functions of f in the space $\mathrm{Diff}^1_\mu(M)$ of C^1 diffeomorphisms preserving a given probability measure μ, endowed with the C^1 topology. We also define

$$\Lambda_i(f,\mu) = \lambda_1(f,\mu) + \cdots + \lambda_i(f,\mu) \quad \text{and} \quad V_i(f,\mu) = \lambda_i(f,\mu) + \cdots + \lambda_d(f,\mu)$$

Proposition 12.2. *Let μ be any Borel measure on M. For every $1 \leq i \leq d$ the functions*

$$g \mapsto \Lambda_i(g,\mu) \quad \text{and} \quad g \mapsto V_i(g,\mu)$$

are, respectively, upper semi-continuous and lower semi-continuous on the space $\mathrm{Diff}^1_\mu(M)$.

Upper semi-continuity is proved as follows. For $i = 1$ one uses the relation (see [254])

$$\Lambda_1(g,\mu) = \lambda_1(g,\mu) = \inf_{n\geq 1} \frac{1}{n} \int \log \|Dg^n\|\, d\mu. \tag{12.2}$$

The infimum of continuous functions being upper semi-continuous, the conclusion follows. For $i \geq 2$ one considers the action $Dg^{\wedge i} : W^i \to W^i$ induced by Dg on the vector bundle W^i over M whose fiber W_x^i is the space of i-forms on the co-tangent space $(T_x M)^*$. A natural choice of a norm is such that $\|(Dg^{\wedge i}(x))^n\|$ is the supremum of the i-volume of $Df^n(x)(P)$ over i-dimensional parallelepipeds $P \subset T_x M$ of unit i-dimensional volume. The largest Lyapunov exponent of $Dg^{\wedge i}$ is precisely $\Lambda_i(g, \mu)$. Thus we have the relation corresponding to (12.2) for $Dg^{\wedge i}$:

$$\Lambda_i(g, \mu) = \inf_{n \geq 1} \frac{1}{n} \int \log \|(Dg^{\wedge i})^n\| \, d\mu \,, \tag{12.3}$$

and semi-continuity follows as before. A dual argument gives the lower semi-continuity statement in the proposition.

The following theorem characterizes continuity of Lyapunov exponents for volume preserving C^1 diffeomorphisms:

Theorem 12.3 (Bochi, Viana [63, 67, 66]). *Let μ be normalized Lebesgue measure on a compact manifold M. If $f \in \mathrm{Diff}^1_\mu(M)$ is a continuity point for the map*

$$\mathrm{Diff}^1_\mu(M) \to \mathbb{R}^d, \quad g \mapsto (\lambda_1(g, \mu), \cdots, \lambda_d(g, \mu))$$

then, for almost every point $x \in M$,

(a) either all Lyapunov exponents $\lambda_i(f, x) = 0$ for $1 \leq i \leq d$,
(b) or the Oseledets splitting of f is dominated on the orbit of x.

The second alternative means that there are at least two distinct Oseledets subspaces and there exists $m \in \mathbb{N}$ such that

$$\frac{\|Df^n(y)u\|}{\|u\|} \geq 2 \frac{\|Df^n(y)v\|}{\|v\|} \quad \text{for all} \quad n \geq m \tag{12.4}$$

and all non-zero $u \in E_y^{1,i} = E_y^1 \oplus \cdots \oplus E_y^i$ and $v \in E_y^{i+1,k} = E_y^{i+1} \oplus \cdots \oplus E_y^k$, for any $1 \leq i < k$, and every iterate $y = f^l(x)$. In other words, the fact that Df^n will eventually expand $E^{1,i}$ more than $E^{i+1,k}$ can be observed in a finite number m of iterates, uniform over the orbit of x. This implies that *the Oseledets splitting extends continuously to the closure of the orbit* and, in particular, the angles between the Oseledets subspaces are bounded from zero over the orbit. Compare Remark C.2 in Appendix C.

Problem 12.4. Does the theorem remain valid if the conclusion is replaced by

(a') either $\lambda_j(f, x) = 0$ for all $1 \leq j \leq d$ and μ-almost every point $x \in M$,
(b') or the Oseledets splitting of f extends continuously to a dominated splitting on the whole ambient space M (in particular, m may be chosen independent of x).

This more global version of the conclusion does hold when $d = 2$: Bochi [63] proves that continuity points either have vanishing Lyapunov exponents almost everywhere or else they are uniformly hyperbolic (Anosov). Notice also that the two versions (a)-(b) and (a')-(b') coincide if f is ergodic: the number and dimensions of the Oseledets subspaces are constant almost everywhere, and so either (a) or (b) must hold on a full measure subset. In the second case, the Oseledets splitting extends to a dominated splitting on the whole ambient (see Section B.1).

Remark 12.5. Conversely, if f satisfies either (a') or (b') then it is a continuity point for all Lyapunov exponents. In case (a') this is a direct consequence of the fact that the largest Lyapunov exponent is an upper semi-continuous function of the dynamics, by Proposition 12.2. In case (b') one applies a similar argument, and the dual fact that the smallest exponent is lower semi-continuous, to each subbundle of the dominated splitting: by domination, every nearby diffeomorphism has an invariant splitting to the same number of subbundles, which are uniformly close to the initial ones.

Let us say a few words about the proof of Theorem 12.3. It is clear that $(\lambda_1, \ldots, \lambda_d)$ is continuous at some f if and only $(\Lambda_1, \ldots, \Lambda_d)$ is. Suppose the Oseledets splitting of f is neither trivial nor dominated, over a positive measure set of orbits: for some i and for arbitrarily large m there exist iterates y for which (12.4) does not hold. The strategy is to take advantage of this fact to, by a small perturbation of the map, cause a vector originally in $E_y^{1,i}$ to move to $E_z^{i+1,k}$, $z = f^m(y)$ thus "blending" different expansion rates.

More precisely, given a perturbation size $\epsilon > 0$ one takes m sufficiently large with respect to ϵ. Then for $n \gg m$ one chooses a convenient $y = f^l(x)$, with $l \approx n/2$, where domination fails. By composing Df with small rotations near the first m iterates of y, one causes the orbit of some $Df^l(x)v \in E_y^{1,i}$ to move to $E_{f^m(y)}^{i+1,k}$. Thus, one constructs a perturbation g preserving the orbit segment $\{x, \ldots, f^n(x)\}$ such that $Dg^n(x)v \in E_{f^n(x)}^{i+1,k}$. As a result, it is possible to show that

$$\|(Dg^{\wedge j}(x))^n\| \lesssim \exp\left[n\left(\Lambda_{j-1} + \frac{\lambda_j + \lambda_{j+1}}{2}\right)\right] \ll \exp(n\Lambda_j),$$

where $j = \dim E^{1,i}$ and $\lambda_j = \lambda_j(f, x)$, $\Lambda_j = \Lambda_j(f, x)$.

This local procedure is then repeated for several points $x \in M$. Using (12.3) and a Kakutani tower argument, one proves that Λ_j drops under such arbitrarily small perturbations, contradicting continuity.

Problem 12.6. 1. Characterize the continuity points of Lyapunov exponents relative to more general Borel measures μ on M.
 2. Is there a non-conservative version of Theorem 12.3, that is, in the space $\mathrm{Diff}^1(M)$ of all C^1 diffeomorphisms?

Even more important, one would like to have a corresponding understanding for higher topologies (see also Section 12.4):

Problem 12.7. Let μ be Lebesgue measure on M. What are the continuity points of $\mathrm{Diff}_\mu^r(M) \ni g \mapsto (\lambda_1(g, \mu), \cdots, \lambda_d(g, \mu))$ when $r > 1$?

Now we state a version of Theorem 12.3 for symplectic diffeomorphisms. Let (M, ω) be a compact symplectic manifold of dimension $\dim M = 2q$, that is ω is a differentiable non-degenerate 2-form on M. A diffeomorphism is *symplectic* if it preserves the symplectic form ω. The space $\mathrm{Symp}^\nu_\omega(M)$ of C^ν symplectic diffeomorphisms is a subspace of $\mathrm{Diff}^\nu_\mu(M)$, where μ is the volume measure associated to the volume form $\omega^q = \omega \wedge \cdots \wedge \omega$. The Lyapunov exponents of symplectic diffeomorphisms satisfy

$$\lambda_i(f, x) + \lambda_{2q-i+1}(f, x) = 0 \quad \text{for all } 1 \leq i \leq q.$$

In particular, $\lambda_q(x) \geq 0$ and $\Lambda_q(f, \mu)$ is the integral of the sum of all non-negative exponents. Moreover, let E_x^+, E_x^0, and E_x^- be the sums of all Oseledets spaces associated to positive, zero, and negative Lyapunov exponents, respectively. Then $\dim E_x^+ = \dim E_x^-$ and $\dim E_x^0$ is even.

Theorem 12.8 (Bochi, Viana [66]). *Suppose $f \in \mathrm{Symp}^1_\omega(M)$ is a continuity point for the function*

$$\mathrm{Symp}^1_\omega(M) \ni g \mapsto \Lambda_q(g, \mu).$$

Then for μ-almost every $x \in M$ either $\dim E_x^0 \geq 2$ or the splitting $T_x M = E_x^+ \oplus E_x^-$ is uniformly hyperbolic along the orbit of x.

In the second alternative, one just needs to prove that the splitting is dominated, because for symplectic diffeomorphisms dominated splittings into two subspaces of the same dimension are uniformly hyperbolic. See [65, 66].

12.2 A dichotomy for conservative systems

From Theorem 12.3 one gets the following dichotomy for generic volume preserving diffeomorphisms:

Theorem 12.9 (Bochi, Viana [63, 67, 66]). *Let μ be normalized Lebesgue measure on a compact manifold M. There exists a residual subset $\mathcal{R}$ of $\mathrm{Diff}^1_\mu(M)$ such that, for every $f \in \mathcal{R}$ and μ-almost every point x,*

(a) either all Lyapunov exponents $\lambda_i(f, x) = 0$ for $1 \leq i \leq d$,
(b) or the Oseledets splitting of f is dominated on the orbit of x.

This result was first proved for $d = 2$ by Bochi [63], partially based on an outline of proof by Mañé [284]. The statement had been announced by Mañé in the early eighties [279]: *for a residual subset of C^1 area preserving diffeomorphisms on any surface, either the Lyapunov exponents vanish almost everywhere or the diffeomorphism is Anosov and the surface is the torus.*
Theorem 12.9 is a direct consequence of Proposition 12.2 and Theorem 12.3. The proposition gives that every $g \mapsto \Lambda_i(g, \mu)$ is upper semi-continuous. Since the set of continuity points of a semi-continuous function on

a Baire space always contains a residual subset, we find that the elements of a residual subset of $\mathrm{Diff}^1_\mu(M)$ are continuity points for $g \mapsto (\Lambda_1, \ldots, \Lambda_d)$ and, hence, also for $g \mapsto (\lambda_1, \ldots, \lambda_d)$. By Theorem 12.3, they satisfy the conclusion of Theorem 12.9.

A similar argument, starting from Theorem 12.8, gives the following generic dichotomy for symplectic diffeomorphisms, which implies that generic symplectic diffeomorphisms on most manifolds (not supporting Anosov diffeomorphisms) are *not* hyperbolic!

Theorem 12.10 (Bochi, Viana [66]). *There exists a residual set* $\mathcal{R} \subset \mathrm{Symp}^1_\omega(M)$ *such that for every* $f \in \mathcal{R}$ *either the diffeomorphism* f *is Anosov or Lebesgue almost every point has zero as Lyapunov exponent, with multiplicity* ≥ 2.

A very useful ingredient in deducing Theorem 12.10 is the fact that generic C^1 diffeomorphisms (including all C^2 diffeomorphisms) either are Anosov or all their hyperbolic sets have zero Lebesgue measure. See [65] for a proof. This is true also in the volume-preserving category. On the other hand, no analogue is known for invariant sets with a dominated splitting. In fact, it is easy to find examples of C^∞ diffeomorphisms for which both alternatives (a) and (b) in Theorem 12.9 occur simultaneously:

Example 12.11. Let $\{f_t : t \in S^1\}$ be a family of volume preserving diffeomorphisms of $\mathbb{T}^3$ such that $f_t = \mathrm{id}$ for t in some interval and f_t is partially hyperbolic for t in another interval. One such a family can be obtained is using the construction of partially hyperbolic maps isotopic to the identity given in [71]. Then $F : \mathbb{T}^4 \to \mathbb{T}^4$, $F(t, x) = (t, f_t(x))$ satisfies both (a) and (b) in Theorem 12.9 on positive measure sets.

It is unclear whether such examples can be made generic. In this direction we ask the following strong form of Problem 12.4:

Problem 12.12. Is there a residual subset of C^r volume preserving or symplectic diffeomorphisms, $r \geq 1$, for which the compact invariant sets with a dominated splitting either have zero volume or coincide with the ambient space? Same in the space of all diffeomorphisms.

Remark 12.13. For r sufficiently large, KAM theory yields C^r open sets of symplectic maps which are not hyperbolic, due to the presence of invariant Lagrangian tori restricted to which the map is conjugate to rotations and whose union has positive volume. Moreover, Herman has constructed C^∞ open sets of volume preserving diffeomorphisms having invariant sets with positive volume, which are unions of codimension-1 invariant tori. See [454, § 4.6]. In all these examples where hyperbolicity fails, *all* Lyapunov exponents vanish. Question 1 in [412] asks whether for generic volume preserving diffeomorphisms either all Lyapunov exponents are zero or none is zero, almost everywhere.

The next almost-example describes a possible surprising consequence of Theorem 12.9: Lyapunov exponents with multiplicity persistently larger than 1. The argument would be complete if we knew the answer to

Problem 12.14. Are C^2 transformations C^1 dense among volume preserving C^1 diffeomorphisms?

Zehnder [466] gives a complete solution in the symplectic category: C^∞ maps are C^k dense among C^k symplectic diffeomorphisms, for any $k \geq 1$. He also proves that C^∞ maps are C^k dense among volume preserving diffeomorphisms of class $C^{k,\gamma}$ (the derivative of order k is γ-Hölder), for any $k \geq 1$ and $0 < \gamma \leq 1$.

The following example corresponds to Sections 7.1.4 and 8.1.3:

Example 12.15 (Bonatti,Viana [84], Tahzibi [424]). For $M = \mathbb{T}^4$ and μ = Lebesgue measure on M, there exists an open subset $\mathcal{U}$ of $\mathrm{Diff}^1_\mu(M)$ such that, for every $f \in \mathcal{U}$,

1. f admits a dominated splitting $TM = E \oplus F$ with $\dim E = \dim F = 2$;
2. there are no other continuous invariant subbundles;
3. f is transitive and if f is C^2 then it is even ergodic.
4. f is not hyperbolic: $Df \mid E$ is not expanding and $Df \mid F$ is not contracting.

Assume C^2 diffeomorphisms are dense in $\mathcal{U}$ (this could be true even if the global density statement fails). The set of ergodic maps is always a G_δ (countable intersection of open subsets) in $\mathrm{Diff}^r_\mu(M)$, $r \geq 0$, by [335, § 8]. Thus property (3) implies that the set of ergodic diffeomorphisms contains a residual (*dense* G_δ) subset $\mathcal{E}$ of $\mathcal{U}$. Take $\mathcal{S} = \mathcal{R} \cap \mathcal{E}$, where $\mathcal{R}$ is the residual set in Theorem 12.9. Because of (1), the Lyapunov exponents of f can not be all zero. So, using Theorem 12.9 and ergodicity, the Oseledets splitting must extend to a dominated splitting on the whole M. By (2), this implies that the Oseledets splitting has the form $E^1_x \oplus E^2_x$ with $\dim E^i_x = 2$. In other words, there exists a residual subset $\mathcal{S}$ of $\mathcal{U}$ such that every $f \in \mathcal{S}$ is ergodic, and its Lyapunov exponents satisfy

$$\lambda_1(f,\mu) = \lambda_2(f,\mu) > \lambda_3(f,\mu) = \lambda_4(f,\mu).$$

By Remark 12.5, every $f \in \mathcal{S}$ is a continuity point for all Lyapunov exponents.

A classical theorem of Oxtoby, Ulam [335] states that ergodic maps are generic among all volume preserving *homeomorphisms*. The following question, posed to us by A. Katok, asks for a C^1 version of this result:

Problem 12.16 (Katok). Do ergodic diffeomorphisms contain a residual subset of all area preserving C^1 diffeomorphisms on any surface? Same question in higher dimensions.

12.3 Deterministic products of matrices

Lyapunov exponents have an important role in several other areas, such as the theory of random matrices and the spectral theory of Schrödinger operators. applications are usually formulated in the more general setting of linear cocycles over a dynamical system. The questions we have been discussing extend in a natural way to this setting.

Let $f : M \to M$ be a continuous transformation on a compact metric space M. A *linear cocycle* over f is a vector bundle automorphism $F : \mathcal{E} \to \mathcal{E}$ covering f, where $\pi : \mathcal{E} \to M$ is a finite-dimensional (real or complex) vector bundle over M. This means that

$$\pi \circ F = f \circ \pi$$

and F acts as a linear isomorphism on every fiber. The prototype is the derivative $F = Df$ of a diffeomorphism on a manifold (*dynamical cocycle*). A linear cocycle over a flow $f^t : M \to M$, $t \in \mathbb{R}$, is a flow $F^t : \mathcal{E} \to \mathcal{E}$, $t \in \mathbb{R}$, such that F^t is a linear cocycle over f^t for every $t \in \mathbb{R}$.

For simplicity, we focus on the case when the vector bundle is trivial $\mathcal{E} = M \times \mathbb{R}^d$ (the complex case $\mathcal{E} = M \times \mathbb{C}^d$ is entirely similar). Then the cocycle has the form $F(x, v) = (f(x), A(x)v)$ with $A : M \to \mathrm{GL}(d, \mathbb{R})$. The iterates are given by

$$F^n(x, v) = (f^n(x), A^n(x)v) \quad \text{with } A^n(x) = A(f^{n-1}(x)) \cdots A(f(x)) \, A(x).$$

When f is invertible, $A^{-n}(x)$ is the inverse of $A^n(f^{-n}(x))$. In the continuous time case $F^t(x, v) = (f^t(x), A^t(x)v)$ for $t \in \mathbb{R}$.

Given any measurable $A : M \to \mathrm{GL}(d, \mathbb{R})$ such that $\log^+ \|A^{\pm 1}\|$ are integrable relative to some invariant probability μ, the largest Lyapunov exponent

$$\lambda(A, x) = \lim_{n \to \infty} \frac{1}{n} \log \|A^n(x)\|$$

exists μ-almost everywhere. In fact, the theorem of Oseledets holds in this setting of linear cocycles, see Appendix C. The symbols $\hat{\lambda}_j(A, x)$, $\lambda_i(A, x)$, $\lambda_i(A, \mu)$, $\Lambda_i(A, \mu)$, $V_i(A, \mu)$ will have the same meaning as in the dynamical case. Except where stated otherwise, we assume cocycles to be continuous. Then the integrability condition is automatic, by compactness.

Example 12.17. (Products of i.i.d. [1] random matrices) Let G be a matrix group, $M = G^{\mathbb{N}}$ be the space of sequences $(\alpha_j)_{j \geq 0}$ in G, and $f : M \to M$ be the shift map. Let $A : M \to G$ be the projection onto the zero:th coordinate: $A((\alpha_j)_{j \geq 0}) = \alpha_0$. Then

$$A^n((\alpha_j)_{j \geq 0}) = \alpha_{n-1} \cdots \alpha_1 \, \alpha_0. \tag{12.5}$$

---

[1]Independent and identically distributed.

Let ν be a probability in G and $\mu = \nu^{\mathbb{N}}$. The cocycle defined by A over (f, μ) models the product of random matrices chosen in G independently according to the distribution ν.

A classical theory initiated by Furstenberg and Kesten [186, 188] states, in its simplest form, that *for the majority of choices of the probability ν in $G = \mathrm{SL}(d, \mathbb{R})$ the product* (12.5) *grows exponentially fast.* More precisely, as long as there is no probability on the projective space $\mathbb{RP}^{d-1}$ invariant under the action of *all* the matrices in $\mathrm{supp}\,\nu$, the largest Lyapunov exponent $\lambda(A, \mu)$ of the corresponding cocycle is positive. Also in great generality, one even has that the Lyapunov spectrum is simple: all Oseledets subspaces have dimension 1. See Guivarch, Raugi [202] and Gold'sheid, Margulis [192].

Quite in contrast, the following version of the dichotomy Theorem 12.9 shows that vanishing Lyapunov exponents are quite common for linear cocycles. We give the statement in the ergodic invertible case. Recall that a group action is *transitive* if for any pair of points on the manifold there is a group element sending one point to the other.

Theorem 12.18 (Bochi, Viana [63, 67, 66]). *Assume $f : (M, \mu) \to (M, \mu)$ is ergodic and invertible. Let $G \subset \mathrm{SL}(d, \mathbb{R})$ be any closed subgroup acting transitively on the projective space $\mathbb{RP}^{d-1}$. Then there exists a residual subset $\mathcal{R}$ of continuous maps $A : M \to G$ for which either the Lyapunov exponents $\lambda_i(A, \mu)$ are all zero or the Oseledets splitting of A extends to a dominated splitting over the support of μ.*

An important point to be noted is that existence of a dominated splitting is a strong property, which can often be excluded *a priori* for topological or geometric reasons. The next couple of examples illustrate this point. In such cases we are left with the first alternative in the dichotomy: vanishing Lyapunov exponents for generic cocycles.

Example 12.19. Let $f : M \to M$ and $A : M \to \mathrm{SL}(d, \mathbb{R})$ be such that for every $1 \leq i < d$ there exists a periodic point p_i in the support of μ, with period q_i, such that the eigenvalues $\{\beta_j^i : 1 \leq j \leq d\}$ of $A^{q_i}(p_i)$ satisfy

$$|\beta_1^i| \geq \cdots \geq |\beta_{i-1}^i| > |\beta_i^i| = |\beta_{i+1}^i| > |\beta_{i+2}^i| \geq \cdots \geq |\beta_d^i| \qquad (12.6)$$

and β_i^i, β_{i+1}^i are complex conjugate (not real). Such an A may be found, for instance, starting with a constant cocycle and deforming it on disjoint neighborhoods of the periodic orbits. Property (12.6) remains valid for every B in a C^0 neighborhood $\mathcal{U}$ of A. It implies that no B admits an invariant dominated splitting over the support of μ: if such a splitting $E \oplus F$ existed then, at every periodic point, the $\dim E$ largest eigenvalues would be strictly larger than the other eigenvalues, which is incompatible with (12.6). It follows, by Theorem 12.18, that every cocycle in a residual subset $\mathcal{U} \cap \mathcal{R}$ of the neighborhood has all the Lyapunov exponents equal to zero.

Example 12.20. Let $f : S^1 \to S^1$ be a homeomorphism and μ be any invariant ergodic measure with $\operatorname{supp} \mu = S^1$. Let $\mathcal{N}$ be the set of all continuous $A : S^1 \to \mathrm{SL}(2, \mathbb{R})$ non-homotopic to a constant. For a residual subset of $\mathcal{N}$, the Lyapunov exponents of the corresponding cocycle over (f, μ) are zero. That is a direct consequence of Theorem 12.18, together with the fact that the cocycle has no invariant continuous subbundle. The latter can be shown as follows.

A continuous subbundle defines a continuous map from S^1 to $\mathbb{RP}^1$ and, hence, an element of $\pi_1(\mathbb{RP}^1) = \mathbb{Z}$. The space $\mathrm{SL}(2, \mathbb{R})$ has the homotopy type of the circle, and the map A defines an element $\deg A$ of its fundamental group $\mathbb{Z}$. The cocycle associated to A induces an action in the space of continuous sub-bundles. This action is defined at the level of the fundamental group $\pi_1(\mathbb{RP}^1)$, and is given by a translation: $i \mapsto i + \deg A$. An invariant subbundle yields a fixed point for this action, and so it can only exist if $\deg A = 0$ or, in other words, if A is homotopic to a constant.

Let us say a few words about the proof of Theorem 12.18. Firstly, Proposition 12.2 extends to linear cocycles: the sum $\Lambda_i(A, \mu)$ of the largest Lyapunov exponents is an upper semi-continuous function and the sum $V_a(A, \mu)$ of the smallest Lyapunov exponents is a lower semi-continuous function of A in the space $C^0(M, G)$ of continuous maps from M to G endowed with the C^0 topology. Then Theorem 12.18 follows from a corresponding version of Theorem 12.3: if A is a continuity point for the map

$$C^0(M, G) \ni B \mapsto (\lambda_1(B, \mu), \dots, \lambda_d(B, \mu)),$$

then either the Lyapunov exponents are all zero or the Oseledets splitting extends to a dominated splitting over the whole support. The converse of this latter result is also true, and follows from the arguments in Remark 12.5.

Theorem 12.18 also carries over to the space $L^\infty(M, \mathrm{SL}(d, \mathbb{R}))$ of measurable bounded cocycles, still with the uniform topology. Let us point out that in weaker topologies cocycles having a dominated splitting may cease to constitute an open set. In fact, in the L^p topology, $p < \infty$, uniform hyperbolicity has empty interior!

Theorem 12.21 (Arnold, Cong [31], Arbieto, Bochi [21]). *For every $1 \le p < \infty$, there is a residual subset of all L^p linear cocycles over any ergodic transformation have one-point Lyapunov spectra, that is, all the Lyapunov exponents coincide.*

First, Arnold, Cong [31] proved the conclusion on a dense subset. Then Arbieto, Bochi [21] proved the rather surprising fact that semi-continuity of Lyapunov exponents (Proposition 12.2) is true in any L^p topology, and they deduced the full statement of the theorem.

12.4 Abundance of non-zero exponents

We are going to see that the conclusions of the previous section change radically if one considers cocycles which are better than just continuous: assuming the base dynamics is hyperbolic (uniformly or not), almost all Hölder continuous or even differentiable cocycles have non-zero Lyapunov exponents.

Let G be any closed subgroup of $\mathrm{SL}(d, \mathbb{R})$. For $0 < \nu \leq \infty$ denote by $C^\nu(M, G)$ the space of C^ν maps from M to G endowed with the C^ν topology. For $\nu \geq 1$ it is implicitly assumed that M has a smooth structure. When ν is not an integer, C^ν means the map is $[\nu]$ differentiable and the derivative of order $[\nu]$ is $(\nu - [\nu])$-Hölder continuous. In the integer case, C^ν means either that f is ν times differentiable with continuous derivative of order ν, or that it is $\nu - 1$ times differentiable with Lipschitz continuous derivative of order $\nu - 1$: all statements are meant for both interpretations.

Definition 12.22. A matrix Lie group G *acts independently* on the projective space if the map

$$G \to (\mathbb{RP}^{d-1})^d, \quad B \mapsto (B\xi_1, \ldots, B\xi_d)$$

is a submersion, for any linearly independent set $\{\xi_1, \ldots, \xi_d\} \subset \mathbb{RP}^{d-1}$.

This holds, for instance, for the special linear group $\mathrm{SL}(d, \mathbb{R})$ and for the symplectic group.

Let μ be a hyperbolic measure for a diffeomorphism $f : M \to M$. We assume the derivative Df to be Hölder continuous, so that Pesin theory applies (Appendix C). The notion of measure with local product structure is recalled in Section C.5. Most interesting hyperbolic measures have local product structure, e.g. Lebesgue measure, measures absolutely continuous along stable or along unstable manifolds, and the equilibrium states of Axiom diffeomorphisms relative to Hölder potentials [86].

Theorem 12.23 (Viana [437]). *Assume (f, μ) is ergodic, non-uniformly hyperbolic, with local product structure. Let $G \subset \mathrm{SL}(d, \mathbb{R})$ be a Lie subgroup acting independently on the projective space. Then, for any $\nu > 0$ the set of cocycles A with largest Lyapunov exponent $\lambda(A, x) > 0$ at μ-almost every point contains an open dense subset $\mathcal{A}$ of $C^\nu(M, G)$. Moreover, its complement has infinite codimension.*

The last property means that the set of cocycles with vanishing exponents is locally contained inside finite unions of closed submanifolds of $C^\nu(M, G)$ with arbitrary codimension. Thus, generic parametrized families of cocycles do not intersect this exceptional set at all!

We may drop the ergodicity assumption in Theorem 12.24, at the price of replacing open and dense by residual in the statement and allowing for countable unions in the definition of infinite codimension. This is because any measure with local product structure has countably many ergodic components, which also have local product structure.

When $f : M \to M$ is uniformly hyperbolic, for instance, a two-sided shift of finite type or an Axiom A diffeomorphism restricted to some hyperbolic basic set, the ergodic components are finitely many, and so one recovers the full statement of the theorem even in the non-ergodic case. Clearly, in this case every invariant measure is hyperbolic.

Theorem 12.24 (Bonatti, Gomez-Mont, Viana [79, 437]). *Assume the map* $f : M \to M$ *is uniformly hyperbolic, and let* $G \subset \mathrm{SL}(d, \mathbb{R})$ *be a Lie subgroup acting independently on the projective space. Assume also that* μ *has local product structure. Then, for every* $\nu > 0$, *the set of cocycles* A *having largest Lyapunov exponent* $\lambda(A, x) > 0$ *at* μ-*almost every point contains an open dense subset* $\mathcal{A}$ *of* $C^\nu(M, G)$. *Moreover, its complement has infinite codimension.*

Theorem 12.24 was first proved in [79] under an additional hypothesis called domination: the cocycle induced on the projective bundle $M \times \mathbb{RP}^{d-1}$ is partially hyperbolic, with the projective fibers as central leaves. In this case the set $\mathcal{A}$ may be taken the same for all invariant measures with local product structure, ergodic or not. Moreover, still in the dominated case, Bonatti, Viana [83] get a stronger conclusion: for the majority of cocycles all Lyapunov exponents have multiplicity 1, in other words, the Oseledets subspaces E^i are one-dimensional. We expect this to be true in full generality:

Conjecture 12.25. Theorems 12.23 and 12.24 remain true if one replaces $\lambda(A, x) > 0$ by all Lyapunov exponents $\lambda_i(A, x)$ having multiplicity 1.

Problem 12.26. Can the open dense set $\mathcal{A}$ in Theorem 12.24 always be taken the same for every invariant measure with local product structure?

Theorems 12.23 and 12.24 extend to cocycles over non-invertible transformations, respectively, local diffeomorphisms equipped with invariant non-uniformly expanding probabilities (all Lyapunov exponents positive), and uniformly expanding continuous maps, like one-sided shifts of finite type, or smooth expanding maps.

Problem 12.27. What are the continuity points of Lyapunov exponents as functions of the cocycle in $C^\nu(M, \mathrm{SL}(d, \mathbb{R}))$ when $\nu > 0$? Analogously, assuming the base system (f, μ) is hyperbolic. Is every dominated cocycle a continuity point?

Lyapunov exponents may vary discontinuously in the C^ν topology if ν is small. See [437]. The currently known examples are not dominated.

In the remainder of this section we outline some main ingredients in the proofs of Theorems 12.23 and 12.24 that are of independent interest. For clearness, we focus on a particular situation: we suppose f is uniformly expanding,

$G = \mathrm{SL}(2,\mathbb{R})$, and μ is ergodic with $\operatorname{supp}\mu = M$. It is no restriction to suppose $\nu \geq 1$, because the cases $0 < \nu < 1$ can be transformed to $\nu = 1$ by replacing the distance $d(x,y)$ in M by $d(x,y)^{\nu}$.

12.4.1 Bundle-free cocycles

We call a cocycle bundle-free if it admits no Lipschitz continuous invariant sub-bundle, not even up to finite covering. More precisely:

Definition 12.28. We say that $A : M \to \mathrm{SL}(2,\mathbb{R})$ is *bundle-free* if, given any $\eta \geq 1$, there exists *no* Lipschitz continuous map $\psi : x \mapsto \{v_1(x), \ldots, v_\eta(x)\}$ assigning to each $x \in M$ a subset of $\mathbb{RP}^1$ with exactly η elements, invariant under the cocycle in the sense that

$$A(x)\big(\{v_1(x), \ldots, v_\eta(x)\}\big) = \{v_1(f(x)), \ldots, v_\eta(f(x))\} \quad \text{for all } x \in M.$$

Notice that the definition is independent of the invariant measure μ. We also point out that regularity is crucial here: existence of Lipschitz invariant subbundles is rare (most cocycles are bundle-free) whereas existence of Hölder invariant subbundles with poor Hölder constants is often robust. The following construction illustrates these two assertions, in a simplified situation:

Example 12.29. Let $F : S^1 \times \mathbb{R} \to S^1 \times \mathbb{R}$, $F(\theta,x) = (f(\theta), g(\theta,x))$ be a smooth map with

$$\sigma_1 \geq |f'| \geq \sigma_2 > \sigma_3 > |\partial_x g| > \sigma_4 > 1.$$

Let θ_0 be a fixed point of f and x_0 be the fixed point of $g(\theta_0, \cdot)$. Then

1. The set of points whose forward orbit is bounded is the graph of a continuous function $u : S^1 \to \mathbb{R}$ with $u(\theta_0) = x_0$. This function is ν-Hölder for any $\nu < \log \sigma_4 / \log \sigma_1$. Typically it is not Lipschitz:
2. The fixed point $p_0 = (\theta_0, x_0)$ has a strong-unstable set $W^{uu}(p_0)$ invariant under F and which is locally a Lipschitz graph over S^1. If u is Lipschitz then its graph must coincide with $W^{uu}(p_0)$.
3. However, for an open dense subset of choices of g the strong-unstable set is not globally a graph: it intersects vertical lines at infinitely many points.

Theorem 12.30. *Suppose $A \in C^\nu(M, \mathrm{SL}(2,\mathbb{R}))$ has $\lambda(A,\mu) = 0$. Then A is approximated in $C^\nu(M, \mathrm{SL}(2,\mathbb{R}))$ by stably bundle-free maps.*

Here is a sketch of the proof, which is inspired in the previous example. Using the hypothesis one finds periodic points p of f, such that the matrix $A^q(p)$, $q = \mathrm{per}(p)$ is hyperbolic and the normalized eigenvalues are close to 1. Then the eigenspaces, seen as periodic points of the cocycle acting in the projective space, have strong-unstable sets that are *locally* Lipschitz graphs over M. The graph of any Lipschitz map ψ as in Definition 12.28 has to be contained in the strong-unstable sets. But a simple transversality argument shows that *globally* the strong-unstable sets are not graphs, not even up to finite covering, if certain configurations with positive codimension are avoided.

12.4.2 A geometric criterium for non-zero exponents

Another main step is

Theorem 12.31. *Suppose* $A \in C^\nu(M, \mathrm{SL}(2,\mathbb{R}))$ *is bundle-free and there exists some periodic point* $p \in M$ *of* f *such that* A *is hyperbolic over the orbit of* p. *Then* $\lambda(A, \mu) > 0$ *for every invariant probability* μ *with local product structure.*

The second condition, existence of some periodic point over which the cocycle is hyperbolic, is satisfied by an open and dense subset of $C^\nu(M, \mathrm{SL}(2,\mathbb{R}))$, that we denote HP. See the last section of [84] for a proof, and note that the same statement for $\mathrm{SL}(2,\mathbb{C})$ is easy. We also denote by BF the subset of bundle-free maps. The next example helps understand the significance of Theorem 12.31.

Example 12.32. Let $M = S^1$, $f : M \to M$ be given by $f(x) = kx \bmod \mathbb{Z}$, for some $k \geq 2$, and μ be Lebesgue measure on M. Let

$$A : M \to \mathrm{SL}(2,\mathbb{R}), \qquad A(x) = \begin{pmatrix} \beta(x) & 0 \\ 0 & 1/\beta(x) \end{pmatrix}$$

for some smooth function β such that $\int \log \beta \, d\mu = 0$. It is easy to ensure that the set $\beta^{-1}(1)$ is finite and does not contain $x = 0$. Then $A \in \mathrm{HP}$ and indeed the matrix A "looks hyperbolic" at most points. Nevertheless, the Lyapunov exponent $\lambda(A, \mu) = \int \log \beta \, d\mu = 0$. Notice that A is not bundle-free.

This suggests the following heuristic principle: assuming there is some source of hyperbolicity (here the fact that $A \in \mathrm{HP}$), the only way Lyapunov exponents may happen to vanish is by having expanding directions mapped *exactly* onto contracting directions, thus causing hyperbolic behavior to be "wasted way".

12.4.3 Conclusion and an application

It is now easy to finish the argument using Theorems 12.30 and 12.31. Let ZE be the subset of $A \in C^\nu(M, \mathrm{SL}(2,\mathbb{R}))$ such that $\lambda(A, \mu) = 0$. Theorem 12.30 says that any $A \in \mathrm{ZE}$ is approximated by the interior of BF. Since HP is open and dense, A is also approximated by the interior of $\mathrm{BF} \cap \mathrm{HP}$. By Theorem 12.31, the latter is contained in the complement of ZE. This proves that the interior of $C^\nu \setminus \mathrm{ZE}$ is dense in ZE, and so it is dense in the whole $C^\nu(M, \mathrm{SL}(2,\mathbb{R}))$, as claimed. To get the infinite codimension statement observe that it suffices to avoid the positive codimension configurations mentioned at the end of Section 12.4.1, for *some* of infinitely many periodic points of f. This completes our outline of the proof of Theorems 12.23 and 12.24, in a particular case.

We conclude with an application that puts Theorems 12.18 and 12.31 together to give a very complete description of Lyapunov exponents for a whole open set of continuous cocycles. It also shows that these results are, in some sense, optimal.

Example 12.33. Let $M = S^1$, $f : M \to M$ be a uniformly expanding map, and μ be any ergodic invariant measure supported on the whole M. Given any continuous map $\alpha : S^1 \to S^1$, define

$$A : M \to \mathrm{SL}(2, \mathbb{R}), \qquad A(x) = R_{\alpha(x)} \cdot A_0$$

where R_α denotes the rotation of angle α and A_0 is some constant matrix.

Corollary 12.34. *Assume $2 \deg(\alpha)$ is not a multiple of $\deg(f) - 1$ and the matrix $A(p)$ is hyperbolic at some fixed point p. Then there exists a C^0 neighborhood $\mathcal{U}$ of A such that*

1. *for B in a residual subset $\mathcal{R} \cap \mathcal{U}$ we have $\lambda(B, \mu) = 0$;*
2. *for every $B \in C^\nu(M, \mathrm{SL}(2, \mathbb{R}))$, $\nu > 0$ we have $\lambda(B, \mu) > 0$.*

Proof. Start by taking $\mathcal{U}$ to be the isotopy class of A in the space of continuous maps from M to $\mathrm{SL}(2, \mathbb{R})$. We claim that, given any $B \in \mathcal{U}$, there is no *continuous* B-invariant map

$$\psi : M \ni x \mapsto \{\psi_1(x), \dots, \psi_\eta(x)\}$$

assigning a constant number $\eta \geq 1$ of elements of $\mathbb{RP}^1$ to each point $x \in M$. The proof is by contradiction. Suppose there exists such a map and the graph

$$G = \{(x, \psi_i(x)) \in S^1 \times \mathbb{RP}^1 : x \in S^1 \text{ and } 1 \leq i \leq \eta\}$$

is connected. Then G represents some element (η, ζ) of the fundamental group $\pi_1(S^1 \times \mathbb{RP}^1) = \mathbb{Z} \oplus \mathbb{Z}$. Because B is isotopic to A, the image of G under the cocycle must represent $(\eta \deg(f), \zeta + 2 \deg(\alpha)) \in \pi_1(S^1 \times \mathbb{RP}^1)$; here the factor 2 comes from the fact that S^1 is the 2-fold covering of $\mathbb{RP}^1$. By the invariance of ψ we get

$$\zeta + 2 \deg(\alpha) = \deg(f)\zeta$$

which contradicts the hypothesis that $\deg(f) - 1$ does not divide $2 \deg(\alpha)$. If the graph G is not connected, consider the connected components instead. Since connected components are pairwise disjoint, they all represent elements with the same direction in the fundamental group. Then the same type of argument as before proves the claim in full generality.

Now let $\mathcal{R}$ be the residual subset in Theorem 12.18. The previous observation implies that no $B \in \mathcal{R} \cap \mathcal{U}$ may have an invariant dominated splitting. Then B must have all Lyapunov exponents equal to zero as claimed in (1). Similarly, that observation ensures that every $B \in \mathcal{U} \cap C^\nu$ is bundle-free. Our assumptions ensure that A is in HP, and so is any map C^0 close to it. Thus, reducing $\mathcal{U}$ if necessary, we may apply Theorem 12.31 to conclude that $\lambda(B, \mu) > 0$. This proves (2). ☐

12.5 Looking for non-zero Lyapunov exponents

In this section we discuss a number of very useful techniques for finding dynamical systems and linear cocycles with Lyapunov exponents different from zero.

12.5.1 Removing zero Lyapunov exponents

Since existence of an Anosov diffeomorphism imposes very strong restrictions on the manifold, it is natural to ask whether all manifolds support non-uniformly hyperbolic diffeomorphisms. This problem has a fairly long history. Katok [230] constructed non-uniformly hyperbolic Bernoulli diffeomorphisms on any surface. Then Brin, Feldman, Katok [96] showed that every compact surface supports Bernoulli diffeomorphisms. Their examples had 2 Lyapunov exponents equal to zero, but Brin [94] could bring that number down to 1. Very recently, Dolgopyat, Pesin used a perturbation argument going back to [412] to get rid of this last vanishing exponent, thus settling the matter completely:

Theorem 12.35 (Dolgopyat, Pesin [169]). *Every compact manifold supports volume preserving C^∞ diffeomorphisms which are Bernoulli and non-uniformly hyperbolic relative to Lebesgue measure.*

Even more recently, the perturbation argument of [412] was given a local form by Baraviera, Bonatti [45], and this allowed them to prove

Theorem 12.36 (Baraviera, Bonatti [45]). *Every volume preserving partially hyperbolic diffeomorphism $f : M \to M$ on a compact manifold may be approximated, in the C^1 topology, by another for which the integrated sum of all central Lyapunov exponents is different from zero.*

The integrated sum of all central Lyapunov exponents is given by

$$\lambda^c(f) - \int_M \log |\det(Df \mid E^c)| \, dm$$

where E^c is the central sub-bundle of f. Observe that this number varies continuously with f, within the set of partially hyperbolic C^1 diffeomorphisms, because the central sub-bundle does.

Related to all this, we recall

Conjecture 12.37 (Viana [442]). Let a diffeomorphism $f : M \to M$ be non-uniformly hyperbolic at Lebesgue almost every point: there exists a measurable splitting of the tangent space $T_x M = E^u_x \oplus E^s_x$, defined Lebesgue almost everywhere, and such that

$$\liminf_{n \to \pm\infty} \frac{1}{n} \log \|Df^n(x)v_u\| > 0 > \limsup_{n \to \pm\infty} \frac{1}{n} \log \|Df^n(x)v_s\|$$

for every $v_u \in E^u_x$ and $v_s \in E^s_x$. Then f admits some SRB measure.

The results of [12, 84] contain special cases of this conjecture, recall Theorem 11.34. Another special case is Keller's theorem [238] (a simpler proof is given in [147]): *for unimodal maps* with negative Schwarzian derivative, there exists an absolutely continuous invariant measure if and only if there exists a positive Lebesgue measure set of points with positive Lyapunov exponent. If such a measure exists, it is ergodic and, hence, an SRB measure.

12.5.2 Lower bounds for Lyapunov exponents

We close with a brief discussion of Herman's [213] sub-harmonicity argument, and some recent related results.

Let f be a holomorphic function defined on a neighborhood of $0 \in \mathbb{C}^k$, such that $f(0) = 0$ and there is $r > 0$ such that $f(\mathbb{D}_r^k) \subset \mathbb{D}_r^k$ and $f(\mathbb{T}_r^k) = \mathbb{T}_r^k$,

$$\mathbb{D}_r^k = \{(z_1, \ldots, z_k) : |z_j| \leq r\} \qquad \mathbb{T}_r^k = \{(z_1, \ldots, z_k) : |z_j| = r\},$$

and f preserves the Haar measure m on the torus $\mathbb{T}_r^k$. For example, let $f(z_1, \ldots, z_k) = (e^{i\theta_1} z_1, \ldots, e^{i\theta_n} z_k)$. Let $A : \mathbb{T}_r^k \to SL(d, \mathbb{R})$ be a continuous map, seen as a cocycle over (f, m), and

$$\lambda(A, m) = \int \lambda(A, x) \, dm(x) = \inf_{n \geq 1} \frac{1}{n} \int_{\mathbb{T}_r^k} \log \|A^n(z)\| \, dm(z)$$

denote its largest Lyapunov exponent. The spectral radius of a matrix B is

$$\rho(B) = \inf_{n \geq 1} \|B^n\|^{1/n}.$$

Theorem 12.38 (Herman [213]). *We have $\lambda(A, m) \geq \log \rho(A(0))$.*

Herman simply observes that every function $\mathbb{D}_r^k \ni z \mapsto \log \|A^n(z)\|$ is plurisubharmonic. Consequently, it satisfies the maximum principle:

$$\frac{1}{n} \int_{\mathbb{T}_r^k} \log \|A^n(z)\| \, dm(z) \geq \frac{1}{n} \log \|A^n(0)\|.$$

Taking the infimum over n one immediately gets the conclusion of the theorem!

An extension of this ingenious idea leads to explicit lower bounds for the average Lyapunov exponents on certain parametrized families of cocycles. The main technical ingredient is the following result of Herman [213] (see also Knill [245]). *Let $f : M \to M$ be a transformation on a compact metric space M, preserving some probability μ, and let $\mathcal{A} : \mathbb{D}_r^p \times M \to \mathcal{L}(\mathbb{R}^d, \mathbb{R}^d)$ be continuous and such that $z \mapsto \mathcal{A}(z, x)$ is holomorphic on the interior of $\mathbb{D}_r^p$ for every $x \in M$. Then the function $z \mapsto \lambda(\mathcal{A}(z, \cdot), \mu)$ is plurisubharmonic.*

The best known application is the following lower bound for average Lyapunov exponents of $SL(2, \mathbb{R})$ cocycles. Given a continuous $A : M \to SL(2, \mathbb{R})$ let AR_θ be the cocycle obtained multiplying $A(x)$ by the rotation of (constant) angle θ at every point $x \in M$. Then (Herman [213, Section 6.2])

$$\frac{1}{2\pi} \int_0^{2\pi} \lambda(AR_\theta, \mu) \, d\theta \geq \int_M N(A(x)) \, d\mu(x), \tag{12.7}$$

where, following [35], we write

$$N(B) = \log \frac{\|B\| + \|B\|^{-1}}{2}, \quad \text{for } B \in \mathrm{SL}(2, \mathbb{R}).$$

Note that $N(B) \geq 0$ and $N(B) = 0$ if and only B is a rotation. So, except for the exceptional case when A acts by rotation at every point in the support of μ, the right hand side of (12.7) is positive, and so the inequality implies that the Lyapunov exponent of the cocycle AR_θ is positive for many values of θ (with explicit estimates, see [213, 245]).

Knill [245] used this fact to prove that a dense subset of bounded $\mathrm{SL}(2, \mathbb{R})$ cocycles have non-zero Lyapunov exponents. His result is optimal since, as mentioned before, Bochi [63] proves that the cocycles with vanishing exponents contain a residual subset.

The relation (12.7) has recently been improved by Avila, Bochi [35], who show that the equality actually holds:

$$\frac{1}{2\pi} \int_0^{2\pi} \lambda(AR_\theta, \mu) \, d\theta = \int_M N(A(x)) \, d\mu(x). \tag{12.8}$$

This is a corollary of an elegant general formula for $\mathrm{SL}(2, \mathbb{R})$ matrices, which has a number of other applications:

Theorem 12.39 (Avila, Bochi [35]). *Given any $A_1, \ldots, A_n \in \mathrm{SL}(2, \mathbb{R})$,*

$$\frac{1}{2\pi} \int_0^{2\pi} N(A_n R_\theta \cdots A_1 R_\theta) \, d\theta = \sum_{j=1}^n N(A_j) = \frac{1}{2\pi} \int_0^{2\pi} \rho(A_n R_\theta \cdots A_1 R_\theta) \, d\theta.$$

To deduce (12.8), begin by noting that $N(B) \leq \log \|B\| \leq \log 2 + N(B)$. Therefore, taking $A_{j+1} = A(f^j(x))$ in the theorem,

$$\sum_{j=0}^{n-1} N(A(f^j(x))) \leq \frac{1}{2\pi} \int_0^{2\pi} \log \|(AR_\theta)^n(x)\| \, d\theta \leq \log 2 + \sum_{j=0}^{n-1} N(A(f^j(x))).$$

So, by the ergodic theorem and the dominated convergence theorem,

$$\int_M N(A) \, d\mu = \lim_{n \to \infty} \frac{1}{2\pi} \int_0^{2\pi} \frac{1}{n} \log \|(AR_\theta)^n\| \, d\theta = \frac{1}{2\pi} \int_0^{2\pi} \lambda(AR_\theta, \mu) \, d\theta$$

(μ-almost everywhere) as claimed.

While Lyapunov exponents of individual cocycles are often be very unstable under perturbations (recall e.g. Corollary 12.34), relations like (12.7) or (12.8) show that the average behavior in parametrized families of cocycles may be surprisingly robust. Moreover, as we have just seen, such relations may be used to deduce that global properties of a family yield good properties for many of its elements.

This general philosophy has also been much pursued by Shub and collaborators, in an effort to obtain lower bounds for Lyapunov exponents of deterministic cocycles in terms of the exponents of certain random counterparts (which are often easier to estimate from below). A concrete implementation has been provided by Dedieu, Shub [148] for linear systems. We just quote a 2-dimensional version of their result:

Theorem 12.40 (Dedieu, Shub [148]). *Let ν be a probability on $\mathrm{SL}(2,\mathbb{R})$ invariant under all rotations and such that the function $\log \|B\|$, $B \in \mathrm{SL}(2,\mathbb{R})$ is ν-integrable. Then*

$$\int \log \rho(B)\, d\nu(B) = \lambda(\nu)$$

where $\lambda(\nu)$ is the largest Lyapunov exponent of the cocycle defined by the product of i.i.d. random matrices with probability law ν (Example 12.17).

See [35, Section 6] for a quick proof of this statement from Theorem 12.39. Ledrappier *et al* [255] have been investigating similar ideas in a non-linear setting. They consider a monotone twist map of the unit 2-sphere in $\mathbb{R}^3$: the dynamics on each parallel circle is a rotation, with angle of rotation that varies monotonically from one pole to the other. Next, they consider the parametrized family of dynamical systems obtained by composing this twist map with the group of orthogonal orientation preserving transformations of $\mathbb{R}^3$. They compare the Lyapunov exponents for random products of maps in this family with the mean of the Lyapunov exponents for random products restricted to a small neighborhood of the individual elements. They prove that these two numbers are asymptotic as the twist (that is, the variation of the rotation angle with the latitude) tends to infinity. In addition, there is numerical evidence that the same should be true for the Lyapunov exponents of the deterministic systems, averaged over the parameter. If that is really the case, these would be the first examples of analytic families of area preserving maps of the 2-sphere containing twist maps and maps with positive entropy. Compare the comments following Problem 4.39.

12.5.3 Genericity of non-uniform hyperbolicity

A very nice application of methods we have been presenting was given, very recently, by Bochi, Fayad, Pujals, who announce that *non-uniform hyperbolicity is generic among stably ergodic diffeomorphisms*. Let $\mathrm{Diff}_m^{1+}(M)$ denote the space of volume preserving diffeomorphisms with Hölder continuous derivative on a compact manifold M, endowed with the C^1 topology, and let SE denote the subset of stably ergodic diffeomorphisms.

Theorem 12.41 (Bochi, Fayad, Pujals [64]). *There exists an open and dense subset of diffeomorphisms $f \in \mathrm{SE}$ admitting a dominated splitting $TM = E^+ \oplus E^-$ such that the Lyapunov exponents along E^+ are all positive and those along E^- are all negative. In particular, Lebesgue measure m is a hyperbolic measure for f.*

The argument has three main steps. The first one is to show that stable ergodicity implies the existence of a dominated splitting. The idea is that in the absence of dominated splitting one may create periodic points where the map is tangent to the identity [2] (Bonatti, Díaz, Pujals [73]) and, after a new small perturbation, this breaks ergodicity (Arbieto, Matheus [22]).

We want to prove that every $f \in \mathrm{SE}$ is approximated by an open set consisting of non-uniformly hyperbolic diffeomorphisms. Let $TM = E^1 \oplus \cdots \oplus E^k$ be the finest dominated splitting for f (see Appendix B). This has a continuation $TM = E_g^1 \oplus \cdots \oplus E_g^k$ as a dominated splitting for any nearby diffeomorphism g. Replacing f by an arbitrarily close C^{1+} diffeomorphism if necessary, we may and do suppose from the start that this extension remains the finest dominated splitting in a neighborhood of f inside $\mathrm{Diff}_m^{1+}(M)$. The second step is to use the perturbation Theorem 12.36, to approximate f by some $g \in \mathrm{Diff}_m^{1+}(M)$ such that the sum

$$\int \log |\det(Dg \mid E_g^i)| \, dm$$

of all Lyapunov exponents along E_g^i is non-zero, for every $i = 1, \ldots, k$. By continuity, the same is true for every diffeomorphism in a C^1 neighborhood U of g, with integrals uniformly bounded from zero.

The last step is based on Theorem 12.9. The theorem says that there exists a residual subset $\mathcal{R}$ of diffeomorphisms $f \in \mathrm{Diff}_m^1(M)$ for which the Oseledets splitting is either dominated or trivial. A variation of the proof gives a more quantitative version: for any $\delta > 0$ there is an open and dense subset $\mathcal{R}_\delta$ of diffeomorphisms $f \in \mathrm{Diff}_m^{1+}(M)$ (or $\mathrm{Diff}_m^1(M)$) admitting a dominated splitting such that the Lyapunov exponents along each of the subbundles differ by less than δ. Fixing δ sufficiently small, this splitting must coincide with the previous one, $TM = E_h^1 \oplus \cdots \oplus E_h^k$, for every $h \in U \cap \mathcal{R}_\delta$. Thus, for such maps h all Lyapunov exponents along every E_h^i are almost equal, while their sum is uniformly bounded from zero. This implies that all the Lyapunov exponents of $h \in U \cap \mathcal{R}_\delta$ are non-zero, and that gives the claimed density.

This result and the arguments in the proof motivate the following question, which also appeared in [412]:

Problem 12.42. Is it true that for generic $f \in \mathrm{Diff}_m^1(M)$ either all Lyapunov exponents are non-zero m-almost everywhere or they are all equal to zero m-almost everywhere?

---

[2]That is, $f^n(p) = p$ and $Df^n(p) = \mathrm{id}$.

In view of Theorem 12.9, this would follow from an affirmative answer to the following natural question:

Problem 12.43. Are ergodic diffeomorphisms generic in $\mathrm{Diff}_m^1(M)$?

Recall that, according to Theorem 8.14, transitive diffeomorphisms are generic in $\mathrm{Diff}_m^1(M)$.

12.6 Hyperbolic measures are exact dimensional

As an illustration of the role of Lyapunov exponents in connection with other aspects of dynamics, we briefly discuss a recent result of Barreira, Pesin, Schmeling [48] stating that hyperbolic invariant measures have well-defined dimension. This settled an old question of Eckmann, Ruelle [176].

Given a Borel measure μ on a complete metric space M, the *pointwise dimension* of μ at $x \in M$ is defined by

$$d(\mu, x) = \lim_{r \to 0} \frac{\log \mu(B_r(x))}{\log r}. \tag{12.9}$$

In other words, if the pointwise dimension does exist then small balls around x have

$$\mu(B_r(x)) \approx r^{d(\mu,x)}.$$

In general, one speaks of upper and lower dimensions, $\overline{d}(\mu, x)$ and $\underline{d}(\mu, x)$, defined by the $\limsup$ and the $\liminf$ in (12.9). If μ is invariant under a diffeomorphism $f : M \to M$ then $\overline{d}$ and $\underline{d}$ are constant on trajectories; hence, they are constant almost everywhere if μ is ergodic.

Eckmann, Ruelle [176] highlighted the importance of this concept for invariant measures of smooth systems, and thus raised the question whether the pointwise dimension always exists at almost every point if μ is a hyperbolic invariant measure. Young [456] proved that if $d(\mu, x) = d$ at μ-almost every point then several natural definitions of dimension of the measure μ coincide with d. Then μ is said to be *exact dimensional*. She also proved that ergodic hyperbolic measures of a *surface* $C^{1+\alpha}$ diffeomorphism f are exact dimensional, in fact,

$$d(\mu) = d(\mu, x) = h_\mu(f)\big(\lambda_1^{-1} - \lambda_2^{-1}\big)$$

where $h_\mu(f)$ is the entropy of (f, μ) and $\lambda_1 > 0 > \lambda_2$ are the Lyapunov exponents. Ledrappier, Young [257, 258] introduced stable and unstable pointwise dimensions of an invariant probability, denoted $d^s(\mu, x)$ and $d^u(\mu, x)$. Avoiding technicalities, let us just say that these are the pointwise dimensions of μ conditioned (see Appendix C) to stable manifolds and to unstable manifolds, respectively. They proved that stable and unstable pointwise dimensions always exist almost everywhere, and

$$\overline{d}(\mu, x) \leq d^s(\mu, x) + d^u(\mu, x),$$

if μ is hyperbolic. Finally, the matter was settled by the following remarkable

Theorem 12.44 (Barreira, Pesin, Schmeling [48]). *Let μ be a hyperbolic invariant measure of $C^{1+\alpha}$ diffeomorphism $f : M \to M$ on any compact manifold M. Then $d(\mu, x)$ exists and is equal to $d^s(\mu, x) + d^u(\mu, x)$ at almost every point. In particular, if μ is ergodic then it is exact dimensional and $d(\mu) = d^s(\mu) + d^u(\mu)$.*

The main ingredient in the proof is a property of *local quasi product structure* for arbitrary invariant probability measures: for any δ there exists a subset Λ with $\mu(M \setminus \Lambda) < \delta$ and a constant $\kappa > 1$ such that

$$r^\delta \mu_x^s(B_{r\kappa^{-1}}(x)) \mu_x^u(B_{r\kappa^{-1}}(x)) \leq \mu(B_r(x)) \leq r^{-\delta} \mu_x^s(B_{r\kappa}(x)) \mu_x^u(B_{r\kappa}(x))$$

for every $x \in \Lambda$ and $r < r(x)$, where μ_x^s and μ_x^u are conditional measures of μ along the stable manifold and the unstable manifold of x. Compare Definition C.6.

The reader is referred to the supplementary chapter by Katok, Mendoza in [232], and the book of Barreira, Pesin [47] for broad presentations of nonuniform hyperbolicity theory.

A

Perturbation Lemmas

In this appendix we review certain general perturbation statements that are at the basis of most genericity results mentioned in the text. So far these perturbation lemmas have been proved in the C^1-topology only. That is the reason why, at this stage, a nice description of generic diffeomorphisms is available only in that topology.

Pugh's closing lemma (Section A.1) has a very intuitive statement which gives no hint of the complexity of its proof: any orbit going back arbitrarily close to itself can be closed by a C^1-small perturbation of the map. The proof contains two main ingredients. The first one is to estimate the effect of small perturbations of the diffeomorphism on a small neighborhood of a finite segment of orbit. The second one is to select two returns of the orbit close to itself which are, in some sense, closer to one another than to any intermediary returns. See also Section A.4.

Each of this arguments has been generalized, leading important new perturbations lemmas. For the ergodic closing lemma (Section A.2), Mañé considered how frequently, one may take the initial point as one of the selected returns. The reason why this question is important is that the periodic orbit one creates by closing an orbit segment in this way has derivative close to that of the original orbit segment.

Hayashi's connecting lemma (Section A.3) was originally designed as a tool for creating intersections between invariant manifolds of periodic points. For that purpose, the strategy was to shorten in a systematic way (Pugh's perturbations) orbit segments almost joining the two invariant manifolds. See Section A.4 for further comments on the proof

In Section A.6 we state a very useful simple result of Franks, allowing to realize perturbations of the derivative along of periodic obits by C^1 perturbations of the actual dynamical systems. This argument turns certain perturbations problems into problems about how eigenspaces and eigenvalues of products of matrices depend on small perturbations the factors.

A.1 Closing lemmas

A typical question is: *Given a recurrent point of a system, is there a C^r nearby system for which that point is periodic?* The difficulty of the question depends, dramatically, on the topology and, to some extent, on the dimension of the ambient space and the class of systems one considers.

For $r = 0$ the answer is positive and easy. For $r = 1$ it is still positive, as stated by Pugh's closing lemma:

Theorem A.1 (Pugh [363, 364]). *Let* $\mathbf{f} = \{f^t : t \in \mathbb{R}\}$ *be a* C^1 *flow on a compact manifold* M, *and* $x \in M$ *be a non-wandering point for this flow: there exist* $x_j \to x$ *and times* $t_j \to +\infty$ *such that* $f^{t_j}(x_j) \to x$. *Then there exist flows* $\mathbf{g} = \{g^t : t \in \mathbb{R}\}$ *arbitrarily close to* $\mathbf{f}$ *in the* C^1 *topology, for which* x *is a periodic point.*

Partial results had been obtained by Peixoto [351], for flows on orientable surfaces, and Anosov [16], for the class of systems named after him. Actually, in the special case of hyperbolic systems this problem has a complete answer, known as the shadowing lemma: a closing orbit exists in the system itself, no perturbation is needed.

Theorem A.1 was extended by Pugh, Robinson [365] to several other classes of systems, including C^1 diffeomorphisms, Hamiltonian flows, symplectic diffeomorphisms, and volume preserving diffeomorphisms and flows. See also [23] for recent proofs based on a different approach. Let us mention that the C^1 closing lemma remains open in the setting of geodesic flows (C^2 perturbation of the Riemannian metric).

For $r > 1$ the question of the closing lemma remains very much open. Gutierrez [203] showed that local perturbations as used by Pugh in the C^1 case do not suffice if $r \geq 2$. Most important, Herman [214, 215] constructed counterexamples to the C^∞ closing lemma in the setting of conservative systems, symplectic diffeomorphisms or Hamiltonian vector fields, in sufficiently high dimensions. More precisely, give any $r > 2n \geq 4$, in [214] he exhibits C^r open sets of Hamiltonian functions on $\mathbb{T}^{2n}$, such that the corresponding Hamiltonian vector fields [1] on the $2n$-torus, endowed with a constant symplectic form, exhibit open sets formed by energy surfaces that contain no periodic orbit. In fact, on each component of these energy surfaces the dynamics is C^1 conjugate to a linear irrational (Diophantine) flow. See also [454]. On the other hand, no counterexamples are known in the most interesting *mechanical case*, that is, when the ambient space is the cotangent bundle T^*M of some symplectic manifold (M, ω), endowed with the natural symplectic form.

In view of these negative results, it is convenient to look for weaker useful versions of the closing lemma. The result of Peixoto mentioned above holds for any $0 \leq r \leq \infty$. It suggests that a perhaps more realistic goal could be

---

[1]The Hamiltonian vector field X corresponding to a function H on a symplectic manifold (M, ω) is defined by $\Omega(X, \cdot) = \nabla H$. If H is C^r then X is C^{r-1}.

Problem A.2. Given a C^r system with a recurrent point x, is there an arbitrarily C^r close system for which either x is periodic or else there are no recurrent points in a neighborhood of x?

On the other hand, the closing lemma has been refined in two important directions, always in the C^1 topology:

A.2 Ergodic closing lemma

In many situations one would like to know: *Does the periodic orbit obtained after perturbation remain close, up to the period, to a segment of the original orbit?* Mañé's ergodic closing lemma asserts that this is true for a *full probability* set of points, that is, having full measure with respect to any invariant probability of the unperturbed map. For the precise statement, let $\Sigma(f)$ be the set of points $x \in M$ such that given any C^1 neighborhood $\mathcal{U}$ of f and any $\varepsilon > 0$ there exist $g \in \mathcal{U}$, coinciding with f outside the ε-neighborhood of the f-orbit of x, and there exists a g-periodic point $y \in M$ such that

$$d(f^i(x), g^i(y)) \leq \varepsilon \quad \text{for all } 0 \leq i \leq \mathrm{per}(y).$$

Theorem A.3 (Mañé [278]). *For any C^1 diffeomorphism f of a compact manifold, the set $\Sigma(f)$ has full probability for f.*

A major application of this statement is to express in terms of periodic orbits dynamical behaviors that are detected by invariant measures. The following argument, which is part of Mañé's proof of the 2-dimensional version of Theorem 7.5 in [278], is a good illustration:

Let f be a surface diffeomorphism with a dominated splitting $E \oplus F$ and suppose the sub-bundle F is *not* uniformly expanding. Then there are arbitrarily long orbit segments over which expansion along F is arbitrarily weak. Any weak accumulation point of Dirac averages along such orbit segments is an invariant measure for which the sub-bundle F is not expanding. Pugh's closing lemma permits to close a typical orbit of such a measure. However, not knowing whether the periodic orbit remains close to the original one, one loses control on the dilation along F. On the contrary, the ergodic closing lemma ensures that the derivative of the perturbed map g is, at best, weakly expanding along the continuation F_g of F for g over the orbit of the periodic point y. By another small perturbation (this is Franks' lemma, that we shall discuss in a while) we can now turn any slight expansion into a contraction, which means that y becomes a sink.

A.3 Connecting lemmas

The connecting lemma is motivated by the following question. Assume that the unstable manifold of a periodic point accumulates on the stable manifold of another (possibly the same) periodic point. *Is there a small perturbation of the dynamical system for which the two invariant manifolds corresponding to the continuations of the two periodic points actually intersect each other?*

While this is closely related to the problem of the closing lemma, there is one subtle difference. In the context of the closing lemma it suffices to create a periodic orbit passing close to the initial recurrent point x: once that is done, a change of coordinates close to the identity allows us to suppose that the periodic point is x itself. In other words, points near x are indistinguishable from each other. This is clearly no longer true in the setting of the connecting lemma, because now one wants to create orbits connecting invariant (stable and unstable) manifolds that are given a priori.

Even in the simplest case, flows on orientable surfaces, this problem remained open for more than three decades after the corresponding closing problem had been settled. This was also a major difficulty in the proof of the stability conjecture for diffeomorphisms [283], where a main step consists in showing that if the map is not hyperbolic then homoclinic connections may be created by small C^1 perturbations. In [282], Mañé bypassed the lack, at the time, of a connecting lemma by proving several connecting results with additional assumptions: in a few words, he supposed that orbits accumulating the two invariant manifolds spent a positive fraction of time near the associated periodic point.

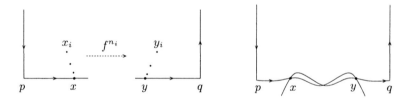

Fig. A.1. Connecting invariant manifolds

Eventually, the problem was solved by Hayashi [208] for C^1 diffeomorphisms and flows. In fact he proves a stronger statement, cf. Figure A.1:

Theorem A.4 (Hayashi [208]). *Given two periodic points p and q, if there are orbits visiting the neighborhoods of some point in the unstable manifold of p and some point in the stable manifold of q then, by an arbitrarily C^1-small perturbation, one can make the two invariant manifolds intersect each other.*

Using this result, Hayashi could simplify Mañé's proof of the stability conjecture substantially, as well as extend it to C^1 flows. Moreover, these results were extended to the conservative setting by Xia, Wen [450, 453].

It is fair to say that the C^1 connecting lemma was a decisive ingredient in the recent progress towards understanding C^1 generic dynamical systems. On the other hand, many of these applications require variations of the original statement. In particular, the following question posed itself quite naturally: Assume the unstable manifold of p and the stable manifold of q accumulate on some point x. *Is there a C^1 perturbation of the system for which the two invariant manifolds intersect each other?* This has been answered recently, assuming x is not periodic itself, independently by Arnaud [25], Hayashi [209], and Wen, Xia [450]. Their statements are slightly different, here we quote [25]:

Theorem A.5. *Let f be a diffeomorphism on a compact manifold, $x \in M$ be a non-periodic point, and $\mathcal{U}$ be a C^1 neighborhood of f. Then there exist $N \geq 1$ such that for any neighborhood V of x there exists another neighborhood $W \subset V$ such that given any points y and z outside $\cup_{0 \leq n < N} f^n(V)$ and such that W contains some forward iterate of y and some backward iterate of z, there exists $g \in \mathcal{U}$ coinciding with f outside $\cup_{0 \leq n < N} f^n(V)$ and for which z is in the forward orbit of y, passing through V.*

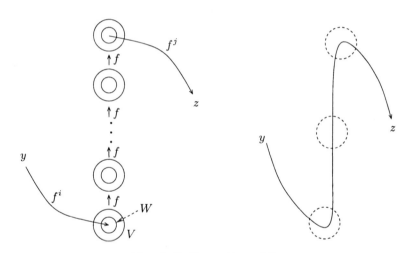

Fig. A.2. Connecting orbits

See Figure A.2. The statement remains true for non-compact manifolds, as long as the orbit of x has some accumulation point in M.

A.4 Some ideas of the proofs

We are going to give a sketch of the proof of the closing lemma and the connecting lemma.

Consider a diffeomorphism f on a compact manifold M, and a point $x \in M$ having a positive iterate $f^n(x)$ very close to x. The naive strategy for "closing"

282   A Perturbation Lemmas

the orbit of x is to compose f with a diffeomorphism h close to the identity, such that $h(f^n(x)) = x$. If *the support of h*, that is, the region where h differs from the identity, does not contain any intermediate iterate $f^i(x)$, $i \in \{1, \ldots, n-1\}$ then the diffeomorphism $g = h \circ f$ satisfies $g^n(x) = x$. This simple idea actually gives the proof of the C^0 closing lemma: if $\dim M \geq 2$ and the distance $d(x, f^n(x))$ is less than some number δ then there is a topological disk Δ of diameter δ containing x and $f^n(x)$ and disjoint from $f^i(x)$, $i \in \{1, \ldots, n-1\}$. Now there is a diffeomorphism h supported in Δ, and so δ-close to identity in the C^0 topology, such that $h(f^n(x)) = x$.

This idea is no longer sufficient in the C^1 topology. To explain this, consider a diffeomorphism h such that $h(y) = x$ for some point y with $d(x, y) \leq \delta$. In order that the C^1 distance from h to the identity be bounded by ε we need some "safety space": the support of h must contain the geometric (round) disk of radius $d(x, y)/\varepsilon$ around y. Taking $y = f^n(x)$, there is no reason why this disk of prescribed radius would be disjoint from the orbit segment $f^i(x), i \in \{1, \ldots, n-1\}$. So, while trying to close the orbit by sending $f^n(x)$ to x one might be opening it at some intermediate place.

Pugh's idea for solving this difficulty was to spread the perturbation in time along some segment of the orbit of x. Since we deal with local perturbations, we may think of the dynamics as being essentially linear, so that the key ingredient for reducing the safety zone is the following lemma from linear algebra:

Lemma A.6 ([365]). *Let $\mathbb{R}^d$ be endowed with some Euclidean metric $d(\cdot, \cdot)$, and let M_i be any sequence of linear isomorphisms of $\mathbb{R}^d$. Given $\eta > 0$ and $\varepsilon > 0$ there exists an integer $N \geq 1$ and some basis of $\mathbb{R}^d$ such that the following holds; given any pair of points z, y in the cube $[-1, 1]^d$ of radius 1 relative to this basis, there exist points z_i in the cube of radius $1 + \eta$ such that $z_0 = z$, $z_N = y$, and*

$$d\big(M_i(z_{i-1}), M_i(z_i)\big) \leq \varepsilon d\big(M_i(z_{i-1}), \partial M_i([-(1+2\eta), (1+2\eta)]^d)\big)$$

for all $i = 1, \ldots, N$.

The case $d = 1$ is easy: it suffices to take for z_i an equally spaced sequence between z and y, and to use the simple fact that one-dimensional linear maps preserve ratios of distances. Thus, in this case by spreading the perturbation along N iterates one reduces the safety zone by a factor N.

To finish the proof of the closing lemma, one considers those iterates of x in the time interval $\{1, \ldots, n\}$ that return to some local chart around x. Among them, one chooses a pair $f^j(x)$ and $f^k(x)$, with $j > k$, which are closest in the following sense. If C represents the smallest cube, in local coordinates, containing $f^j(x)$ and $f^k(x)$ then there is no other iterate $f^l(x)$, $1 \leq l \leq n$ in the cube C_η with the same center as C and radius $(1 + 2\eta)$ as large. Taking the local chart sufficiently small, we may consider the N first iterates of f to be roughly linear restricted to it, so that the previous lemma may be applied with

$z = f^j(x)$, $y = f^k(x)$, and $M_i = Df^i(x)$. The factor ε permits to construct a perturbation h supported inside slightly smaller domains: h is the identity outside the union of $f^i(C_\eta)$ over $1 \leq i \leq N$, and satisfies $h(f^i(z_{i-1})) = f^i(z_i)$, where z_i is the sequence given by the lemma. Taking $g = h \circ f$ one has that $g^N(f^j(x)) = f^{k+N}(x)$, and $f^j(x)$ is a periodic orbit for g. Notice that $f^j(x)$ is close to x.

In summary, first one selects a convenient place to close the orbit (iterates $k < j$), then one carries out small perturbations during N iterates, so that for the new diffeomorphism the N:th iterate of $f^j(x)$ coincides with $f^{k+N}(x)$ and all the iterates in the time interval $[k+N, j)$ are not affected. In this way, $f^j(x)$ becomes a periodic point.

A key difference in the context of the connecting lemma is that the initial and final points must be involved in the new orbit, and so one can not afford to retain only an intermediate orbit segment. More precisely, given two orbit segments $f^i(y)$, $0 \leq i \leq m$ and $f^{-i}(z)$, $0 \leq i \leq n$ such that $f^m(y)$ is close to $f^{-n}(z)$, one wants to find a perturbation for which y and z themselves belong to the same orbit.

To go around this difficulty, Hayashi renounced trying to exhibit a perturbation that would leave intermediate iterates unchanged. His strategy was to make perturbations at several places in order to shorten the two segments every time one of them comes close to itself or to the other: at each visit close to some point of a segment, instead of following the original trajectory one jumps directly to last visit close to that point, discarding all intermediate iterates.

Should "closeness" be an equivalence relation, which of course it is not, this strategy could be neatly implemented. As it is, one must replace this relation by more technical versions which, on the other hand, are transitive. The first step is to consider a small cube contained in a local chart around $f^m(y)$ and $f^{-n}(z)$ as in the closing lemma (using Lemma A.6). Next one tiles this cube into sub-cubes such that the size of each tile is comparable, up to a bounded factor, to the size of every one of its neighbors, and so the number of neighbors is uniformly bounded. Moreover, $f^m(y)$ and $f^{-n}(z)$ belong to the same tile. We consider two points as being "close" if they belong to the same tile, and then we carry out the shortening procedure as explained before: orbit segments in between the first and the last visit to each tile are discarded.

We still have to deal with the possibility of having points in neighboring tiles which are actually very close to each other. This is where having a bounded number of neighbors, all with comparable sizes, becomes crucial. We consider another closeness relation for such points, where we say that they are close if the distance is much smaller than the size of the tile and all its neighbors. As the number of neighbors is uniformly bounded, this is an equivalence relation, and the equivalence classes have small diameter. Then, we carry out a second shortening of the orbit segments as before.

At this point local perturbations as in the proof of the closing lemma, with disjoint supports, allow us to close all the jumps introduced by the various shortenings performed in the previous steps, and to complete the proof.

A.5 A connecting lemma for pseudo-orbits

Notice that while the connecting lemma aims at closing a pseudo-orbit with a single jump, the proof creates and then closes pseudo-orbits with any number of jumps, under the following condition: each jump is contained in a tile of a cube, itself tiled by Pugh's cubes, and is disjoint from its N first iterates, where N is given by the linear algebra argument in Lemma A.6. One calls such a tiled cube a *perturbation box* and the union of the its N first iterates the *support of the perturbation box*. This suggests that one should look for a more general *connecting lemma for arbitrary pseudo-orbits*. Such a statement has indeed been obtained recently:

Theorem A.7 (Bonatti, Crovisier [70, 69]). *Let f be a C^1 diffeomorphism whose periodic points are all hyperbolic, and y, z be any two points which may be joined by pseudo-orbits with arbitrarily small jumps. Then there exists a C^1 nearby diffeomorphism g for which y and z are in the same orbit.*

This statement remains true within volume-preserving and within symplectic maps. Moreover, in the symplectic case one may relax the hypothesis to include elliptic points.

The fundamental idea for proving Theorem A.7 is to cover the space of all (pseudo-)orbits by perturbation boxes as in Hayashi's argument.

Let us first present a very simple (and not realistic) situation: assume that there is a finite family of perturbations boxes with disjoint support, such that any orbit meets the interior of some tile of these perturbation boxes. Then, by compactness of the ambient manifold, there is a finite family of such tiles and some time $n > 0$ such that any segment of orbit with length larger than n meets the interior of one of these tiles, at distance larger than some $\delta > 0$ from the tiles's boundary. Up to reducing δ, the same occurs for any ε-pseudo-orbit, for ε small enough. As the time between two visits to the interior of a tile is bounded by n, one can replace the pseudo-orbit by another one having jumps only at the visit times: if ε is small enough, the new pseudo-orbit will have jumps bounded by δ, that is, inside the same tiles. Now, Hayashi's argument permits to perturb the diffeomorphism in the support of a perturbation box, obtaining a (shorter) pseudo-orbit without jumps in this box. Proceeding inductively, in all the perturbation boxes of the family, one gets a perturbation of the original diffeomorphism having a genuine orbit joining the extremities of the initial pseudo-orbit.

The main difficulty for building such a family of perturbation boxes with disjoint supports is that, if it existed, the union of such a family would be an open set disjoint from its N first iterates and meeting every orbit of the diffeomorphism (such an open set is called a *topological tower of length N*). This is clearly impossible if there are periodic orbits with period less than N. It turns out that the existence of periodic orbits with low periods is the unique obstruction to the existence of a topological tower: there is a constant k, depending only on the ambient dimension, such that any invariant compact set without periodic orbits of period kN admits a topological tower of length N (over that compact set).

These topological towers allow us to build perturbations boxes with disjoint supports, far from the periodic orbits with low periods. The possible visits of a pseudo-orbit to the neighborhood of such a periodic orbit needs a separate analysis of the local dynamics: it is for this reason that Theorem A.7 requires all periodic orbits to be hyperbolic; in the symplectic case, where this would too strong a requirement, one uses an additional argument to include the case of elliptic periodic orbits without non-trivial resonances. In fact the proof in [28] uses an abstract condition called *uniform avoidability* of periodic orbits, which is satisfied in the two cases we have just mentioned.

A.6 Realizing perturbations of the derivative

The following result permits to perform dynamically perturbations of the linear co-cycle defined by the derivative of a diffeomorphism f. It may also be stated uniformly for any g in a neighborhood of f.

Theorem A.8 (Franks [185]). *Let f be a C^1-diffeomorphism defined on a closed manifold M and consider any $\delta > 0$. Then there is $\varepsilon > 0$ such that, given any finite set Σ, any neighborhood U of Σ, and any linear maps*

$$A_x \colon T_x M \to T_{f(x)} M \,, \quad x \in \Sigma$$

such that every A_x is ε-close to the derivative of f at x, there is a diffeomorphism g which is δ-close to f in the C^1 topology, coinciding with f on $M \setminus U$ and on Σ, and such that the derivative of g at any point $x \in \Sigma$ is A_x.

The key point in the proof is that the C^1 norm of a map is not affected by conjugacy under a homothetic transformation. This allows one to perform perturbations of the derivative in arbitrarily small neighborhoods of the point without changing the C^1 size of the perturbation.

This argument clearly fails in any C^r topology for any $r \geq 2$, and so does the conclusion. Since one often uses Franks' lemma to perturb the derivative along periodic orbits, let us give a C^2 counterexample in that context. Consider a linear Anosov map f of the 2-torus and fix some $\delta > 0$. Then for any $\varepsilon > 0$ there is a periodic orbit Σ which is "sufficiently dense" in the torus such that there are ε-perturbations of the derivative at the points of Σ that cannot be realized as δ-perturbations of f in the C^2 topology: they can be realized as C^1 perturbations but the variation of the derivative must be very fast, and so the second derivative is necessarily large.

The previous counter-example is rather artificial, but a natural dynamical situation where Franks' lemma fails in the C^2 category was exhibited by Pujals, Sambarino [370]. They consider a diffeomorphism f having a non-trivial homoclinic class Λ which is locally maximal, admits a nontrivial dominated splitting, and contains a non-hyperbolic point p (for instance, a saddle-node). A class of examples are the saddle-node horseshoes obtained by collapsing a sink into a hyperbolic horseshoe, as considered in [425].

Clearly, there are periodic points $p_n \in \Lambda$ or arbitrarily large period having a Lyapunov exponent which is arbitrarily close to one: it suffices to consider points passing close enough to p and having only finitely many iterates outside a neighborhood of p. Then, by small perturbations of the derivatives along these orbits one destroys their hyperbolicity. However, the results in [370] ensure that the periods of the non-hyperbolic periodic points of diffeomorphisms C^2 close to f are uniformly bounded.

B

Normal Hyperbolicity and Foliations

There are several notions extending uniform hyperbolicity in the following spirit: one still requires the existence of an invariant splitting of the tangent bundle, but one allows one or more of the subbundles involved to be such that the derivative exhibits mixed contracting/neutral/expanding behavior along them. Typically, these central (non-hyperbolic) subbundles concentrate the main dynamical difficulties: often one can project the dynamics along the remaining hyperbolic directions thus reducing the study to the central part.

Brin, Pesin [97] and Hirsch, Pugh, Shub [216] considered *partially hyperbolic systems*, that is, admitting an invariant splitting of the tangent bundle as the sum $E^s \oplus E^c \oplus E^u$ of three subbundles, where E^s is uniformly contracting, E^u is uniformly expanding, and the central direction E^c is "in between" the other two (one of the subbundles may be trivial). Examples include the time-1 map of an Anosov flow and the product of an Anosov diffeomorphism by the identity. In the context of the stability conjecture, Liao [261], Mañé [278], Pliss [359] were led to the more general notion of *dominated splitting* into two subbundles: one of them is definitely more contracted (or less expanded) than the other, after a uniform number of iterates.

We begin (Section B.1) by giving the precise definition of dominated splitting (or dominated decomposition) with any number of subbundles, and discussing some basic properties. Then we review some classical results from [97, 216] on the existence and regularity of the invariant foliations of partially hyperbolic systems, and discuss the possible existence of central foliations. Finally, in Section B.3 we adapt the definitions of these notions for flows.

B.1 Dominated splittings

B.1.1 Definition and elementary properties

Definition B.1. Let $f : M \to M$ be a diffeomorphism on a closed manifold M and K be any f-invariant set. A splitting $T_x M = E_1(x) \oplus \cdots \oplus E_k(x)$,

$x \in K$ of the tangent bundle over K is *dominated* if it is invariant under the derivative Df and there exists $\ell \in \mathbb{N}$ such that for every $i < j$, every $x \in K$, and every pair of unit vectors $u \in E_i(x)$ and $v \in E_j(x)$, one has

$$\frac{\|Df^\ell(x)u\|}{\|Df^\ell(x)v\|} < \frac{1}{2} \tag{B.1}$$

and the dimension of $E_i(x)$ is independent of $x \in K$ for every $i \in \{1, \ldots, k\}$.

We write $E_i \prec E_j$. In some contexts we need to specify the strength of the domination, measured by the number ℓ of iterates needed to get the inequality in the definition. Then we write $E_i \prec_\ell E_j$ and say that the splitting is ℓ-*dominated*.

The previous definition may be formulated, equivalently, as follows: there exist $C > 0$ and $\lambda < 1$ such that

$$\frac{\|Df^n(x)u\|}{\|Df^n(x)v\|} < c\lambda^n \quad \text{for all } n \geq 1. \tag{B.2}$$

When using this notion, λ and c may be taken as measures of the strength of the domination.

Let us list some useful elementary properties of dominated splittings, referring the reader to the end of this section for proofs. An important ingredient in the proofs of the properties *extension to a neighborhood* and *persistence* is a characterization of existence of a dominated splitting in terms of invariant cone-fields.

Uniqueness: The dominated splitting is unique if one fixes the dimensions of the subbundles: Let d be the dimension of the ambient manifold and let

$$d = n_1 + n_2 + \cdots + n_k .$$

Then there is at most one dominated splitting $T_K M = E_1 \oplus E_2 \oplus \cdots \oplus E_k$ over K with dimension of E_i equal to n_i for all i. It is clear that without the dimension requirement, there may be more than one dominated splitting. For instance, if $E^s \oplus E^c \oplus E^u$ is dominated then so are $E \oplus E^u$ and $E^s \oplus F$, where $E = E^s \oplus E^c$ and $F = E^c \oplus E^u$.

Transversality: The angles between any two subbundles of a dominated splitting are uniformly bounded from zero. In fact, the angle between the spaces $\oplus_{i \in I} E_i$ and $\oplus_{j \in J} E_j$ is uniformly bounded from zero, for any disjoint sets $I, J \subset \{1, \ldots, k\}$.

Continuity: Every dominated splitting is continuous, meaning that the subbundles $E_i(x)$ depend continuously on the point x.

Extension to the closure: Every ℓ-dominated splitting over a set K may be extended to an ℓ-dominated splitting over the closure of K. Even more, if g_n is a sequence of diffeomorphisms converging to f and K_n is a g_n-invariant set with an ℓ-dominated splitting, with dimensions independent of n, then the f-invariant set

$$K = \limsup_n K_n = \bigcap_N \overline{\bigcup_{n \geq N} K_n}$$

admits an ℓ-dominated splitting.

Extension to a neighborhood: Let K be an f-invariant set with a dominated splitting. Then such a splitting can be extended in a dominated way to the maximal invariant set of f in a neighborhood of K.

Persistence: Every dominated splitting persists under C^1-perturbations. More precisely, consider an f-invariant set K with an ℓ-dominated splitting. Then, for every $\varepsilon > 0$ there is a neighborhood U of K and a C^1-neighborhood $\mathcal{U}$ of f such that, for every $g \in \mathcal{U}$, the maximal g-invariant set in the closure of U admits an $(\ell - \varepsilon)$-dominated splitting [1], having the same dimensions of the initial dominated splitting over K.

Clustering: Suppose that $E_1 \oplus \cdots \oplus E_k$ is a dominated splitting and let $1 \leq i < j \leq k$, with either $i > 1$ or $j < k$. Then

$$E_1 \oplus \cdots \oplus E_{i-1} \oplus F \oplus E_{j+1} \oplus \cdots \oplus E_k\,,$$

where $F = E_i \oplus \cdots \oplus E_j$, is a dominated splitting.

Sub-splittings: Let $E_1 \oplus \cdots \oplus E_k$ be a dominated splitting and suppose that $E_j = F_1 \oplus \cdots \oplus F_m$ is a dominated splitting. Then

$$E_1 \oplus \cdots \oplus E_{j-1} \oplus F_1 \oplus \cdots \oplus F_m \oplus E_{j+1} \oplus \cdots \oplus E_k$$

also is a dominated splitting.

These properties motivate the following notion introduced in [73]:

Proposition B.2. *Let K be an f-invariant set admitting a dominated splitting. Then there is a unique* finest *dominated splitting $T_K M = E_1 \oplus \cdots \oplus E_k$ on K such that no E_i admits a dominated splitting. Any other invariant splitting is obtained by considering a partition of $\{1, \ldots, k\}$ into sub-intervals, and clustering the corresponding subbundles.*

The following notions were used in our study of robustly transitive dynamics given in Chapter 7:

Definition B.3. Let f be a diffeomorphism and K be an f-invariant set having a dominated splitting $T_K M = E_1 \oplus \cdots \oplus E_k$. We say that the splitting and the set K are

- *partially hyperbolic* if there is $n \geq 1$ such that the restriction of Df^n either contracts uniformly E_1 or expands uniformly E_k.

---

[1] By $(\ell - \varepsilon)$-dominated splitting we mean that $\|Df^\ell(x)u\|/\|Df^\ell(x)v\| < \frac{1}{2} + \varepsilon$.

- *volume hyperbolic* if there is $n \geq 1$ such that the Jacobian of $Df^n \mid E_1$ is strictly less than 1 (uniform volume contraction) and the Jacobian of $Df^n \mid E_k$ is strictly greater than 1 (uniform volume expansion).

We say that the diffeomorphism f is partially/volume hyperbolic if the ambient space M is a partially/volume hyperbolic set for f.

Remark B.4. An invariant set is volume hyperbolic if and only if its finest dominated splitting is volume hyperbolic. If a volume hyperbolic splitting has at least one of the extremal bundles E_1 or E_k of dimension 1 then it is partially hyperbolic.

Remark B.5. The notion of partial hyperbolicity is sometimes used in a much more restrictive sense. Many authors (e.g. Pugh, Shub [367]) require both extremal subbundles to be hyperbolic. Moreover, when defining domination one may ask that the maximal expansion of $Df^\ell \mid E_i$ at any point x of K be less than the minimal expansion of $Df^\ell \mid E_{i+1}$ at any point y of K. This corresponds to what Hirsch, Pugh, Shub [216] call *absolute* normal hyperbolicity, whereas the more general notion we use here corresponds to their *relative* normal hyperbolicity.

Just to give some flavor of the dynamical consequences of these properties, let us quote a result on bifurcations of invariant sets with a dominated splitting, from [74]: *Let f be a diffeomorphism having a dominated splitting into one-dimensional subbundles, defined on the whole ambient manifold. Then no diffeomorphism close to f exhibits homoclinic tangencies.* Section 7.2.3 and [74] explore other consequences.

Curiously, the following basic question, which had been raised already in [216, page 5], remains open in general [2]:

Problem B.6. Given a dominated splitting $T_K M = E \oplus F$ over an invariant set K, is there a continuous Riemannian metric over K which is adapted to the splitting in the sense that, for this norm, condition (B.2) holds with $c = 1$?

B.1.2 Proofs of the elementary properties:

We close this section with the proofs of the properties of dominated splittings stated near the beginning. Consider a diffeomorphism $f : M \to M$ on a closed manifold and let K be an f-invariant set.

---

[2]Nicolas Gourmelon has just announced a positive answer: *any dominated splitting admits adapted metrics.*

Transversality (two subbundles):

Assume that the tangent bundle over K contains two subbundles E and F, not necessarily continuous, and some $\ell \geq 1$ such that $E \prec_\ell F$ on K. We first show that the angle between E and F is uniformly bounded from zero. For doing that it is enough to prove that the difference $\|u - v\|$, $u \in E(x)$, $v \in F(x)$, $\|u\| = \|v\| = 1$, is larger than some constant independent of $x \in K$, u and v. Assume otherwise. Then there are sequences $u_n \in E(x_n)$ $v_n \in F(x_n)$ of unit vector with $u_n - v_n$ going to 0. As the derivative of f is bounded, one may find a sequence $m_n \to +\infty$ such that

$$\frac{1}{2} < \frac{\|Df^{m_n}(u_n)\|}{\|Df^{m_n}(v_n)\|} < 2,$$

which contradicts the ℓ-domination. This proves the first part of the transversality property.

Clustering:

Assume now that K contains 3 subbundles E, F, and G, and there exits $\ell \geq 1$ such that $E \prec_\ell F$ and $F \prec_\ell G$. As the angle between E and F is uniformly bounded from zero, by the previous argument, there is $\alpha > 0$ such that any unit vector $u \in E(x) \oplus F(x)$, $x \in K$ may be written as $u = u_E + u_F$ with $\|u_E\| + \|u_F\| < \alpha$. Fix $k \geq 1$ such that $2^k > 2\alpha$. Then, for any unit vector $v \in G(x)$,

$$\frac{\|Df^{k\ell}(u)\|}{\|Df^{k\ell}(v)\|} < \frac{1}{2^k}\alpha < \frac{1}{2}.$$

So we proved $(E \oplus F) \prec_{k\ell} G$. The same proof give that $E \prec (F \oplus G)$. Now a simple induction argument gives the complete clustering property.

Uniqueness:

Assume that $E \prec F$ and $G \prec H$ are two dominated splittings over K such that $\dim E \leq \dim G$. We argue, by contradiction, that $E \subset G$. Actually, this is a stronger result than what we claimed, useful for proving uniqueness of the finest dominated splitting. A simpler proof of the uniqueness statement we claimed is contained in the proof of the continuity and extension properties below.

So, assume $E(x) \not\subset G(x)$ for some $x \in K$ and consider some unit vector $u \in E(x) \setminus G(x)$. Write $u = u_G + u_H$ with $u_H \neq 0$. Then the positive iterates of u grow at the same rate as those of u_H. Write also $u_H = v_E + v_F$. If v_F did not vanish then the positive iterates of v_F would grow at the same rate as those of u_H, that is, at the same rate as the iterates of u, which would contradict $E \prec F$. Therefore, $v_F = 0$, which means that $u_H \in E(x) \cap H(x)$. Since $\dim E \leq \dim G$ and we are assuming that they do not coincide, there is some unit vector $w \in G(x) \setminus E(x)$. Write $w = w_E + w_F$ with $w_F \neq 0$. Then the positive iterates of $w \in G$ grow at the same rate as those of w_F, and so they grow exponentially faster than the iterates of u_H (because $u_H \in E(x)$). Since u_H is also in $H(x)$, this contradicts the domination $G \prec H$, and completes our argument.

This gives the uniqueness of a dominated splitting into two subbundles, once dimensions are fixed. To extend this to any dominated splitting $E_1 \prec E_2 \prec \cdots \prec E_k$ just use the clustering property: all the splittings $(E_1 \oplus \cdots \oplus E_i) \oplus (E_{i+1} \oplus \cdots \oplus E_k)$ are dominated and so they are unique, once the dimensions are fixed. By intersecting their subbundles, one recovers the initial splitting $E_1 \prec E_2 \prec \cdots, \prec E_k$ which is, therefore, unique.

Transversality (any number of subbundles):

Now we prove that the angle between $\oplus_{i \in I} E_i$ and $\oplus_{j \in J} E_j$ is uniformly bounded from zero, for any disjoint sets $I, J \subset \{1, \ldots, k\}$. In the case when I and J are intervals (of integers) this follows easily from the case of two subbundles, using the clustering property. In the general case, it is enough to prove that there are uniformly bounded linear maps sending the spaces E_i to spaces mutually orthogonal.

For that one uses that, by clustering, $E_1 \oplus \cdots \oplus E_i$ is dominated by E_{i+1} and $E_1 \oplus \cdots \oplus E_{i+1}$ is dominated by $E_{i+2} \oplus \cdots \oplus E_k$. Since the angles between these subbundles are bounded from 0, there is a bounded linear map P_i given by the identity on $E_1 \oplus \cdots \oplus E_i$ and on $E_{i+2} \oplus \cdots \oplus E_k$, and which maps E_i to the orthogonal space of $E_1 \oplus \cdots \oplus E_i$ in $E_1 \oplus \cdots \oplus E_{i+1}$. The composition $P_1 \circ \cdots \circ P_n$ has the announced property.

Continuity and extension to the closure:

Let $E \oplus F$ be an ℓ-dominated splitting over K. Let $x_n \in K$ be a sequence converging to some point x. Up to choosing subsequences one may assume that the spaces $E(x_n)$ and $F(x_n)$ converge to subspaces $E(x)$ and $F(x)$. A simple continuity argument show that, for any $k \in \mathbb{N}$, for any unit vectors $u \in E_x$ $v \in F(x)$ one has

$$\frac{\|Df^{k\ell}(u)\|}{\|Df^{k\ell}(v)\|} < \frac{1}{2^\ell} \quad \text{and} \quad \frac{\|Df^{-k\ell}(v)\|}{\|Df^{-k\ell}(u)\|} < \frac{1}{2^\ell}. \tag{B.3}$$

This directly proves that $E(x) \oplus F(x) = T_x M$. The first part of (B.3) characterizes $E(x)$ uniquely (once its dimension is fixed): the iterates of any unit vector u in $E(x)$ grow slower than those of any unit vector $w \notin E(x)$. In the same way, the second part of (B.3) characterizes $F(x)$ uniquely.

This now implies that, for any other sequence y_n converging to x, any accumulation point of $E(y_n)$ coincides with $E(x)$ and, in the same way, any accumulation point of $F(y_n)$ coincides with $F(x)$. As a consequence, one gets that the splitting on K extends by continuity to the closure of K. At the same time, this argument proves that the subbundles are continuous on K.

The assumption that the splitting is invariant on K implies that invariance on the closure as well. Finally, (B.3) shows that this invariant extension is still dominated.

The argument extends easily to any dominated splitting with any number of subbundles, using clustering: the previous argument shows that each splitting $(E_1 \oplus \cdots \oplus E_i) \oplus (E_{i+1} \oplus \cdots \oplus E_k)$, and by intersecting the subbundles in all these extensions one obtains the extension of the original splitting to the closure of K.

Extension to a neighborhood and persistence

Consider a ℓ-dominated splitting $E \oplus F$ on K. Extend it to a continuous splitting $TM = E \oplus F$ in a neighborhood U of K, not necessarily invariant. Then consider the cone fields $\mathcal{C}_\alpha$ on U consisting of all the vectors such that the norm of the component in the F-direction is larger than α times the norm of the component in the E-direction. The domination property implies that, for any $x \in K$, $Df^\ell(\mathcal{C}_1(x)) \subset \mathcal{C}_2(f^\ell(x))$. As a consequence, there is a neighborhood $V \subset U$ of K and a C^1-neighborhood $\mathcal{U}$ of f such that for any $g \in \mathcal{U}$ and any $x \in V$ one has $Dg^\ell(\mathcal{C}_1(x)) \subset \mathcal{C}_{2-\varepsilon}(g^\ell(x))$. This implies that the maximal invariant set of g in V has a dominated splitting with the same dimensions of the initial one, and with almost the same strength.

B.2 Invariant foliations

Let f be a C^r diffeomorphism on a closed manifold M admitting a dominated splitting $TM = E^u \oplus E^c \oplus E^s$ where Df is uniformly expanding along E^u and uniformly contracting along E^s, that is, f is partially hyperbolic.

Theorem B.7 (Brin, Pesin [97], Hirsch, Pugh, Shub [216]). *Assuming* $\dim E^u > 0$, *there exists a unique family* $\mathcal{F}^u$ *of injectively* C^r *immersed submanifolds* $\{\mathcal{F}^u(x) : x \in M\}$ *such that* $x \in \mathcal{F}^u(x)$ *and* $\mathcal{F}^u(x)$ *is tangent to* E^u_x *at every* $x \in M$. *This family is invariant, in the sense that* $f(\mathcal{F}^u(x)) = \mathcal{F}^u(f(x))$ *for all* $x \in K$, *and the leaves* $\mathcal{F}^u(x)$ *are uniformly contracted by some negative iterate of* f. *Moreover,* $\mathcal{F}^u$ *is a continuous lamination of* M, *meaning that every point has a continuous local chart which trivializes the leaves and is uniformly* C^r *along each leaf.*

$\mathcal{F}^u$ is the *strong unstable foliation* and its leaves $\mathcal{F}^u(x)$ are called strong unstable manifolds. Local strong unstable manifolds are neighborhoods of the point inside the corresponding leaf. By reversing time one gets strong stable foliations and (local) strong stable manifolds, assuming $\dim E^s > 0$.

More generally, strong stable and strong unstable foliations exist for any invariant subset $K \subset M$ admitting a dominated splitting as above. The family $\mathcal{F}^u$ is indexed by $x \in K$. These strong invariant foliations are continuous laminations, in the same sense as before.

Our knowledge of the properties of the central bundle E^c is quite different. Some of the most basic questions remain open, including the existence of a tangent foliation, in general. On the other hand, when they exist such central foliations usually exhibit strong rigidity. Let us discuss these issues in some more detail.

As we just mentioned, it is not known in general whether there is a foliation tangent to the central bundle E^c. When E^c has dimension 1, there always exists curves tangent to E^c passing through every point. That is because continuous vector fields always have solutions. But these curves need not be unique, because the central bundle is usually not Lipschitz continuous.

Problem B.8. Let f be a partially hyperbolic diffeomorphism on a compact manifold such that the central bundle E^c has dimension equal to 1. Is there some foliation tangent to E^c? Is this foliation unique? Is there some invariant foliation tangent to E^c? (clearly, invariance follows from uniqueness)

When the central bundle has dimension greater than 1 then it is, generally, not integrable. An example of robust non-integrability can be found in [197]. The construction starts from an example, in [418, Section 3.1], of an Anosov diffeomorphism in $\mathbb{T}^6$ whose hyperbolic splitting $TM = E^u \oplus E^s$ admits a refinement $E^{uu} \oplus E^{wu} \oplus E^{ws} \oplus E^{ss}$ where the weak invariant subbundles E^{wu} and E^{ws} are 2-dimensional and non-integrable. The author explains how this diffeomorphism can be deformed to make some of the weak subbundles non-hyperbolic while remaining non-integrable.

Problem B.9. 1. Build a robustly transitive partially hyperbolic diffeomorphism on a compact 3-dimensional manifold having open sets where no local surface is tangent to E^c.
  2. Build a robustly transitive diffeomorphism on a compact manifold with dimension larger than 3 having no invariant foliation.

On the other hand, most central foliations are structurally stable: actually, that is the case for all known examples. By structural stability, one means that given any nearby C^1 diffeomorphism f there exists a globally defined homeomorphism h_g sending leaves of the central foliation of f to leaves of the central foliation of g and such that $h_g \circ f \circ h_g^{-1}$ is isotopic to g along the leaves. This contrasts with the behavior of strong stable and strong unstable foliations, which are generally not robust.

The following general structural stability result assumes that the initial central foliation is *plaque expansive*: given ε there exists δ such that if x_n and y_n are δ-pseudo orbits with all jumps along the central leaves, and they remain within distance ε for all times, then x_n and y_n are in the same local central leaf for all n.

Theorem B.10 (Hirsch, Pugh, Shub [216]). *If f admits a plaque expansive central foliation $\mathcal{F}^c$ then $(f, \mathcal{F}^c)$ is structurally stable. The hypothesis of plaque expansiveness holds, in particular, if $\mathcal{F}^c$ is of class C^1.*

It is not clear whether differentiability is really necessary in the theorem:

Problem B.11. Is every central foliation of a transitive partially hyperbolic diffeomorphism plaque expansive?

The same questions may be asked about *center stable* and *center unstable* foliations, tangent to the sums $E^c \oplus E^s$ and $E^c \oplus E^u$, respectively. When both invariant foliations $\mathcal{F}^{cs}$ and $\mathcal{F}^{cu}$ do exist, the intersections of leaves define an invariant central foliation $\mathcal{F}^c$, and one says that the system is *dynamically coherent*. Recently, Brin, Burago, Ivanov [95] showed that *for partially hyperbolic diffeomorphisms on 3-dimensional manifolds, with 3 invariant subbundles, existence of a unique central foliation implies dynamical coherence.*

The ergodic properties of orbits propagate along stable and unstable sets, since forward time averages are constant on stable sets and backward time averages are constant on unstable sets. In order to use this idea, often called the Hopf argument, one needs the holonomies of foliations to preserve zero measure sets. This *absolute continuity* property does hold for the strong stable and strong unstable foliations of partially hyperbolic C^2 diffeomorphisms [97]. We discuss this issue in much more generality in Appendix C.3.

In contrast, even when there exists a unique central foliation, it needs not be absolutely continuous. Open sets of partially hyperbolic diffeomorphisms with such pathological central foliations were first exhibited by Shub, Wilkinson [412]. See Section 7.3.1. Their construction also shows that neither $\mathcal{F}^{cu}$ nor $\mathcal{F}^{cs}$ are absolutely continuous, in general.

B.3 Linear Poincaré flows

In the context of flows there are two related, but not quite equivalent, notions of dominated splitting. On the one hand, there is the direct adaptation of the corresponding notion for diffeomorphisms: a splitting $T_\Lambda M = E \oplus F$ of the tangent bundle over an invariant set Λ is dominated if it is invariant under the derivative DX^t of every flow map X^t, and there is $T > 0$ such that

$$\|(DX^T \mid E(x))^{-1}\| \, \|DX^T \mid F(x)\| \leq 1/2 \quad \text{for every } x \in \Lambda. \tag{B.4}$$

Recall that we always assume the dimensions of the subbundles to be constant. Then the decomposition is automatically continuous. Partial hyperbolicity is defined by asking, in addition, that one of the subbundles be hyperbolic: either E is uniformly expanding, $\|(DX^T \mid E(x))^{-1}\| \leq 1/2$ at every x, or F is uniformly contracting, $\|DX^T \mid F(x)\| \leq 1/2$ at every x.

These notions share the main properties of its discrete time counterpart, discussed in the first section of this appendix. In particular, a dominated splitting can always be extended to the closure. Recall also that, following [308, 306], we call a compact invariant set Λ of a 3-dimensional flow singular hyperbolic if it partially hyperbolic in the previous sense, and the central bundle is volume expanding or volume contracting.

However, the following example, suggested by Enrique Pujals, shows that robustly transitive flows in high dimensions need not have a dominated splitting (compare Theorem 9.12):

Example B.12. If $f : M \to M$ is a robustly transitive diffeomorphism, then the flow X obtained by suspending f is also robustly transitive: a C^1-perturbation of X induces a Poincaré map on the fiber which is C^1-close to f and, hence, transitive. Theorem 7.5 shows that, since f is robustly transitive, it admits a dominated splitting. However, [84] exhibits robust examples of transitive diffeomorphisms of $\mathbb{T}^4$ whose finest dominated splitting is $E \prec F$, where neither of the subbundles E and F is contracting nor expanding: some periodic orbits have expanding eigendirections contained in E, and other periodic orbits have contracting eigendirections contained in F. Consequently, the suspension of f is a robustly transitive flow which does not admit any dominated splitting: the direction of the vector field is neither contracting nor expanding, and so it is not dominated by E nor by F.

In contrast, Vivier [444] shows that, for vector fields in any dimension, a robustly transitive set, either with or without singularities, always admits a dominated splitting for the linear Poincaré flow (Theorem 9.13).

The *linear Poincaré flow* of a vector field X is defined over the set of regular points $R(X) = \{x \in M : X(x) \neq 0\}$, as follows. Let $N_{R(X)}$ be the normal subbundle of $T_{R(X)}M$, defined by $N_x = \{X(x)\}^{\perp}$, and let $\tilde{T}_{R(X)}M$ be the quotient vector bundle defined by $\tilde{T}_x M = T_x M / \{X(x)\}$. Since the line bundle defined by $\{X(x)\}$ is invariant under the flow DX^t, there is an induced flow

$$\tilde{D}X^t : \tilde{T}_x M \to \tilde{T}_{X^t(x)}M$$

on the quotient bundle. Via the natural isomorphism $\eta_y : N_y \to \tilde{T}_y M$, the flow $\tilde{D}X^t$ corresponds to the *linear Poincaré flow* $P^t : N_x \to N_{X^t(x)}$ on the normal subbundle. The normal subbundle N_x is endowed with the metric [3] induced from the Riemannian metric of M on $X(x)^{\perp}$.

Given an invariant set $\Lambda \subset R$, not necessarily compact, we call *dominated splitting for the linear Poincaré flow* over Λ any decomposition $N_\Lambda = N^1 \oplus N^2$ of the normal subbundle restricted to Λ which is invariant under the flow P^t and such that there exist constants $C > 0$ and $\lambda < 1$ such that

$$\|(P^t \mid N_x^1)^{-1}\|\, \|P^t \mid N_x^2\| \leq C\lambda^t \quad \text{for every } x \in \Lambda \text{ and } t > 0. \tag{B.5}$$

Moreover, we say that Λ is *hyperbolic for the linear Poincaré flow* if the decomposition may be taken such that N^1 is expanding and N^2 is contracting:

---

[3] The choice of the metric is relevant here, because the normal subbundle is only defined in the complement of the singularities, which is in an open set.

$$\|(P^t \mid N_x^1)^{-1}\| \leq C\lambda^t \text{ and } \|P^t \mid N_x^2\| \leq C\lambda^t \quad \text{for } x \in \Lambda \text{ and } t > 0.$$

Proposition 1.1 in [162] states that a *compact* invariant set $\Lambda \subset R(X)$ is hyperbolic for the linear Poincaré flow P^t if and only it is hyperbolic for X^t, in the usual sense, that is, if and only if there exists an invariant splitting $T_\Lambda M = E^u \oplus E_X \oplus E^s$ such that E^u is expanding, E^s is contracting, and E_X has the direction of the flow. Moreover, Proposition 1.8 in [306] shows that if Λ is a (compact) singular hyperbolic set of a 3-dimensional vector field X, then $\Lambda \cap R(X)$ is hyperbolic for the linear Poincaré flow of X. We know from Chapter 9 that Λ needs not be hyperbolic for X, since it may contain singularities. A stronger fact is that a singular attractor (containing some singularity), such as the geometric Lorenz attractors, has no invariant 1-dimensional subbundle corresponding to the unstable direction of its linear Poincaré flow. However, the combination of [306, Proposition 1.8] with [162, Proposition 1.1] does imply the first part of

Corollary B.13. *Let X be a 3-dimensional vector field and Λ be a singular hyperbolic set. If Λ contains no singularities, then Λ is a hyperbolic set for X. In general, any compact invariant subset of Λ containing no singularities (e.g. any regular periodic orbit) is a hyperbolic set.*

The last part follows by noting that, clearly, any compact invariant subset of a singular hyperbolic set is also singular hyperbolic.

C

Non-Uniformly Hyperbolic Theory

Oseledets' theorem (Section C.1) states that, relative to any invariant probability μ, for almost every point x and any tangent vector v the norm of the iterates $Df^n(x)v$ has a well defined rate of exponential growth or decay as $n \to \pm\infty$, called the Lyapunov exponent. This immediately leads to the following notion: the system (f, μ) is *non-uniformly hyperbolic* if all Lyapunov exponents are non-zero almost everywhere. Thus, there exists an invariant splitting $T_x M = E_x^u \oplus E_x^s$ defined at almost every point x, such that E_x^u is expanding and E_x^s is contracting for Df, at non-uniform rates.

A theory initiated by Pesin [355] provides fundamental geometric information on non-uniformly hyperbolic systems, especially existence of stable and unstable manifolds at almost every point (Section C.2) which form absolutely continuous invariant laminations (Section C.3). Invariant measures with absolutely continuous conditional probabilities along unstable (or stable) manifolds play a key role in the theory, as they include most SRB measures as well as Gibbs states in Chapter 11. See Section C.4. The notion of conditional measures is founded on Rokhlin's disintegration theorem, for which we give an elementary proof in Section C.6.

For more information, see Mañé [281, Chapter 4], Pollicott [362], Katok, Hasselblatt [232], or Barreira, Pesin [46].

C.1 The linear theory

Let $f : M \to M$ be an invertible measurable transformation, and μ be an f-invariant probability on M. Let $\pi : \mathcal{E} \to M$ be a finite-dimensional vector bundle over M, endowed with a measurable Riemannian metric $\| \cdot \|$. Let $F : \mathcal{E} \to \mathcal{E}$ be a measurable vector bundle automorphism over f: this means that $\pi \circ F = f \circ \pi$ and the action $F_x : \mathcal{E}_x \to \mathcal{E}_{f(x)}$ of F on each fiber $\mathcal{E}_x = \pi^{-1}(x)$ is a linear isomorphism.

Oseledets ergodic theorem states that, under an integrability condition, almost every fiber splits as a direct sum of subspaces such that the iterates

$F_x^n : \mathcal{E}_x \to \mathcal{E}_{f^n(x)}$ have well-defined rates of exponential growth, in norm, restricted to each subspace:

Theorem C.1 (Oseledets [334], Ruelle [392]). *Assume* $\log^+ \|F_x\|$ *and* $\log^+ \|F_x^{-1}\|$ *are integrable for* μ. *For* μ-*almost all* $x \in M$, *there exist* $k(x) \in \mathbb{N}$, *real numbers* $\hat{\lambda}_1(x) > \cdots > \hat{\lambda}_{k(x)}$, *and a splitting* $\mathcal{E}_x = E_x^1 \oplus \cdots \oplus E_x^{k(x)}$, *such that* $F_x(E_x^i) = E_{f(x)}^i$ *for* $1 \le i \le k(x)$, *and*

$$\lim_{n \to \pm\infty} \frac{1}{n} \log \|F_x^n(v_i)\| = \hat{\lambda}_i(x) \quad \text{for every non-zero } v_i \in E_x^i.$$

The Lyapunov exponents $\hat{\lambda}_i(x)$, *and the Oseledets subspaces* E_x^i *are unique* μ-*almost everywhere, and they depend measurably on* x.

In most situations we are interested in, M is a compact space and the transformation f, the vector bundle $\pi : \mathcal{E} \to M$, the automorphism F, and the norm $\|\cdot\|_x$ are all continuous. Then the integrability hypothesis is automatically satisfied, and the Lyapunov exponents and Oseledets subspaces do not depend on the actual choice of the norm. An important particular case: $f : M \to M$ is a C^1 diffeomorphism on a compact Riemannian manifold M, $\mathcal{E} = TM$ is the tangent bundle, and $F = Df$ is the derivative of f.

Theorem C.1 extends to the case when f is non-invertible: instead of a splitting one gets a filtration $\mathcal{E}_x = E_x^1 > \cdots > E_x^{k(x)} > 0 = E_x^{k(x)+1}$ of the fiber into vector subspaces such that

$$\lim_{n \to +\infty} \frac{1}{n} \log \|F_x^n(v_i)\| = \hat{\lambda}_i(x) \quad \text{for every } v_i \in E_x^i \setminus E_x^{i+1}.$$

There is also a continuous time version of the theorem, where one considers a flow (or semi-flow) $(F^t)_{t \in \mathbb{R}} : \mathcal{E} \to \mathcal{E}$ of vector bundle automorphisms over a flow (or semi-flow) $(f^t)_{t \in \mathbb{R}} : M \to M$.

An outline of Mañé's proof [281] of Theorem C.1 follows. It is no restriction to suppose that μ is ergodic. The largest Lyapunov exponent is the growth rate of the norm over the whole fiber:

$$\hat{\lambda}_1(x) = \limsup_{n \to +\infty} \frac{1}{n} \log \|F_x^n\|$$

and E_x^1 is the subset of vectors $v \in \mathcal{E}_x$ realizing this growth rate, in the sense that

$$\liminf_{n \to -\infty} \frac{1}{n} \log \|F_x^n(v)\| \ge \hat{\lambda}_1(x). \tag{C.1}$$

Then E_x^1 is a vector subspace, the limit actually exists for both $n \to \pm\infty$, and it coincides with $\hat{\lambda}_1(x)$. One key step in the proof is to show that the dimension of E_x^1 is positive at μ-almost every point. This is done as follows.

Fix $\varepsilon > 0$. Given $N \ge 1$, consider the set $Y_{\varepsilon,N}$ of points x such that

$$\log \|F_x^n(u)\| \le n(\hat{\lambda}_1(x) - \varepsilon)$$

for some non-zero vector u and for all $0 > n > -N$. Using the definition of $\hat{\lambda}_1(x)$ and Pliss Lemma 11.5, one finds that $\mu(Y_{\varepsilon,N}) \ge \delta$ where $\delta > 0$ is independent of N. Consequently, $\mu(\cap_N Y_{\varepsilon,N}) \ge \delta$. Thus, there is a positive measure set of points such that

$$\liminf_{n \to -\infty} \frac{1}{n} \log \|F_x^n(u)\| \ge \hat{\lambda}_1(x) - \varepsilon$$

for some non-zero vector u. By ergodicity, this (invariant) set has full measure. Making $\varepsilon \to 0$, it follows that $\dim E_x^1 > 0$ at μ-almost every point.

Another key step in the proof of Theorem C.1 is a graph-transform construction of a measurable subbundle $\mathcal{E}'$, invariant under F and such that $\mathcal{E} = \mathcal{E}' \oplus E^1$. The proof proceeds by induction on the dimension of the vector bundle, considering the restriction of F to $\mathcal{E}'$.

Remark C.2. In general, the angles between Oseledets subspaces are not bounded from zero, not even over individual orbits. However, angles may decrease at most sub-exponentially fast: for μ-almost x and any disjoint subsets I, J of $\{1, \dots, k(x)\}$,

$$\lim_{n \to \pm\infty} \frac{1}{n} \log \text{angle} \left(\oplus_{i \in I} E_{f^n(x)}^i, \oplus_{j \in J} E_{f^n(x)}^j \right) = 0.$$

One consequence is the following useful relation: almost everywhere,

$$\sum_{j=1}^{k(x)} \hat{\lambda}_i(x) \dim E_x^i = \lim_{n \to \pm\infty} \frac{1}{n} \log |\det(F_x^n)|.$$

C.2 Stable manifold theorem

Let $f : M \to M$ be a C^1 diffeomorphism on a compact manifold, and let $\mathcal{E} = TM$ and $F = Df$. The *stable set* of a point $x \in M$ is defined by

$$W^s(x) = \{y \in M : d(f^n(x), f^n(y)) \to 0 \text{ as } n \to +\infty\}$$

The unstable set $W^u(x)$ is defined analogously, with f^{-1} in the place of f. For $x \in M$ as in Theorem C.1, define

$$E_x^s = \oplus_{\hat{\lambda}_j(x) < 0} E_x^j \quad \text{and} \quad E_x^u = \oplus_{\hat{\lambda}_j(x) > 0} E_x^j.$$

Moreover, let $\tau_x > 0$ be smaller than the norm of every non-zero Lyapunov exponent at x. Pesin's stable manifold theorem states that x has a *local stable manifold* which is a smooth disk tangent to E_x^s at x and whose points approach the orbit of x exponentially fast:

Theorem C.3 ([179, 355, 366, 392]). *Assume the derivative of f is Hölder continuous. Then there exists a C^r embedded disk $W^s_{loc}(x)$ through x and there exists $C_x > 0$ such that*

1. *$W^s_{loc}(x)$ is tangent to E^s_x at x;*
2. *$d(f^n(x_1), f^n(x_2)) \leq C_x e^{-n\tau_x} \operatorname{dist}(x_1, x_2)$ for all $x_1, x_2 \in W^s_{loc}(x)$;*
3. *$f(W^s_{loc}(x)) \subset W^u_{loc}(f(x))$;*
4. *if all Lyapunov exponents are non-zero, $W^s(x) = \bigcup_{n=0}^{\infty} f^{-n}(W^s_{loc}(f^n(x)))$.*

Analogously, x has a *local unstable manifold* $W^u_{loc}(x)$ satisfying corresponding properties with f^{-1} in the place of f.

An important difficulty is that these manifolds may be arbitrarily small, and they usually vary discontinuously with the point x. However, the local stable manifold and local unstable manifold do depend measurably on x, as C^1 embedded disks, and the constants C_x, τ_x may also be chosen depending measurably on the point. One way to understand this statement is that one may find compact *hyperbolic blocks* $\mathcal{H}(\kappa)$, $\kappa \in \mathbb{N}$, such that

- $C_x < \kappa$ and $\tau_x > 1/\kappa$ for every $x \in \mathcal{H}(\kappa)$;
- $\mathcal{H}(\kappa) \subset \mathcal{H}(\kappa + 1)$ for all $\kappa \in \mathbb{N}$ and $\mu(\mathcal{H}(\kappa)) \to 1$ as $\kappa \to \infty$.
- the C^1 disks $W^s_{loc}(x)$ and $W^u_{loc}(x)$ vary continuously with x in $\mathcal{H}(\kappa)$.

In particular, the sizes of $W^s_{loc}(x)$ and $W^u_{loc}(x)$ are uniformly bounded from zero on each $x \in \mathcal{H}(\kappa)$, and so is the angle between the two disks.

C.3 Absolute continuity of foliations

Most important for the applications, the holonomy maps associated to the Pesin stable lamination $\mathcal{W}^s = \{W^s_{loc}(x)\}$ preserve zero Lebesgue measure sets. More precisely, let x be in some hyperbolic block $\mathcal{H}(\kappa)$ and x_1, x_2 be points of $W^s_{loc}(x)$ close to x. Let Σ_1 and Σ_2 be small smooth disks transverse to $W^s_{loc}(x)$ at x_1 and x_2:

$$T_{x_i}\Sigma_i \oplus T_{x_i}W^s_{loc}(x) = T_{x_i}M \quad \text{for } i = 1, 2.$$

Given any point $y \in \mathcal{H}(\kappa)$ close to x, the local stable manifold $W^s_{loc}(y)$ intersects each cross section Σ_i in exactly one point y_i. This defines a holonomy map

$$\pi^s : y_1 \to y_2$$

which is a homeomorphism between subsets of Σ_1 and Σ_2.

Theorem C.4 (Pesin [355, 366]). *Assume the derivative of f is Hölder continuous. Then Pesin's stable lamination $\mathcal{W}^s$ is absolutely continuous: every holonomy map π^s as before maps zero Lebesgue measure subsets of Σ_1 to zero Lebesgue measure subsets of Σ_2. A dual statement holds for the unstable lamination $\mathcal{W}^u$.*

This fundamental property was first observed by Anosov [16], for the class of systems named after him. Even in that uniformly hyperbolic setting the statement is not true, in general, if one drops the assumption of Hölder continuity of the derivative [383].

C.4 Conditional measures along invariant foliations

Let κ and $x \in \mathcal{H}(\kappa)$ be fixed. For each small $\delta > 0$, let $\mathcal{B}_\delta^s(\kappa, x)$ be the union of the local stable manifolds through points $y \in \mathcal{H}(\kappa)$ in the δ-neighborhood of x. Let $\nu = \nu(\kappa, x, \delta)$ be the restriction of μ to $\mathcal{B}_\delta^s(\kappa, x)$.

Let $\mathcal{P}$ be the coarsest partition of $\mathcal{B}_\delta^s(\kappa, x)$ such that every local stable manifold $W_{loc}^s(y)$ is entirely contained in some atom of $\mathcal{P}$. Then $\mathcal{P}$ is a *measurable partition* in the sense of Rokhlin [385, 386]: it may be written as the product

$$\mathcal{P} = \bigvee_{n=1}^{\infty} \mathcal{P}_n \tag{C.2}$$

of an increasing sequence of *finite* partitions $\mathcal{P}_n$ of $\mathcal{B}_\delta^s(\kappa, x)$.

Let $\hat{\nu}$ be the projection of ν to $\mathcal{P}$, that is, the measure defined in $\mathcal{P}$ by $\hat{\nu}(Q) = \nu(\pi^{-1}(Q))$, where π is the canonical map $x \mapsto \mathcal{P}(x)$ assigning to each point x the atom of $\mathcal{P}$ that contains it. By the Rokhlin disintegration theorem, there exists a family $\{\nu_P : P \in \mathcal{P}\}$ of *conditional probability measures* satisfying

1. $\nu_P(P) = 1$ for $\hat{\nu}$-almost every $P \in \mathcal{P}$;
2. $\nu(E) = \int \nu_P(E) \, d\hat{\nu}(P)$ for every measurable subset E,

It is not difficult to see that such a family is essentially unique: given any other choice $\{\nu_P' : P \in \mathcal{P}\}$, we have $\nu_P = \nu_P'$ for $\hat{\nu}$-almost every P.

The precise statement and proof of the disintegration theorem are given in Section C.6. Here we just give a quick outline. One way to construct the family of conditional measures is as follows. For each $P \in \mathcal{P}$ and continuous function φ define

$$e_n(\varphi)(P) = \frac{1}{\nu(P_n)} \int_{P_n} \varphi \, d\nu$$

where P_n is the element of the partition $\mathcal{P}_n$ that contains P. We may restrict to the case when the denominator is non-zero, for the union of all $P \in \mathcal{P}$ such that $\nu(P_n) = 0$ for some n has zero measure. The main step is to show that for each functions φ the sequence $e_n(\varphi)(P)$ converges at $\hat{\nu}$-almost every P. Let $e(\varphi)(P)$ be the limit. Using the fact that the space of continuous function has a countable dense subset (see [389]), there is a full ν-measure subset of $P \in \mathcal{P}$ such that $e(\varphi)(P)$ is defined for all continuous functions φ. By the Riesz-Markov theorem, each $e(\cdot)(P)$ corresponds to a probability ν_P on M, and one checks that these probabilities satisfy (1) and (2) above.

Definition C.5. An invariant measure μ is *absolutely continuous along stable manifolds* if, given any $\mathcal{B}_\delta^s(\kappa, x)$ as above, the conditional measure ν_P is absolutely continuous with respect to Lebesgue measure in P for $\hat{\mu}$-almost every $P \in \mathcal{P}$.

This makes sense because the atoms $P \in \mathcal{P}$ are embedded submanifolds. One defines absolute continuity along unstable manifolds in the same way.

Lebesgue measure is always absolutely continuous along stable manifolds and along unstable manifolds. This is a consequence of the absolute continuity Theorem C.4.

C.5 Local product structure

The technical notion of measure with local product structure plays a significant role in the theory of non-uniformly hyperbolic systems, and was used, for instance, in Chapter 12. Let μ be a hyperbolic measure for a diffeomorphism f, that is, an invariant probability such that all Lyapunov exponents $\hat{\lambda}_j(x)$ are non-zero at μ-almost every point x. Assume also that μ is non-atomic. We say that μ has local product structure if the stable holonomy maps (respectively unstable holonomy maps) are absolutely continuous relative to the conditional measures along leaves of the unstable lamination (respectively the stable lamination). Let us state this in a more formal way.

Let $\mathcal{H}(\kappa)$ be a hyperbolic block and $x, y \in \mathcal{H}(\kappa)$ be such that y is in the closed δ-neighborhood $B_\delta(x)$ of x. If δ is small enough, $W_{loc}^s(y)$ intersects $W_{loc}^u(x)$ exactly in one point, and analogously for $W_{loc}^u(y)$ and $W_{loc}^s(x)$. Let

$$\mathcal{H}_\delta^u(\kappa, x) \subset W_{loc}^s(x) \quad \text{and} \quad \mathcal{H}_\delta^s(\kappa, x) \subset W_{loc}^s(x)$$

be the (compact) intersection sets obtained in this way. Reducing δ if necessary, $W_{loc}^s(\xi) \cap W_{loc}^u(\eta)$ consists of exactly one point $[\xi, \eta]$, for every $\xi \in \mathcal{H}_\delta^u(\kappa, x)$ and $\eta \in \mathcal{H}_\delta^s(\kappa, x)$. Let $\mathcal{H}_x(\kappa, \delta)$ be the set of such points $[\xi, \eta]$. By construction, $\mathcal{H}_x(\kappa, \delta)$ contains $\mathcal{H}(\kappa) \cap B_\delta(x)$ and its diameter goes to zero when $\delta \to 0$. Moreover, $\mathcal{H}_x(\kappa, \delta)$ is homeomorphic to $\mathcal{H}_\delta^u(\kappa, x) \times \mathcal{H}_\delta^s(\kappa, x)$ via

$$(\xi, \eta) \mapsto [\xi, \eta] \,.$$

Definition C.6. A hyperbolic non-atomic measure μ has *local product structure* if, for μ-almost every point x and every small $\delta > 0$ as before, the restriction $\nu = \mu \mid \mathcal{H}_\delta(\kappa, x)$ is equivalent to the product measure $\nu^u \times \nu^s$, where ν^u and ν^s are the projections of ν to $\mathcal{H}_\delta^u(\kappa, x)$ and $\mathcal{H}_\delta^s(\kappa, x)$, respectively.

In particular, Lebesgue measure always has local product structure if it is invariant and hyperbolic. More generally, using Theorem C.4, every hyperbolic invariant measure which is absolutely continuous along stable manifolds or along unstable manifolds has local product structure.

C.6 The disintegration theorem

Let Z be a compact metric space, μ be a Borel probability measure on Z, and $\mathcal{P}$ be a partition of Z into measurable subsets. Let $\pi : Z \to \mathcal{P}$ be the map associating to each $z \in Z$ the atom $P \in \mathcal{P}$ that contains it. By definition, $\mathcal{Q}$ is a measurable subset of $\mathcal{P}$ if and only if $\pi^{-1}(\mathcal{Q})$ is a measurable subset of Z. Let $\hat{\mu}$ be the push-forward of μ under π, in other words, $\hat{\mu}$ is the probability measure on $\mathcal{P}$ defined by $\hat{\mu}(\mathcal{Q}) = \mu(\pi^{-1}(\mathcal{Q}))$ for every measurable set $\mathcal{Q} \subset \mathcal{P}$.

Definition C.7. A *system of conditional measures of μ with respect to $\mathcal{P}$* is a family $(\mu_P)_{P \in \mathcal{P}}$ of probability measures on Z such that

1. $\mu_P(P) = 1$ for $\hat{\mu}$-almost every $P \in \mathcal{P}$;
2. given any continuous $\varphi : Z \to \mathbb{R}$, the function $\mathcal{P} \ni P \mapsto \int \varphi \, d\mu_P$ is measurable and $\int \varphi \, d\mu = \int \left(\int \varphi \, d\mu_P \right) d\hat{\mu}(P)$.

Lemma C.8. *If $(\mu_P)_{P \in \mathcal{P}}$ is a system of conditional measures of μ relative to the partition $\mathcal{P}$, then $\mathcal{P} \ni P \mapsto \int \psi \, d\mu_P$ is measurable and $\int \psi \, d\mu = \int \left(\int \psi \, d\mu_P \right) d\hat{\mu}(P)$, for any bounded measurable function $\psi : Z \to \mathbb{R}$.*

Proof. The class of functions that satisfy the conclusion of the lemma contains all the continuous functions, and is closed under dominated pointwise convergence. Therefore, it contains all bounded measurable functions. □

In particular, $P \mapsto \mu_P(E)$ is measurable, and $\mu(E) = \int \mu_P(E) \, d\hat{\mu}(P)$, for any measurable set $E \subset Z$.

Conditional measures, when they exist, are unique almost everywhere:

Proposition C.9. *If $(\mu_P)_{P \in \mathcal{P}}$ and $(\nu_P)_{P \in \mathcal{P}}$ are two systems of conditional measures of μ with respect to $\mathcal{P}$, then $\mu_P = \nu_P$ for $\hat{\mu}$-almost every $P \in \mathcal{P}$.*

Proof. Suppose otherwise, that is, there exists a measurable set $\mathcal{Q}_0 \subset \mathcal{P}$ with $\hat{\mu}(\mathcal{Q}_0) > 0$ such that $\mu_P \neq \nu_P$ for every $P \in \mathcal{Q}_0$. Let $\{\varphi_k : k \in \mathbb{N}\}$ be a countable dense subset of $C^0(Z, \mathbb{R})$, and define

$$A_k = \{P \in \mathcal{Q}_0 : \int \varphi_k \, d\mu_P \neq \int \varphi_k \, d\nu_P\}.$$

Noting that $\cup_k A_k = \mathcal{Q}_0$, there exists $\varphi \in C^0(Z, \mathbb{R})$ and a subset $\mathcal{Q}$ of $\mathcal{Q}_0$ such that $\hat{\mu}(\mathcal{Q}) > 0$ and (interchanging the roles of μ_P and ν_P, if necessary) $\int \varphi \, d\mu_P > \int \varphi \, d\nu_P$ for every $P \in \mathcal{Q}$. Then

$$\int_{\mathcal{Q}} \left(\int \varphi \, d\mu_P \right) d\hat{\mu}(P) > \int_{\mathcal{Q}} \left(\int \varphi \, d\nu_P \right) d\hat{\mu}(P). \tag{C.3}$$

On the other hand, by Lemma C.8,

$$\int (\varphi \mathcal{X}_{\pi^{-1}(\mathcal{Q})}) \, d\mu = \int \left(\int (\varphi \mathcal{X}_{\pi^{-1}(\mathcal{Q})}) \, d\mu_P \right) d\hat{\mu}(P).$$

By assumption $\mu_P(P) = 1$ for $\hat{\mu}$-almost every $P \in \mathcal{P}$. For any such P, we have

$$\int (\varphi \mathcal{X}_{\pi^{-1}(\mathcal{Q})}) \, d\mu_P = \mathcal{X}_{\mathcal{Q}}(P) \int \varphi \, d\mu_P.$$

Therefore,

$$\int (\varphi \mathcal{X}_{\pi^{-1}(\mathcal{Q})}) \, d\mu = \int \left(\mathcal{X}_{\mathcal{Q}}(P) \int \varphi \, d\mu_P \right) d\hat{\mu}(P) = \int_{\mathcal{Q}} \left(\int \varphi \, d\mu_P \right) d\hat{\mu}(P).$$

Analogously, we find

$$\int (\varphi \mathcal{X}_{\pi^{-1}(\mathcal{Q})}) \, d\mu = \int_{\mathcal{Q}} \left(\int \varphi \, d\nu_P \right) d\hat{\mu}(P).$$

These two last equalities contradict (C.3). Therefore, $\mu_P = \nu_P$ for $\hat{\mu}$-almost every P, as claimed.   □

Definition C.10. $\mathcal{P}$ is a *measurable partition* if there exist measurable subsets $E_1, E_2, \ldots, E_n, \ldots$ of Z such that

$$\mathcal{P} = \{E_1, Z \setminus E_1\} \vee \{E_2, Z \setminus E_2\} \vee \cdots \vee \{E_n, Z \setminus E_n\} \vee \cdots \qquad \text{mod } 0.$$

In other words, there exists some full μ-measure subset $F_0 \subset Z$ such that, given any atom P of $\mathcal{P}$ we may write

$$P \cap F_0 = E_1^* \cap E_2^* \cap \cdots \cap E_n^* \cap \cdots \cap F_0 \qquad (C.4)$$

where E_j^* is either E_j or its complement $Z \setminus E_j$, for every $j \geq 1$.

Example C.11. Every finite or countable partition is a measurable partition. In fact, $\mathcal{P}$ is measurable if and only if there exists a non-decreasing sequence of finite or countable partitions $\mathcal{P}_1 \prec \mathcal{P}_2 \prec \cdots \prec \mathcal{P}_n \prec \cdots$ such that $\mathcal{P} = \vee_{n=1}^{\infty} \mathcal{P}_n$ mod 0.

Example C.12. Let $Z = X \times Y$, where X and Y are compact metric spaces, and $\mathcal{P}$ be the partition of Z into horizontal lines $X \times \{y\}$, $y \in Y$. Then $\mathcal{P}$ is a measurable partition of Z.

Theorem C.13 (Rokhlin [385, 386]). *If $\mathcal{P}$ is a measurable partition, then there exists some system of conditional measures of μ relative to $\mathcal{P}$.*

Proof. For the purpose of the conclusion of the theorem, we may replace the space Z by any full measure subset. So, it is no restriction to suppose that the set F_0 in (C.4) actually coincides with Z, and we do so in all that follows. Let ψ be any bounded measurable real function on Z. For each $n \geq 1$ let

$$\mathcal{P}_n = \{E_1, Z \setminus E_1\} \vee \{E_2, Z \setminus E_2\} \vee \cdots \vee \{E_n, Z \setminus E_n\}$$

that is, $\mathcal{P}_n$ is the partition of Z whose atoms are the sets $E_1^* \cap \cdots \cap E_n^*$, with $E_j^* = E_j$ or $E_j^* = Z \setminus E_j$, for each $1 \leq j \leq n$. Define $\tilde{\psi}_n : Z \to \mathbb{R}$ as follows. If the atom $P_n(z)$ of $\mathcal{P}_n$ that contains z has positive μ-measure, then

$$\tilde{\psi}_n(z) = \frac{1}{\mu(P_n(z))} \int_{P_n(z)} \psi \, d\mu. \qquad (C.5)$$

Otherwise, $\tilde{\psi}_n(z) = 0$. Clearly, the second case in the definition of $\tilde{\psi}_n$ applies only to a zero μ-measure set of points.

Lemma C.14. *Given any bounded measurable function $\psi : Z \to \mathbb{R}$, there exists a full μ-measure subset $F = F(\psi)$ of Z such that $\tilde{\psi}_n(z)$, $n \geq 1$, converges to some real number $\tilde{\psi}(z)$, for every $z \in F$.*

Proof. We may always write $\psi = \psi^+ - \psi^-$, where $\psi^\pm$ are measurable, bounded, and non-negative: for instance, $\psi^\pm = (|\psi| \pm \psi)/2$. Then $\tilde{\psi}_n = \tilde{\psi}_n^+ - \tilde{\psi}_n^-$ for $n \geq 1$, and so the conclusion holds for ψ if it holds for ψ^+ and ψ^-. This shows that it is no restriction to assume that ψ is non-negative. We do so from now on.

For any $\alpha < \beta$, let $S(\alpha, \beta)$ be the set of points $z \in Z$ such that

$$\liminf \tilde{\psi}_n(z) < \alpha < \beta < \limsup \tilde{\psi}_n(z).$$

Clearly, given $z \in Z$, the sequence $\tilde{\psi}_n(z)$ diverges if and only if $z \in S(\alpha, \beta)$ for some pair of rational numbers α and β. So, the lemma will follow if we show that $S = S(\alpha, \beta)$ has zero μ-measure for all α and β.

For each $z \in S$, fix some sequence of integers $1 \leq a_1^z < b_1^z < \cdots < a_i^z < b_i^z < \cdots$ such that

$$\tilde{\psi}_{a_i^z}(z) < \alpha \quad \text{and} \quad \tilde{\psi}_{b_i^z}(z) > \beta \quad \text{for every } i \geq 1.$$

Define A_i to be the union of the partition sets $P_{a_i^z}(z)$, and B_i to be the union of the partition sets $P_{b_i^z}(z)$ obtained in this way, for all the points $z \in S$. By construction,

$$S \subset A_{i+1} \subset B_i \subset A_i \quad \text{for every } i \geq 1.$$

In particular, S is contained in the set

$$\tilde{S} = \bigcap_{i=1}^{\infty} B_i = \bigcap_{i=1}^{\infty} A_i.$$

Given any two of the sets $P_{a_i^z}(z)$ that form A_i, either they are disjoint or else they coincide. This is because $\mathcal{P}_n$, $n \geq 1$, is a non-decreasing sequence of partitions. Consequently, A_i may be written as a pairwise disjoint union of such sets $P_{a_i^z}(z)$. Hence,

$$\int_{A_i} \psi \, d\mu = \sum_{P_{a_i^z}(z)} \int_{P_{a_i^z}} \psi \, d\mu < \sum_{P_{a_i^z}(z)} \alpha \mu(P_{a_i^z}) = \alpha \mu(A_i),$$

for any $i \geq 1$ (the sums are over that disjoint union). Analogously,

$$\int_{B_i} \psi \, d\mu = \sum_{P_{b_i^z}(z)} \int_{P_{b_i^z}} \psi \, d\mu > \sum_{P_{b_i^z}(z)} \beta\mu(P_{b_i^z}) = \beta\mu(B_i).$$

Since $A_i \supset B_i$ and we are assuming $\psi \geq 0$, it follows that

$$\alpha\mu(A_i) > \int_{A_i} \psi \, d\mu \geq \int_{B_i} \psi \, d\mu > \beta\mu(B_i),$$

for every $i \geq 1$. Taking the limit as $i \to \infty$, we find

$$\alpha\mu(\widetilde{S}) \geq \beta\mu(\widetilde{S}).$$

This implies that $\mu(\widetilde{S}) = 0$, and so $S \subset \widetilde{S}$ also has zero μ-measure. $\quad\square$

Given any bounded measurable function $\psi : Z \to \mathbb{R}$, we shall represent as $e_n(\psi)$, $e(\psi)$, respectively, the functions $\tilde{\psi}_n$, $\tilde{\psi}$ defined by (C.5) and Lemma C.14.

Let $\{\varphi_k : k \in \mathbb{N}\}$ be some countable dense subset of $C^0(Z, \mathbb{R})$, and let

$$F_* = \bigcap_{k=1}^{\infty} F(\varphi_k),$$

where $F(\varphi_k)$ is as given by Lemma C.14.

Lemma C.15. *Given any continuous function $\varphi : Z \to \mathbb{R}$, the sequence $e_n(\varphi)(z)$ converges to $e(\varphi)(z)$ as $n \to \infty$, for every $z \in F_*$.*

Proof. Fix $z \in F_*$. It is clear that $\psi \mapsto e_n(\psi)(z)$ is a bounded linear functional on $C^0(Z, \mathbb{R})$, with norm 1, and the same is true for $\psi \mapsto e(\psi)(z)$. For any $\varepsilon > 0$, choose k such that $\|\varphi - \varphi_k\|_0 < \varepsilon/3$. Then, if n is large enough,

$$|e_n(\varphi)(z) - e(\varphi)(z)| \leq$$
$$\leq |e_n(\varphi)(z) - e_n(\varphi_k)(z)| + |e_n(\varphi_k)(z) - e(\varphi_k)(z)| + |e(\varphi_k)(z) - e(\varphi)(z)|$$
$$\leq 2\|\varphi - \varphi_k\|_0 + \varepsilon/3 < \varepsilon.$$

This proves the claim. $\quad\square$

Let $\varphi : Z \to \mathbb{R}$ be continuous. By construction, $e_n(\varphi)$ is constant on each $P_n \in \mathcal{P}_n$, and so it is also constant on each atom P of $\mathcal{P}$, for every $n \geq 1$. Therefore, $e(\varphi)$ is constant on $P \cap F_*$ for every $P \in \mathcal{P}$. Let $e_n(\varphi)(P_n)$ represent the value of $e_n(\varphi)$ on each $P_n \in \mathcal{P}_n$. Similarly, $e(\varphi)(P)$ represents the value of $e(\varphi)$ on $P \cap F_*$ whenever the latter set is non-empty. Then, since (C.5) defines $e_n(\varphi)$ on a full μ-measure subset of Z,

$$\int \varphi \, d\mu = \sum_{\mu(P_n)>0} \int_{P_n} \varphi \, d\mu = \sum_{\mu(P_n)>0} \mu(P_n) \, e_n(\varphi)(P_n) = \int e_n(\varphi) \, d\mu.$$

Observe also that $|e_n(\varphi)| \le \sup |\varphi| < \infty$ for every $n \ge 1$. Therefore, we may use the dominated convergence theorem to conclude that

$$\int \varphi \, d\mu = \int e(\varphi) \, d\mu. \tag{C.6}$$

Now we are in a position to construct a system of conditional measures of μ. Let P be any atom of $\mathcal{P}$ such that $P \cap F_*$ is non-empty. It is easy to see that

$$C^0(Z,\mathbb{R}) \ni \varphi \to e(\varphi)(P) \in \mathbb{R}$$

is a non-negative linear functional on $C^0(Z,\mathbb{R})$. From the fact that $e_n(1)(P) = 1$ and the Riesz-Markov theorem, there exists a unique probability measure μ_P on Z such that

$$\int \varphi \, d\mu_P = e(\varphi)(P). \tag{C.7}$$

For completeness, we should define μ_P also when P does not intersect F_*. In this case we let μ_P be any probability measure on Z: since the set of all these atoms P has zero $\hat{\mu}$-measure in $\mathcal{P}$ (in other words, their union has zero μ-measure in Z), the choice is not relevant. In view of these definitions, (C.6) may be rewritten as

$$\int \varphi \, d\mu = \int \left(\int \varphi \, d\mu_P \right) d\hat{\mu}(P),$$

the fact that $\mathcal{P} \ni P \mapsto \int \varphi \, d\mu_P$ is a measurable function being a direct consequence of (C.7). Therefore, to conclude that $(\mu_P)_{P \in \mathcal{P}}$ do form a system of conditional measures of μ with respect to $\mathcal{P}$ we only have to prove

Lemma C.16. $\mu_P(P) = 1$ *for $\hat{\mu}$-almost every $P \in \mathcal{P}$.*

We use the following auxiliary result.

Lemma C.17. *Given any bounded measurable function $\psi : Z \to \mathbb{R}$ there exists a full $\hat{\mu}$-measure set $\mathcal{F}(\psi) \subset \mathcal{P}$ such that the set $P \cap F_*$ is non-empty, and so $\int \psi \, d\mu_P = e(\psi)(P)$, for any $P \in \mathcal{F}(\psi)$.*

Proof. The class of functions that satisfy the conclusion of the lemma contains all the continuous functions, and is closed under dominated pointwise convergence. Therefore, it contains all bounded measurable functions.   □

Now we can prove Lemma C.16:

Proof. Define $\mathcal{F}_* = \cap_{k,P_k} \mathcal{F}(\mathcal{X}_{P_k})$, where the intersection is over the set of all the atoms $P_k \in \mathcal{P}_k$, and every $k \geq 1$. Since this is a countable set, $\mathcal{F}_*$ has full $\hat{\mu}$-measure. We claim that the conclusion of the lemma holds for every $P \in \mathcal{F}_*$. Indeed, let $k \geq 1$ and P_k be the element of $\mathcal{P}_k$ that contains P. By the definition of $\mathcal{F}_*$

$$\mu_P(P_k) = \int \mathcal{X}_{P_k} \, d\mu_P = e(\mathcal{X}_{P_k})(P). \tag{C.8}$$

For each $n \geq 1$, let P_n be the atom of $\mathcal{P}_n$ that contains P. Given any $z \in P \cap F_*$,

$$e_n(\mathcal{X}_{P_k})(z) = \frac{1}{\mu(P_n)} \int_{P_n} \mathcal{X}_{P_k} \, d\mu.$$

Now, for any $n \geq k$ we have $P_n \subset P_k$, and then the last term is equal to 1. Therefore,

$$e(\mathcal{X}_{P_k})(P) = e(\mathcal{X}_{P_k})(z) = \lim_{n \to \infty} e_n(\mathcal{X}_{P_k})(z) = 1.$$

Replacing this in (C.8) we get $\mu_P(P_k) = 1$ for every $k \geq 1$. Finally,

$$\mu_P(P) = \lim_{k \to \infty} \mu_P(P_k) = 1$$

because the P_k, $k \geq 1$, are a decreasing sequence whose intersection is P. □

The proof of Theorem C.13 is complete. □

Example C.18. Let Z be the 2-dimensional torus, α be some irrational number, and $\mathcal{P}$ be the partition of Z into the straight lines of slope α. Then $\mathcal{P}$ is not a measurable partition. One way to see this is to observe that the Haar (Lebesgue) measure on Z admits no system of conditional measures with respect to $\mathcal{P}$. Indeed, by uniqueness of the disintegration, almost every conditional measure would have to be invariant under translations along the straight lines, and it is clear that translations have no invariant probabilities.

Example C.19. (ergodic decomposition) Let $f : Z \to Z$ be a continuous transformation on a compact metric space Z, and B_f be the subset of points $z \in Z$ such that time averages are well-defined on the orbit of z: given any continuous function $\varphi : Z \to \mathbb{R}$, the sequence

$$\frac{1}{n} \sum_{j=0}^{n-1} \varphi(f^j(z))$$

converges to some $\tilde{\varphi}(z) \in \mathbb{R}$ when $n \to \infty$. Let $\mathcal{P}$ be the partition of Z defined by (i) $Z \setminus B_f$ is an atom of $\mathcal{P}$ and (ii) two points $z_{1,\,2}$ in B_f are in the same atom of $\mathcal{P}$ if and only if they have the same time averages: $\tilde{\varphi}(z_1) = \tilde{\varphi}(z_2)$ for every continuous function φ. Then $\mathcal{P}$ is a measurable partition, with respect to any probability measure μ in Z. If μ is f-invariant, then $\mu(Z \setminus B_f) = 0$ and any system of conditional measures $(\mu_P)_P$ of μ relative to $\mathcal{P}$ is such that μ_P is f-invariant and ergodic for $\hat{\mu}$-almost every $P \in \mathcal{P}$.

D

Random Perturbations

Interest in random perturbations of dynamical systems goes back to Kolmogorov's point of view that, since observations of natural phenomena always involve some level of noise, physical perception corresponds to some sort of stochastic fluctuation rather than a purely deterministic law. This can be modelled through a Markov chain with fairly localized transitions (small noise), its stationary measure representing the "physical" observations. Stochastic stability amounts to this being consistent, for a given dynamical system, with the other interpretation of "observable" data we have been considering, namely, in terms of time averages: stochastic stability means that for small random noise the stationary measures are close to the physical (SRB) measure of the unperturbed system. See Section D.3.

In this appendix we introduce two main random perturbation models for maps (Sections D.1 and D.2) and discuss some of the relations between them (Section D.4). Most of the times we focus on discrete time systems, for which the theory is more elementary, but we do discuss the case of flows in Section D.6. Stochastic stability is conceptually related to shadowing properties, but this connections has yet to be exploited; see Section D.5 for some quick comments.

Broad presentations of random dynamical systems can be found in the books of Arnold [30] and Kifer [241, 242].

D.1 Markov chain model

First, we consider the case of discrete-time systems $f : M \to M$. Let U be an open subset of M such that $f(U)$ is relatively compact in U. The noise level $\varepsilon > 0$ will always be smaller than $\mathrm{dist}(f(U), M \setminus U)$, so that random orbits can not escape from U.

The Markov chain perturbation model is defined by a family $\{p_\varepsilon(\,\cdot\mid z) : z \in U,\ \varepsilon > 0\}$ of Borel probability measures, such that every $p_\varepsilon(\,\cdot\mid z)$ is supported inside the ε-neighborhood of $f(z)$. The random orbits are the sequences $\{z_j\}$ where each z_{j+1} is a random variable with probability distribution $p_\varepsilon(\,\cdot\mid z_j)$.

A probability measure μ_ε is *stationary* for the Markov chain if

$$\mu_\varepsilon(E) = \int p_\varepsilon(E\mid z)\,d\mu_\varepsilon(z) \qquad (\text{D.1})$$

for every Borel set $E \subset U$. Equivalently, the skew-product measure $\mu_\varepsilon \times p_\varepsilon^{\mathbb{N}}$

$$d(\mu_\varepsilon \times p_\varepsilon^{\mathbb{N}})(z_0, z_1, \ldots, z_n, \ldots) = \mu_\varepsilon(dz_0)\,p_\varepsilon(dz_1\mid z_0)\cdots p_\varepsilon(dz_n\mid z_{n-1})\cdots$$

is invariant under the shift map $\mathcal{F} : U \times U^{\mathbb{N}} \to U \times U^{\mathbb{N}}$ in the space of random orbits $\{z_j : j \geq 0\}$. By the ergodic theorem, the time average of every continuous function $\varphi : U \to \mathbb{R}$

$$\tilde{\varphi}(\mathbf{z}) = \lim_{n\to\infty} \frac{1}{n}\sum_{j=0}^{n-1}\varphi(z_j) = \lim_{n\to\infty} \frac{1}{n}\sum_{j=0}^{n-1}(\varphi\circ\pi_0)(\mathcal{F}^j(\mathbf{z}))$$

exists over a full $\mu_\varepsilon \times p_\varepsilon^{\mathbb{N}}$-measure subset of random orbits $\mathbf{z} = \{z_j\}$. Stationary measures always exist, if the transition probabilities $p_\varepsilon(\,\cdot\mid z)$ depend continuously on the point z. Consider the operator $\mathcal{T}_\varepsilon$ acting on the space of Borel measures by

$$\mathcal{T}_\varepsilon\eta(E) = \int p_\varepsilon(E\mid z)\,d\eta(z).$$

Lemma D.1. *Suppose $z \mapsto p_\varepsilon(\,\cdot\mid z)$ is continuous, relative to the weak* topology in the space of Borel probabilities. Then every weak* accumulation point μ_ε of the sequence $n^{-1}\sum_{j=0}^{n-1}\mathcal{T}_\varepsilon^j\eta$ is a stationary measure, for any probability measure η supported in U.*

Proof. The space of probability measures on the closure of $f(U)$ is weak* compact, and so accumulation points do exist. The assumption on the transition probabilities ensures that the operator $\mathcal{T}_\varepsilon$ is continuous, relative to the weak* topology. It follows that every accumulation point is a fixed point for $\mathcal{T}_\varepsilon$. This is equivalent to being a stationary measure. $\qquad\square$

A function $\phi : U \to \mathbb{R}$ is *invariant* for the Markov chain if its average with respect to the probability $p_\varepsilon(\,\cdot\mid x)$ coincides with $\phi(x)$

$$\phi(x) = \int \phi(y)\,p_\varepsilon(dy\mid x)\ ^1$$

for μ_ε-almost every x. A stationary measure μ_ε is *ergodic* if every invariant function is constant μ_ε-almost everywhere.

Lemma D.2. *If μ_ε is ergodic then $\tilde{\varphi}(\mathbf{z}) = \int \varphi\,d\mu_\varepsilon$ for every continuous function $\varphi : U \to \mathbb{R}$ and $\mu_\varepsilon \times p_\varepsilon^{\mathbb{N}}$-almost every random orbit $\mathbf{z}$.*

---

[1]Interpret the right hand side as $\int \phi\,dp_\varepsilon(\,\cdot\mid x)$.

Proof. For each $k \geq 0$, define

$$\tilde{\varphi}_k(z_0, \dots, z_k) = \int \tilde{\varphi}(\mathbf{z}) \, p_\varepsilon(dz_{k+1} \mid z_k) \cdots p_\varepsilon(dz_{n+1} \mid z_n) \cdots$$

Using $\tilde{\varphi} = \tilde{\varphi} \circ \mathcal{F}$ almost everywhere, we get that $\tilde{\varphi}_0$ is an invariant function:

$$\begin{aligned}
\tilde{\varphi}_0(z_0) &= \int \tilde{\varphi}(\mathbf{z}) \, p_\varepsilon(dz_1 \mid z_0) \cdots p_\varepsilon(dz_{n+1} \mid z_n) \cdots \\
&= \int \tilde{\varphi}(\mathcal{F}(\mathbf{z})) \, p_\varepsilon(dz_1 \mid z_0) \, p_\varepsilon(dz_2 \mid z_1) \cdots p_\varepsilon(dz_{n+1} \mid z_n) \cdots \\
&= \int p_\varepsilon(dz_1 \mid z_0) \int \tilde{\varphi}(\mathcal{F}(\mathbf{z})) \, p_\varepsilon(dz_2 \mid z_1) \cdots p_\varepsilon(dz_{n+1} \mid z_n) \cdots \\
&= \int \tilde{\varphi}_0(z_1) \, p_\varepsilon(dz_1 \mid z_0).
\end{aligned}$$

So, by hypothesis, $\tilde{\varphi}_0$ is constant almost everywhere. Moreover, $\tilde{\varphi}_k = \tilde{\varphi}_{k-1} \circ \mathcal{F}$ for every $k \geq 1$ (at this point, we think of $\tilde{\varphi}_k$ as a function of the whole $\mathbf{z}$ depending only on the first $k+1$ coordinates):

$$\begin{aligned}
\tilde{\varphi}_k(z_0, z_1, \dots, z_k) &= \int \tilde{\varphi}(\mathbf{z}) \, p_\varepsilon(dz_{k+1} \mid z_k) \cdots p_\varepsilon(dz_{n+1} \mid z_n) \cdots \\
&= \int \tilde{\varphi}(\mathcal{F}(\mathbf{z})) \, p_\varepsilon(dz_{k+1} \mid z_k) \cdots p_\varepsilon(z_{n+1} \mid z_n) \cdots \\
&= \tilde{\varphi}_{k-1}(z_1, \dots, z_k).
\end{aligned}$$

Consequently, by induction, every $\tilde{\varphi}_k$ is constant almost everywhere. Using that $\tilde{\varphi}(\mathbf{z}) = \lim \tilde{\varphi}_k(z_0, \dots, z_k)$ almost everywhere, we conclude that the same is true for $\tilde{\varphi}$, and that implies the lemma. $\qquad\square$

By Kifer [241, Proposition 2.1], every stationary measure is a convex combination of ergodic ones. Araújo [18] proves that, in great generality, the set of all stationary measures is a finite-dimensional simplex, whose extreme points are the ergodic measures. Moreover, the time average $\tilde{\varphi}(\mathbf{z})$ exists for every initial point z_0 and almost every choice of z_j, $j \geq 1$, and it is given by one of the ergodic measures.

D.2 Iterations of random maps

In this model one considers sequences obtained by iteration $z_j = g_j \circ \cdots \circ g_1(z_0)$ of maps g_j chosen at random ε-close to the original f. Consider a family $\{\nu_\varepsilon : \varepsilon > 0\}$ of probabilities in the space of C^r maps, some $r \geq 0$, such that each supp ν_ε is contained in the ε-neighborhood of f. The random orbit associated to a $(z_0, \mathbf{g})$ is the sequence $z_j = g_j \cdots g_1(z_0)$, where the g_j are independent random variables with probability distribution ν_ε.

There is an associated Markov chain, defined by

$$p_\varepsilon(E \mid z) = \nu_\varepsilon(\{g : g(z) \in E\}).\tag{D.2}$$

Observe that, for any Borel subsets $A_0, A_1, \ldots, A_m$ of U,

$$
\begin{aligned}
(\mu_\varepsilon \times \nu_\varepsilon^{\mathbb{N}})&(\{(z_0, \mathbf{g}) : z_0 \in A_0, g_1(z_0) \in A_1, \ldots, g_m \cdots g_1(z_0) \in A_m\}) = \\
&= \int_{A_0} d\mu_\varepsilon(z_0) \int \mathcal{X}_{\{g_1 : g_1(z_0) \in A_1\}} d\nu_\varepsilon(g_1) \cdots \int \mathcal{X}_{\{g_m : (g_m \cdots g_1)(z_0) \in A_m\}} d\nu_\varepsilon(g_m) \\
&= \int_{A_0} d\mu_\varepsilon(z_0) \int_{A_1} p_\varepsilon(dy_1 \mid z_0) \cdots \int_{A_m} p_\varepsilon(dy_m \mid y_{m-1}) \\
&= (\mu_\varepsilon \times p_\varepsilon^{\mathbb{N}})(A_0 \times A_1 \times \cdots \times A_m).
\end{aligned}
$$

This means that the statistics of the random orbits obtained from randomly perturbing the dynamical system are faithfully reproduced by the Markov chain. A probability measure μ_ε is stationary, in the sense of (D.1), if and only if

$$\mu_\varepsilon(E) = \int \mu_\varepsilon(g^{-1}(E)) \, d\nu_\varepsilon(g)\tag{D.3}$$

for every Borel set $E \subset U$. Most of the time we deal with continuous maps. Then the transition probabilities $p_\varepsilon(\cdot \mid z)$ given by (D.2) vary continuously with z, and so stationary measures do exist.

Example D.3. (Additive noise) Let $f : M \to M$ be a diffeomorphism on some Lie group M, e.g., a d-dimensional torus. Let $\{\theta_\varepsilon : \varepsilon > 0\}$ be a family of probability measures on M such that every $\operatorname{supp} \theta_\varepsilon$ is contained in the ε-neighborhood of unity $e \in M$. Let ν_ε be the measure induced by θ_ε in the space of diffeomorphisms, via left multiplication $M \ni \xi \mapsto \xi \cdot f$. The corresponding transition probabilities are

$$p_\varepsilon(E \mid z) = \theta_\varepsilon(\{\xi \in M : \xi \cdot f(z) \in E\}) = \theta_\varepsilon(E \cdot f(z)^{-1}),$$

where $f(z)^{-1}$ represents the algebraic inverse of $f(z)$.

Example D.4. (Noise in parameter space) Let $F : \mathbb{R}^k \times U \to U$ be a C^r map such that every $F(0, \cdot) = f$ and $f_t(U)$ is relatively compact in U for every $f_t = F(t, \cdot)$. Let θ_ε be the normalized Lebesgue measure on the ε-ball around 0, and ν_ε be the probability induced by θ_ε in the space of diffeomorphisms via the parametrization $t \mapsto f_t$. Clearly, $\operatorname{supp} \nu_\varepsilon \to f$ in the C^r topology as $\varepsilon \to 0$.

D.3 Stochastic stability

Let μ be an invariant probability measure for $f : U \to U$. The main case we have in mind is when μ is an SRB measure supported in an attractor Λ with the basin property (1.6), and U is contained in the basin of attraction of Λ. In that case,

$$\frac{1}{n}\sum_{j=0}^{n-1}\varphi(f^j(z)) \to \int \varphi\, d\mu$$

for Lebesgue almost every $z \in U$ and every continuous φ.

First, suppose there is a unique stationary probability measure μ_ε, for every small $\varepsilon > 0$. Then

$$\frac{1}{n}\sum_{j=0}^{n-1}\varphi(z_j) \to \int \varphi\, d\mu_\varepsilon$$

for almost every random orbit $\{z_j\}$ and every continuous φ. Uniqueness holds, for instance, if the attractor $\cap_{n\geq1} f^n(U)$ is transitive, assuming $p_\varepsilon(\,\cdot\mid z)$ is absolutely continuous and the support contains a $\rho(\varepsilon)$-neighborhood of $f(z)$. See Araújo [19, Proposition 2.1].

Definition D.5. The system (f,μ) is *stochastically stable* under the perturbation scheme $\{p_\varepsilon(\,\cdot\mid z) : z \in U,\ \varepsilon > 0\}$ (or $\{\nu_\varepsilon : \varepsilon > 0\}$) if μ_ε converges to μ in the weak* sense:

$$\lim_{\varepsilon\to0} \int \varphi\, d\mu_\varepsilon = \int \varphi\, d\mu \quad \text{for every continuous } \varphi : U \to \mathbb{R}.$$

The definition extends naturally to the case where the stationary measure is not unique: the whole simplex should converge to μ when $\varepsilon \to 0$.

Remark D.6. It is useful to extend the definition to domains containing several relevant invariant measures, including the whole manifold M. For instance, an attractor may support more than one SRB measure (although it is not yet clear how general this phenomenon is). Also, if the basin of attraction is just a positive Lebesgue measure set, or if one considers random noise which is not supported on small neighborhoods, then random orbits may escape from the basin of attraction. In such cases, a more global notion of stochastic stability can be applied: the simplex of stationary measures should be weak* close to the simplex generated by the SRB measures of the unperturbed system, when ϵ is small. This is substantiated by the following important result of Araújo [19] for absolutely continuous random noise: if there is stochastic stability restricted to each basin of attraction, and the union of the basins contains Lebesgue almost every point, then the system is stochastically stable in the whole ambient space $U = M$, in the previous sense.

In existing results of stochastic stability, the assumptions on the random noise often depend on the class of systems. Even in the simplest uniformly hyperbolic setting, one can not expect stability under *every* Markov chain perturbation. The following example is an improvement of an example due to Keller [237] and told to us by Gary Froyland: the main novelty in our construction is that the random noise varies continuously with the point. This example also highlights the phenomenon of *localization*, pointed out by Blank, Keller [59] as an important mechanism for stochastic instability.

Example D.7. Let $f : S^1 \to S^1$, $f(z) = 2z$ mod $\mathbb{Z}$. Lebesgue measure on $S^1 = \mathbb{R}/\mathbb{Z}$ is the unique SRB measure of f. For small $\varepsilon > 0$ let $f_\varepsilon : S^1 \to S^1$ be a smooth transformation without critical points such that $f_\varepsilon(z) = 0$ for all $|z| \leq \varepsilon$ and $f_\varepsilon(z) = f(z)$ for all $|z| \geq 3\varepsilon/2$. Consider the Markov chain defined by $p_\varepsilon(\,\cdot\mid z) = $ normalized Lebesgue measure on $[f_\varepsilon(z) - \varepsilon, f_\varepsilon(z) + \varepsilon]$. Almost every random orbit eventually enters $|z| \leq 3\varepsilon/2$, and then has probability $\geq 1/2$ of hitting $|z| \leq \varepsilon$. Thus, almost every random orbit enters the domain $|z| \leq \varepsilon$, where it remains trapped forever (localized). This implies that the Markov chain has a unique stationary measure $\mu_\varepsilon = $ normalized Lebesgue measure on $[-\varepsilon, \varepsilon]$. Clearly, μ_ε converges to the Dirac measure δ_0, and not to the SRB measure of f, when $\varepsilon \to 0$.

This construction may be realized as a random maps scheme: just let ν_ε be the image under $[-\varepsilon, \varepsilon] \ni t \mapsto f_\varepsilon + t$ of the normalized Lebesgue measure. However, this is not a small random perturbation in the C^1 category: the maps $f_\varepsilon + t$ are C^0 close but not C^1 close to f.

Problem D.8. Give examples of systems that are stochastically stable in the C^r but not in the C^s category, $s < r$.

In the uniformly hyperbolic case, Kifer [242] assumes absolutely continuity with respect to Lebesgue measure m

$$p_\varepsilon(\,\cdot\mid z) = \rho_{\varepsilon,z}\, m$$

for some integrable function $\rho_{\varepsilon,z}$, and a kind of Lipschitz dependence of the transition probabilities on the point z; see [242, Chapter 2]. In fact, he allows for more general noise, whose support needs not be contained in small balls. On the other hand, C^2 uniformly hyperbolic diffeomorphisms are stochastically stable under any random maps scheme $\{\nu_\varepsilon : \varepsilon > 0\}$: see Young [457] and Sections 3.3 and 4.5 of [441].

Stochastic stability under every random maps scheme $\{\nu_\varepsilon : \varepsilon > 0\}$ includes stability under deterministic perturbations (let $\nu_\varepsilon = $ Dirac measure): the SRB measure must vary continuously with the dynamical system. Alves, Viana [14] call the latter property *statistical stability*. This is a fairly restrictive property: An example of Keller [237, § 6] shows that even uniformly expanding piecewise smooth maps in dimension 1 are not always statistically stable. On the other hand, [14, 436] exhibit open sets of statistical stability outside the uniformly hyperbolic domain.

Next we are going to see that whether a given Markov chain does derive from the iteration of random maps is largely a matter of smoothness of the noise:

D.4 Realizing Markov chains by random maps

We have seen previously, that every random map scheme may be realized as a Markov chain. The present section is devoted to the converse problem. This problem is discussed by Kifer [241, Section 1.1] and Benedicks, Viana [53]. Kifer proves that under a mild condition on the ambient space, and assuming $z \mapsto p_\varepsilon(z \mid E)$ is measurable for every Borel set E, such a realization is possible in the space of measurable maps. When the transition probabilities have positive densities we can say more [53]: *the Markov chain is represented by parametrized families of maps at least as regular as the densities themselves.* An explanation of this statement follows.

For simplicity, we suppose $U \subset \mathbb{R}^d$, but the arguments extend immediately to manifolds, the support being the image of $f(z) + [-\varepsilon, \varepsilon]^d$ under some homeomorphism that depends continuously on the point z.

Proposition D.9. *Suppose every $p_\varepsilon(\cdot \mid z)$, $z \in U$, is absolutely continuous with respect to Lebesgue measure m, with density $\rho_{\varepsilon,z} > 0$ on the support $f(z) + [-\varepsilon, \varepsilon]^d$. Suppose $r_\varepsilon : (\xi, z) \mapsto \rho_{\varepsilon,z}(f(z) + \xi)$ is continuous on the domain $[-\varepsilon, \varepsilon]^d \times U$. Then there exist continuous maps $F_\varepsilon : [0, 1]^d \times U \to U$ such that*

$$p_\varepsilon(E \mid z) = m(\{\omega : F_\varepsilon(\omega, z) \in E\}) \quad \text{for every Borel set } E.$$

Proof. For $z \in U$ and $\xi = (\xi_1, \ldots, \xi_d) \in [-\varepsilon, \varepsilon]^d$, let

$$\omega_i = \frac{p_\varepsilon(f(z) + [-\varepsilon, \xi_1] \times \cdots \times [-\varepsilon, \xi_i] \times [-\varepsilon, \varepsilon]^{d-i} \mid z)}{p_\varepsilon(f(z) + [-\varepsilon, \xi_1] \times \cdots \times [-\varepsilon, \xi_{i-1}] \times [-\varepsilon, \varepsilon]^{d-i+1} \mid z)} \tag{D.4}$$

for $i = 1, \ldots, d$ (for $i = 1$ the denominator is 1). Since the density is positive, this is well-defined if $\xi_1, \ldots, \xi_{i-1}$ are all larger than $-\varepsilon$. Given any $1 \leq s \leq j$, the probability $p_\varepsilon(f(z) + [-\varepsilon, \xi_1] \times \cdots \times [-\varepsilon, \xi_j] \times [-\varepsilon, \varepsilon]^{d-j} \mid z)$ is comparable to

$$(\xi_s + \varepsilon) \int_{-\varepsilon}^{\xi_1} \cdots \int_{-\varepsilon}^{\xi_{s-1}} \int_{-\varepsilon}^{\xi_{s+1}} \cdots \int_{-\varepsilon}^{\xi_j} \int_{-\varepsilon}^{\varepsilon} \cdots \int_{-\varepsilon}^{\varepsilon} (r_\varepsilon \mid \{\xi_s = -\varepsilon\}) \, dm,$$

in the sense that the quotient goes to 1 when $\xi_s + \varepsilon$ goes to zero. There is a similar estimate when several $\xi_s + \varepsilon$ go to zero simultaneously. It follows that (D.4) extends continuously to the points where the denominator is zero.

For fixed $\xi_1, \ldots, \xi_{i-1}$ the map $\xi_i \mapsto \omega_i$ is a homeomorphism from $[-\varepsilon, \varepsilon]$ onto $[0, 1]$. So, $\psi_{\varepsilon,z} : (\xi_1, \ldots, \xi_d) \mapsto (\omega_1, \ldots, \omega_d)$ is a homeomorphism onto $[0, 1]^d$. Moreover, $(\xi, z) \mapsto \psi_{\varepsilon,z}(\xi)$ is continuous, because the densities depend continuously on both variables. Let $\phi_{\varepsilon,z}$ be the inverse of $\psi_{\varepsilon,z}$. Then $(\omega, z) \mapsto \phi_{\varepsilon,z}(\omega)$ and $F_\varepsilon(\omega, z) = f(z) + \phi_{\varepsilon,z}(\omega)$ are continuous. Finally, the definition (D.4) gives that, for every $E = f(z) + [-\varepsilon, \xi_1] \times \cdots \times [-\varepsilon, \xi_d]$,

$$p_\varepsilon(E \mid z) = \omega_1 \cdots \omega_d = m(\{\omega : \phi_{\varepsilon,z}(\omega) \in E - f(z)\}) = m(\{\omega : F_\varepsilon(\omega, z) \in E\}).$$

Since these rectangles generate the σ-algebra on the support, the relation must be true for every Borel set. □

Thus, the Markov chain $\{p_\varepsilon(\,\cdot\mid z) : z \in U\}$ is realized by the probability measure induced by Lebesgue measure in the space of continuous maps through $\omega \mapsto F_\varepsilon(\omega, \cdot)$.

If $(\xi, z) \mapsto r_\varepsilon(\xi, z)$ is C^r, for some $r \geq 1$, then the map $(\xi, z) \mapsto \psi_{\varepsilon,z}(\xi)$ constructed in the proposition is C^r. Moreover, $\psi_{\varepsilon,z}$ is a diffeomorphism, because

$$\frac{\partial \omega_i}{\partial \xi_i} > 0 \quad \text{and} \quad \frac{\partial \omega_i}{\partial \xi_j} = 0 \quad \text{for all } 1 \leq i < j \leq d.$$

It follows that F_ε is C^r, and so ν_ε lives in the space of C^r maps.

For $\{\nu_\varepsilon : \varepsilon > 0\}$ to be a small perturbation of f in the space of C^r maps, we need $\operatorname{supp}\nu_\varepsilon \to f$ in the C^r topology, as $\varepsilon \to 0$. For $r = 0$ this comes for free, because $\sup \|F_\varepsilon(\omega, z) - f(z)\| \leq \varepsilon$. For $r \geq 1$, the following additional conditions suffice (but they are not optimal):

$$c\varepsilon^{-d} \leq r_\varepsilon \leq C\varepsilon^{-d} \quad \text{and} \quad \|\partial^{p+q}r_\varepsilon/\partial\xi^p\partial z^q\| \leq C\varepsilon^{-d-p} \quad \text{for } 1 \leq p+q \leq r.$$

Let us check this, in the case $r = 1$.

Step 1: $\left\|\dfrac{\partial\psi_{\varepsilon,z}}{\partial z}\right\| \leq C$.

Taking derivative with respect to z_j in the expression of ω_i, and then using $c\varepsilon^{-d} \leq r_\varepsilon \leq C\varepsilon^{-d}$ and $\|\partial r_\varepsilon/\partial z_j\| \leq C\varepsilon^{-d}$, we find

$$\left|\frac{\partial\omega_i}{\partial z_j}\right| \leq \frac{2C^2(\xi_i + \varepsilon)}{2c^2\varepsilon} \leq \frac{2C^2}{c^2}$$

for every $1 \leq j \leq i \leq d$. The claim follows immediately.

Step 2: $\left\|\left(\dfrac{\partial\psi_{\varepsilon,z}}{\partial\xi}\right)^{-1}\right\| \leq C$.

Since the derivative is given by a (lower) triangular matrix, to get this bound on the inverse it suffices to prove that

$$\left|\left(\frac{\partial\omega_i}{\partial\xi_i}\right)^{-1}\right| \leq C\varepsilon \quad \text{and} \quad \left|\frac{\partial\omega_i}{\partial\xi_j}\right| \leq C\varepsilon^{-1} \tag{D.5}$$

for every $1 \leq j < i \leq d$. By direct derivation of ω_i, we get

$$\frac{\partial\omega_i}{\partial\xi_i}\mid\{\xi_i = \xi\} = \frac{\int_{-\varepsilon}^{\xi_1}\cdots\int_{-\varepsilon}^{\xi_{i-1}}\int_{-\varepsilon}^{\varepsilon}\cdots\int_{-\varepsilon}^{\varepsilon}r_\varepsilon\mid\{\xi_i = \xi\}}{\int_{-\varepsilon}^{\xi_1}\cdots\int_{-\varepsilon}^{\xi_{i-1}}\int_{-\varepsilon}^{\varepsilon}\cdots\int_{-\varepsilon}^{\varepsilon}r_\varepsilon};$$

the integral in the denominator is with respect to $d\xi_1\cdots d\xi_{i-1}d\xi_{i+1}\cdots d\xi_d$, and the one on the denominator is with respect to $d\xi_i$ too. Using $c\varepsilon^{-d} \leq r_\varepsilon \leq C\varepsilon^{-d}$ we get that

$$\frac{\partial \omega_i}{\partial \xi_i} \geq \frac{c}{C(2\varepsilon)}$$

which proves the first half of (D.5). To prove the second half, take the derivative of ω_i with respect to ξ_j and then use the mean value theorem, to conclude that

$$\left| \frac{\partial \omega_i}{\partial \xi_j} \mid \{\xi_j = \xi\} \right| \leq \frac{\int_{-\varepsilon}^{\xi_1} \cdots \int_{-\varepsilon}^{\xi_{j-1}} \int_{-\varepsilon}^{\xi_{j+1}} \cdots \int_{-\varepsilon}^{\xi_i} \int_{-\varepsilon}^{\varepsilon} \cdots \int_{-\varepsilon}^{\varepsilon} |\partial r_\varepsilon / \partial \xi_j| \mid \{\xi_j = \eta\}}{\int_{-\varepsilon}^{\xi_1} \cdots \int_{-\varepsilon}^{\xi_{j-1}} \int_{-\varepsilon}^{\xi_{j+1}} \cdots \int_{-\varepsilon}^{\xi_{i-1}} \int_{-\varepsilon}^{\varepsilon} \cdots \int_{-\varepsilon}^{\varepsilon} r_\varepsilon / \mid \{\xi_j = \eta\}}$$

for some $\eta \in (-\varepsilon, \xi)$. Using $r_\varepsilon \geq c\varepsilon^{-d}$ and $|\partial r_\varepsilon / \partial \xi_j| \leq C\varepsilon^{d-1}$, we deduce

$$\left| \frac{\partial \omega_i}{\partial \xi_j} \right| \leq \frac{(\xi + \varepsilon) C \varepsilon^{-d-1}}{(2\varepsilon) c \varepsilon^{-d}} \leq \frac{C}{c\varepsilon}.$$

This completes the proof of (D.5) and step 2.

Step 3: Conclusion $\left\| \dfrac{\partial F_\varepsilon(\omega, \cdot)}{\partial z} - \dfrac{\partial f}{\partial z} \right\| \leq C\varepsilon$.

Taking derivatives with respect to z in the equality $\psi_{\varepsilon, z} \circ \phi_{\varepsilon, z}(\omega) \equiv \omega$, and using the conclusions of the previous two steps, we obtain

$$\left\| \frac{\partial F_\varepsilon(\omega, \cdot)}{\partial z} - \frac{\partial f}{\partial z} \right\| = \left\| \frac{\partial \phi_{\varepsilon, z}}{\partial z} \right\| = \left\| \frac{\partial \phi_{\varepsilon, z}}{\partial z} \left(\frac{\partial \phi_{\varepsilon, z}}{\partial z} \right)^{-1} \right\| \leq C\varepsilon.$$

This completes the argument.

D.5 Shadowing versus stochastic stability

The shadowing property of hyperbolic systems means that given every pseudo-orbit, that is, every sequence $(x_n)_n$ such that $f(x_n)$ is close to x_{n+1} for every n, is uniformly approximated (shadowed) by a true orbit. See [86]. An important reformulation of this result for non-uniformly hyperbolic maps was due to Katok [231].

Shadowing is somewhat related to stochastic stability, since random orbits are pseudo-orbits, and proving stability involves showing that the behavior of (most) random orbits mimics that of (typical) true orbits. However, in the strong form stated above, the shadowing property is very rare outside the uniformly hyperbolic world. For instance, it fails already for piecewise smooth uniformly expanding maps of the interval [139]. Smooth counterexamples should occur among quadratic maps of the interval, and Hénon maps, and have been explicitly exhibited in the context of heterodimensional cycles [75, 462]. Namely, [75] gives a C^1 open set of partially hyperbolic diffeomorphisms such that there is $\delta > 0$ such that, for any $\varepsilon > 0$ there are ε-pseudo orbits which are not δ-shadowed by any true orbit.

More than that, it has been recently announced in [3] that non-shadowable diffeomorphisms contain a C^1 open dense subset of all non-hyperbolic robustly transitive diffeomorphisms. Earlier, [401] had shown that the C^1 interior of shadowable diffeomorphisms is formed by the Axiom A diffeomorphisms satisfying the strong transversality condition. However, [357] shows that C^0 generic homeomorphisms are shadowable.

On the other hand, as mentioned in the text, many of these systems have been shown to be stochastically stable, under very general random perturbations. So uniform shadowing is by no means necessary, and a much weaker shadowing property might suffice for stochastic stability:

Problem D.10. Formulate a general criterium for stochastic stability in terms of a shadowing property.

Piecewise smooth uniformly expanding maps of the interval, such as tent maps, are a natural context to start investigating this problem.

D.6 Random perturbations of flows

The main model is *diffusion*: the random orbits are the solutions of stochastic differential equations (for simplicity, pretend $M = \mathbb{R}^d$)

$$dz_t = X(z_t)\, dt + \epsilon A(z_t)\, dw_t \qquad (D.6)$$

where X is the deterministic vector field, $A(\cdot)$ is a matrix-valued function, and dw_t is the standard Brownian motion. See Kifer [242], and also Oksendal [331] or Kunita [249] for general information about stochastic differential equations. A solution of (D.6) with initial condition $z \in M$ is a stochastic process

$$\mathbb{R} \times \Omega \to M, \quad (t, \omega) \mapsto f_\omega^t(z)$$

on an (abstract) probability space Ω, satisfying

$$f_\omega^t(z) = z + \int_0^t X(f_\omega^s(z))\, ds + \varepsilon \int_0^t A(f_\omega^s(z))\, dw_s$$

where the last term is a stochastic integral in the sense of Ito. Under reasonable conditions on $X(\cdot)$ and $A(\cdot)$, there exists exactly one solution *with continuous paths*, that is, such that

$$[0, +\infty) \ni t \mapsto f_\omega^t(z) \qquad (D.7)$$

is continuous for almost $\omega \in \Omega$. See e.g. Theorem 5.5 in [331].

Stochastic stability means that, given any continuous function φ with compact support and any $\delta > 0$, if ε is small enough then

$$\left| \lim_{T \to \infty} \frac{1}{T} \int_0^T \varphi(f_\omega^s(z))\, ds - \int \varphi\, d\mu_{SRB} \right| < \delta$$

for Lebesgue almost every z in the relevant domain U (see the next paragraph) and almost every ω. Here μ_{SRB} is the SRB measure of the deterministic vector field X inside U.

An important difference with respect to the previous models, where noise was supported in small balls, is that diffusion tends to mingle all the attractors together, because the solutions (D.7) tend to spread over the whole ambient M for all $t > 0$. For instance, if A is everywhere nonsingular then the stochastic flow has a unique stationary measure μ_ε, that is, such that

$$\mu_\varepsilon(E) = \int \mu_\varepsilon\left((f_\omega^t)^{-1}(E)\right) d\omega$$

for all $t > 0$ and every measurable set E. Thus, one should use a global notion of stability as in Remark D.6. However, it makes sense to speak of stochastic perturbations restricted to some invariant set U of X, if one considers $A(\cdot)$ supported inside U.

Kifer [242] proved that hyperbolic attractors of flows are stochastically stable, and so are the geometric Lorenz attractors. This was extended by Metzger [292] for the contracting Lorenz-like attractors introduced by Rovella [388].

E

Decay of Correlations

Let $f : M \to M$ be a transformation and μ be an invariant probability. The *correlation sequence* of a pair of functions $\varphi, \psi : M \to \mathbb{R}$ is defined by

$$C_n(\varphi, \psi) = \int \varphi(\psi \circ f^n) \, d\mu - \int \varphi \, d\mu \int \psi \, d\mu$$

when the integrals make sense. The system (f, μ) is *mixing* if $C_n(\varphi, \psi) \to 0$ as $n \to \infty$, for every $\varphi, \psi \in L^2(\mu)$. The definition is not affected if one considers dense subsets; for instance, we may take $\varphi \in L^1(\mu)$ and $\psi \in L^\infty(\mu)$ instead. Notice that

$$C_n(\varphi, \psi) = \int \psi \, d\big(f_*^n(\varphi \mu)\big) - \int \psi \, d\big[\big(\int \varphi \, d\mu\big) \mu\big]$$

so that the mixing property means that the iterates of an initial mass distribution $\varphi \mu$ converge, in the weak topology, to a multiple $(\int \varphi \, d\mu) \, \mu$ of the invariant measure. Hence, it may be seen as a sort of memory dissipation in the system: all the information contained in the initial density, apart from its total mass, is gradually forgotten under iteration.

The problem of decay of correlations corresponds to describing how fast the convergence is. In general, to get non-trivial bounds one must restrict to classes of densities φ that are fairly regular, like Hölder continuous, of bounded variation, or even smooth. We say that (f, μ) has *exponential decay of correlations* in a Banach space $\mathcal{H} \subset L^1(\mu)$ if there exists $\lambda < 1$ such that

$$\big|C_n(\varphi, \psi)\big| \leq C(\varphi, \psi)\lambda^n \quad \text{for every } \varphi \in \mathcal{H}, \ \psi \in L^\infty(\mu), \text{ and } n \geq 1. \quad \text{(E.1)}$$

Sometimes one considers $\psi \in L^p(\mu)$ for some other value of $p \in [1, \infty]$. Changing the sequence on the right-hand side, one obtains notions of slower or faster decay of correlations, like *stretched-exponential* $C(\varphi, \psi)\lambda^{(n^\theta)}$, $0 < \theta < 1$ (*super-exponential* if $\theta > 1$), or *polynomial* $C(\varphi, \psi)n^{-\alpha}$, $\alpha > 0$. In all these

cases, under quite general conditions [1] the factor may be taken of the form
$C(\varphi, \psi) = C\|\varphi\| \, \|\psi\|_p$, where $\|\cdot\|$ represents the norm in $\mathcal{H}$ (see Collet [129]).

In what follows we introduce the main available techniques available for
dealing with this problem: spectral analysis (Section E.1), projective metrics
(Section E.3), and coupling (Section E.7). In parallel, we discuss several applications of these methods, both to hyperbolic systems, in Sections E.2, E.4,
E.5, and to non-hyperbolic ones, in Sections E.6 and E.8. Properties characteristic of independent random variables, such as the central limit theorem,
remain valid for deterministic systems which are sufficiently chaotic, in the
sense that they exhibit sufficiently rapid decay of correlations. This is the
theme of the last section.

A substantial part of this theory is treated in detail by Ruelle [395], Gora,
Boyarsky [91], Viana [441], Baladi [40], Alves [8].

E.1 Transfer operators: spectral gap property

The classical approach to proving exponential decay of correlations, initiated
by Ruelle (see [395] and references therein), is through showing that the transfer operator $\mathcal{L}$ has a spectral gap.

The *transfer operator* $\mathcal{L} : \varphi \mapsto \mathcal{L}\varphi$ of the map f relative to μ is given by

$$f_*(\varphi\mu) = (\mathcal{L}\varphi)\,\mu\,.$$

$\mathcal{L}$ is well defined and a linear operator in $L^1(\mu)$, as long as f is non-singular
with respect to μ, in the sense that f_* preserves the space of measures absolutely continuous with respect to μ. This is automatic in our case, because
we assume μ to be f-invariant. However, transfer operators relative to non-
invariant measures, most especially the Lebesgue measure m, are just as useful. In this latter case, the transfer operator permits to analyze the speed of
convergence to equilibrium

$$f_*^n(\varphi m) \to \left(\int \varphi \, dm\right)\mu\,,$$

which is closely related to, but not quite the same as decay of correlations.

The operator $\mathcal{L}$ never expands the L^1 norm. Indeed, if $\varphi \geq 0$ then $\mathcal{L}\varphi \geq 0$
and

$$\|\mathcal{L}\varphi\|_1 = \int (\mathcal{L}\varphi)\,d\mu = \int 1\,d(f_*(\varphi\mu)) = \int 1\,d(\varphi\mu) = \int \varphi\,d\mu = \|\varphi\|_1\,;$$

---

[1] The space $\mathcal{H}$ should be contained in the dual of $L^\infty(\mu)$, where duality is relative
to integration by μ, with topology at least as fine as the dual norm topology. $L^\infty(\mu)$
may be replaced by any Banach space containing the constant functions and such
that $\varphi \mapsto \int \varphi\,d\mu$ and $\varphi \mapsto \varphi \circ f$ are continuous.

the general case now follows from $|\mathcal{L}\varphi| \leq \mathcal{L}|\varphi|$. When the measure μ is invariant, as we are considering here, the constant functions are fixed points of $\mathcal{L}$. By the previous observation, there can be no eigenvalues of norm larger than 1. Let $\mathcal{H} \subset L^1(\mu)$ be such that $\mathcal{L}(\mathcal{H}) \subset \mathcal{H}$, and let $\|\cdot\|$ be some complete norm in $\mathcal{H}$.

Definition E.1. $\mathcal{L}$ has a *spectral gap* in the Banach space $(\mathcal{H}, \|\cdot\|)$ if

- 1 is a simple eigenvalue of $\mathcal{L} : \mathcal{H} \to \mathcal{H}$ and
- $\mathrm{spec}(\mathcal{L}) \setminus \{1\}$ is contained in some open disk $\mathbb{D}_r$ of radius $r < 1$.

Then $\mathcal{H}$ has an $\mathcal{L}$-invariant splitting $\mathcal{H} = \mathbb{R} \oplus \mathcal{H}_0$ where $\mathcal{H}_0$ is the hyperspace of functions with zero integral relative to μ, and $\mathrm{spec}(\mathcal{L} \mid \mathcal{H}_0) = \mathrm{spec}(\mathcal{L}) \setminus \{1\}$ is contained in $\mathbb{D}_r$. Any $\varphi \in \mathcal{H}$ may be written as $\varphi = \int \varphi \, d\mu + \varphi_0$ with $\varphi_0 \in \mathcal{H}_0$. Then $\|\mathcal{L}^n\varphi_0\| \to 0$ and so $\mathcal{L}^n\varphi \to \int \varphi \, d\mu$, exponentially fast. As long as the norm $\|\cdot\|$ is strong enough, this implies that $f_*^n(\varphi\mu) = (\mathcal{L}^n\varphi)\mu$ converges to $(\int \varphi \, d\mu)\mu$ exponentially fast. Thus, correlations decay faster than $C(\varphi, \psi)r^n$ in the space $\mathcal{H}$.

E.2 Expanding and piecewise expanding maps

This powerful approach was developed and successfully applied to uniformly expanding maps, by Ruelle [395]. See also [40, 91, 441] and references therein. The invariant measure may be any equilibrium state associated to a Hölder potential, including the SRB measure (the unique absolutely continuous invariant probability) if one assumes the derivative of the map to be Hölder. One usually takes $\mathcal{H}$ to be the space of ν-Hölder functions, for any $\nu > 0$. The spectral gap property originates from the fact that the transfer operator tends to improve Hölder constants: there are $C > 0$ and $\rho < 1$ such that

$$\mathrm{Hol}_\nu(\mathcal{L}^n\varphi) \leq C\rho^n \, \mathrm{Hol}_\nu(\varphi) + C \sup |\varphi|, \qquad (\text{E.2})$$

where $\mathrm{Hol}_\nu(\psi)$ is the smallest number a such that ψ is (a, ν)-Hölder. Recall that a function ψ is (a, ν)-Hölder if $|\psi(x) - \psi(y)| \leq a d(x, y)^\nu$ for all x and y, and it is ν-Hölder if it is (a, ν)-Hölder for some a. In this way one proves that (f, μ) has exponential decay of correlations in $\mathcal{H}$, with the radius r in Definition E.1 as a (non-explicit[2]) bound on the rate of decay.

The spectral gap approach was also most successful in the case of piecewise smooth expanding maps in dimension 1, starting from the remarkable work of Lasota, Yorke [252]. Rychlik [398] extended their result on existence of absolutely continuous invariant measures to the case of infinitely many smoothness intervals. See also [40, 91, 441] and references therein. One usually takes $\mathcal{H}$ to

---

[2]We may choose any $\rho \in (\sigma^{-\nu}, 1)$, where $|f'| \geq \sigma > 1$, but r depends also on the norm of the largest eigenvalue of $\mathcal{L}$ in the unit circle and might be larger than ρ.

be the space of functions with bounded variation [3]. The spectral gap relies on the fact that the operator $\mathcal{L}$ tends to improve variations, as expressed by the celebrated Lasota-Yorke inequality: there are $C > 0$ and $\rho < 1$ such that

$$\operatorname{var}(\mathcal{L}^n \varphi) \leq C\rho^n \operatorname{var}(\varphi) + C \int |\varphi| \, d\mu. \tag{E.3}$$

There have been several extensions to piecewise expanding maps in higher dimensions. The pioneer works by Keller [236], Blank [61], and Gora, Boyarsky [194] imposed strong restrictions on the boundaries of smoothness domains, to ensure existence of absolutely continuous invariant measures. Weaker conditions were provided by Buzzi [110], Saussol [406], Tsujii [429]. In fact, Buzzi and Tsujii showed that generic piecewise real-analytic expanding maps on surfaces have absolutely continuous invariant measures. Cowieson [140] proved that the same is true for generic piecewise smooth expanding maps in any dimension; Buzzi [111] had obtained a similar result in the affine setting. Also recently, Buzzi, Maume-Deschamps [112] investigated decay of correlations for piecewise invertible maps in any dimension.

Moreover, Alves [7] extended some of these methods to maps with infinitely many domains of smoothness, as a step toward proving existence and finiteness of absolutely continuous invariant measures in great generality for *non-uniformly* expanding maps in any dimension. Then Alves, Luzzatto, Pinheiro [13] proved super-polynomial decay of correlations for a large class of such higher dimensional systems. Later, Baladi, Gouezel [42, 196] improved this to stretched exponential decay, for a related class of maps also including the ones in Viana [440].

E.3 Invariant cones and projective metrics

Projective metrics associated to convex cones were introduced by Birkhoff [58]. They provide an alternative strategy for proving decay of correlations, initiated by Ferrero, Schmitt [182] and much developed by Liverani [263], Maume [289] and other authors. A strong point is that this approach usually gives explicit bounds for the rates of decay.

A *cone* in a Banach space $\mathcal{H}$ is a subset $\mathcal{C}$ such that $tv \in \mathcal{C}$ for every $v \in \mathcal{C}$ and $t > 0$. The cone is *convex* if $v_1 + v_2 \in \mathcal{C}$ for all $v_1 \in \mathcal{C}$ and $v_2 \in \mathcal{C}$, and it is *proper* if the closure $\bar{\mathcal{C}}$ intersects its symmetric $-\bar{\mathcal{C}}$ only at the origin. Associated to any convex cone $\mathcal{C}$ there is a partial ordering on $\mathcal{H}$

$$v_1 \prec v_2 \Leftrightarrow v_2 - v_1 \in \mathcal{C} \cup \{0\},$$

which is compatible with the algebraic structure: $v_1 \prec v_2 \Leftrightarrow 0 \prec v_2 - v_1$ and the set of "positive" vectors is closed under sum and product by positive scalars. Given $v_1 \in \mathcal{C}$ and $v_2 \in \mathcal{C}$, define

---

[3]That is, such that $\operatorname{var}(\varphi) = \sup \sum_{j=1}^{n} |\varphi(x_j) - \varphi(x_{j-1})|$ is finite, where the supremum is over all choices of $x_0 < x_1 < \cdots < x_n$ in the domain.

$$\Theta(v_1, v_2) = \log \frac{\beta(v_1, v_2)}{\alpha(v_1, v_2)}$$

where (make the convention $\sup \emptyset = 0$ and $\inf \emptyset = \infty$)

$$\alpha(v_1, v_2) = \sup\{t > 0 : tv_1 \prec v_2\} \quad \text{and} \quad \beta(v_1, v_2) = \inf\{t > 0 : v_2 \prec tv_1\}.$$

Assuming $\mathcal{C}$ is a proper convex cone, Θ is a *projective metric*: it is symmetric, it satisfies the triangle inequality, and $\Theta(v_1, v_2) = 0 \Leftrightarrow v_1 = tv_2$ for some $t > 0$. Thus, Θ induces a true distance in the quotient of $\mathcal{C}$ by the projective relation $v_1 \sim v_2 \Leftrightarrow v_1 = tv_2$ for some $t > 0$.

The crucial property is that linear operators that leave a cone invariant tend to contract the associated projective metric. More precisely,

Proposition E.2 (Birkhoff [58]). *Let* $\mathcal{L} : \mathcal{H}_1 \to \mathcal{H}_2$ *be a linear operator and* $\mathcal{C}_i \subset \mathcal{H}_i$ *be proper convex cones such that* $\mathcal{L}(\mathcal{C}_1) \subset \mathcal{C}_2$. *Then*

$$\Theta_2(\mathcal{L}(v_1), \mathcal{L}(v_2)) \leq \tanh\left(\frac{D}{4}\right) \Theta_1(v_1, v_2) \quad \text{for all } v_1, v_2 \in \mathcal{C}_1,$$

where Θ_i *is the projective metric associated to* $\mathcal{C}_i$ *and* D *is the* Θ_2-*diameter of* $\mathcal{L}(\mathcal{C}_1)$, *that is,* $D = \sup\{\Theta_2(\mathcal{L}(v_1), \mathcal{L}(v_2)) : v_1, v_2 \in \mathcal{C}_1\}$.

In particular, the linear operator $\mathcal{L}$ is a contraction relative to Θ_1 and Θ_2 (because $\tanh(D/4) < 1$ if the diameter D is finite). We briefly explain how this result can be used to prove exponential decay of correlations for uniformly expanding maps [182, 263]; details can also be found in [40] and [441]. Liverani [264] used a similar idea in the context of piecewise expanding maps. Other applications will be mentioned later.

Let $f : M \to M$ be a C^1 expanding map with ν-Hölder derivative, and μ be the SRB (absolutely continuous) measure. For each $a > 0$ define $\mathcal{C}(a)$ as the set of functions $\varphi : M \to \mathbb{R}$ such that $\varphi > 0$ and $\log \varphi$ is (a, ν)-Hölder. Then $\mathcal{C}(a)$ is a proper convex cone in the Banach space $\mathcal{H}_\nu$ of ν-Hölder functions endowed with the norm $\|\varphi\| = \sup |\varphi| + \mathrm{Hol}_\nu(\varphi)$. Fix any ρ between $\sigma^{-\nu}$ and 1, where σ is a lower bound for the expansion rate of f. Corresponding to the fact that the transfer operator $\mathcal{L}$ tends to improve Hölder constants, which is also expressed by (E.2) above, we get

$$\mathcal{L}(\mathcal{C}(a)) \subset \mathcal{C}(\rho a) \quad \text{for every large } a. \tag{E.4}$$

From the expression of the projective metric Θ_a associated to $\mathcal{C}(a)$ one sees that $\mathcal{C}(\rho a)$ has finite Θ_a-diameter inside $\mathcal{C}(a)$. Thus, using Proposition E.2, the transfer operator $\mathcal{L}$ is a strict contraction relative to Θ_a. This implies that every $\mathcal{L}^n \varphi$, $\varphi \in \mathcal{C}(a)$ converges exponentially fast to the fixed point 1, relative to the metric Θ_a. Using the expression of the projective metric one deduces that every $\mathcal{L}^n \varphi$, $\varphi \in \mathcal{H}_\nu$ converges exponentially fast to a constant, relative to the sup-norm. This implies that (f, μ) has exponential decay of correlations in the space $\mathcal{H}_\nu$, with rate of decay bounded explicitly in terms of ρ.

E.4 Uniformly hyperbolic diffeomorphisms

The assumption that the system is expanding is crucial for the good behavior of the transfer operator expressed by (E.2) and (E.4). Hence, the situation becomes much more delicate when one considers hyperbolic (rather than purely expanding) transformations: the transfer operator tends to worsen, rather than improve, regularity of functions along the stable direction.

The classical strategy to bypass this difficulty in the case of $C^{1+\nu}$ uniformly hyperbolic maps, going back to Sinai, Ruelle, Bowen [86], is to quotient-out stable directions. More precisely, consider a diffeomorphism f restricted to some hyperbolic basic set Λ. Through a generating Markov partition, one defines a semi-conjugacy $h : \Sigma \to \Lambda$ between $f \mid \Lambda$ and some two-sided shift of finite type. Next, the two-sided shift is reduced to the corresponding one-sided shift, by "forgetting" all negative coordinates, that is to say, by projecting down to the quotient space of local stable leaves.

Since one-sided shifts are uniformly expanding, we already know they have exponential decay of correlations in any space of ν-Hölder functions. Then the same is true for the corresponding two-sided shift. The invariant measure μ may be any equilibrium state associated to a Hölder potential. The semi-conjugacy is Hölder continuous and, for these measures, it is actually a conjugacy (injective) on a full measure subset [4]. Thus, one may use it to translate the conclusion back to the original setting: the uniformly hyperbolic system (f, μ), where μ is an equilibrium state of a Hölder potential, has exponential decay of correlations in the space of Hölder functions.

This very effective approach has, nevertheless, some limitations. For one thing, it relies on the existence of generating Markov partitions, which is known in a few cases only. Moreover, even in the uniformly hyperbolic case, the semi-conjugacy h is only Hölder continuous (this is related to the fact that the stable and unstable foliations of hyperbolic sets are usually not better than Hölder), and so it can not retain other finer aspects of the transfer operator like, for instance, the non-peripheral part of the spectrum. With this in mind, more direct approaches have also been proposed. A general idea is to consider Banach spaces of more complicated objects, that look like a function along the unstable direction and like some dual object along the stable direction, so that their regularity is improved by the transfer operator in both directions.

In the elegant formulation by Liverani [263], one considers a convex cone $\mathcal{C}$ of functions defined through conditions that involve not the functions themselves but rather their averages along local stable leaves. In a few words, on each local stable leaf γ one considers the set $D(\gamma)$ of positive densities ρ such that $\log \rho$ is Lipschitz, for some fixed sufficiently large Lipschitz constant. Then one defines the cone $\mathcal{C}$ to consist of those functions φ on the ambient

---

[4]In general, h may fail to be injective only because different Markov rectangles intersect at the boundaries. For equilibrium states the orbits of the boundaries form a zero measure set.

space such that the average $\int_\gamma \varphi\rho$ is positive, for all $\rho \in D(\gamma)$ and every stable leaf γ, and such that the functions

$$\rho \mapsto \log \int_\gamma \varphi\rho \quad \text{and} \quad \gamma \mapsto \log \int_\gamma \varphi\rho$$

are Hölder, with appropriate Hölder constants (see [263, 441] for precise formulations). The key point is that this cone is strictly invariant under the transfer operator, in the sense of Proposition E.2, so that arguments like those outlined in the previous section may be applied. Thus, Liverani [263] proves exponential decay of correlations in the space of Hölder functions, for uniformly hyperbolic area-preserving surface maps, either smooth (Anosov) or piecewise smooth. This was then extended to solenoid-type attractors in any dimension by Viana [441].

In a similar spirit, and also most interesting, let us mention the works of Rugh [397], focusing on the relations between the dynamical zeta function and the correlation spectrum, and of Blank, Keller, Liverani [60], who propose a Banach space of "generalized functions" (the norm along the stable direction is defined through averaging with respect to certain Hölder densities), and prove the transfer operator has a spectral gap in this space.

E.5 Uniformly hyperbolic flows

For continuous time systems $(f^t)_{t\in\mathbb{R}}$ the correlation sequence is naturally replaced by the *correlation function*

$$C_t(\varphi,\psi) = \int \varphi(\psi \circ f^t)\, d\mu - \int \varphi\, d\mu \int \psi\, d\mu\,.$$

The problem of decay of correlations turns out to be even more subtle than for discrete time systems, already in the uniformly hyperbolic case: some transitive Anosov flows are not even mixing with respect to their SRB measures, or other relevant invariant measures, such as equilibrium states.

Indeed, consider a constant time suspension (or special flow) f^t over a transitive Anosov diffeomorphism $f : M \to M$. That is,

$$f^t : (M \times \mathbb{R}/\sim) \to (M \times \mathbb{R}/\sim), \quad (x,s) \mapsto (x, s+t)$$

where the equivalence relation $\sim$ is generated by $(x,c) \sim (f(x),0)$ for some positive constant c. This is a transitive Anosov flow, with the special feature that the stable and unstable subbundles are *jointly integrable*: the foliation $\{M \times \{s\} : s \in \mathbb{R}/c\mathbb{Z}\}$ is everywhere tangent to $E^s \oplus E^u$. Since this foliation is invariant, the information that two points are in the same leaf, or in nearby leaves, is never dissipated by the flow.

A natural conjecture would then be that every *mixing* uniformly hyperbolic flow is exponentially mixing. However, this is also not true: Ruelle [393]

Pollicott [361] gave examples whose correlations decay more slowly than any predetermined speed. Thus, the problem of characterizing the speed of mixing for uniformly hyperbolic flows, and especially Anosov flows, has remained an outstanding challenge:

Problem E.3. Do generic transitive Anosov flows have exponential decay of correlations?

The examples in [361, 393] show that some kind of geometric condition must be imposed, in order to prove fast decay of correlations. Geodesic flows on surfaces with constant negative curvature were the first case to be dealt with and shown to be exponentially mixing on each energy surface, see [132, 375]. About a decade later, Chernov [125] was able to prove stretched exponential mixing in the space of Hölder functions

$$|C_t(\varphi, \psi)| \leq C(\varphi, \psi)e^{-c\sqrt{t}} \quad \text{for all } t > 0, \tag{E.5}$$

for a large class of Anosov flows in dimension 3 and the corresponding SRB measures. He introduced the notion of uniform non-integrability, a kind of non-infinitesimal version of the Fröbenius non-integrability condition (the subbundles E^s and E^u are generally not differentiable, and so the usual Fröbenius condition does not make sense here). Let $x \in M$ and take $x_s \in W^{ss}(x)$ and $x_u \in W^u(x)$ at distance δ from x. Then consider $z_s \in W^{uu}_{loc}(x_s) \cap W^s_{loc}(x_u)$ and $z_u \in W^u_{loc}(x_s) \cap W^{ss}_{loc}(x_u)$. See Figure E.1. Then $z_u = f^\tau(z_s)$ for some $\tau = \tau(x_s, x_u)$. *Uniform non-integrability* means that $|\tau| \approx \delta^2$, in the sense that the quotient is uniformly bounded from zero and infinity. Chernov proved that C^2 transitive Anosov flows with uniform non-integrability have stretched exponential decay as stated in (E.5). Furthermore, stretched-exponential decay also holds for the geodesic flow on any surface with (variable) negative curvature. Notice, however, that (E.5) is an upper bound only, and so this breakthrough did not answer the question of whether mixing is actually exponential in most cases.

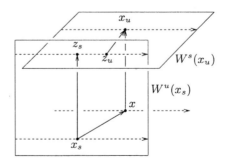

Fig. E.1. Uniform non-integrability of E^s and E^u

Exponential decay of correlations for a broad class of Anosov flows was proved for the first time by Dolgopyat [165], in a remarkable paper that corresponds to his doctoral thesis. Indeed, he showed that given any transitive Anosov $C^{2+\nu}$ flow on a compact manifold whose stable and unstable foliations are of class C^1 and which is uniformly non-integrable, and given the corresponding SRB measure, one has

$$|C_t(\varphi, \psi)| \leq C \, \|\varphi\|_5 \, \|\psi\|_5 \, e^{-ct}$$

for any pair of C^5 functions. Moreover, the same conclusion holds for the geodesic flow of any negatively curved surface. In a subsequent paper [166], Dolgopyat went on to characterize when a suspension of a uniformly hyperbolic diffeomorphism has fast (super-polynomial) decay of correlations, and to prove that this does hold in most cases, e.g. for Lebesgue almost every parameter value in generic one-parameter families.

The best current result is due to Liverani [262], who proves that every C^4 contact Anosov flow has exponential decay of correlations in the space of ν-Hölder functions (any $\nu > 0$):

$$|C_t(\varphi, \psi)| \leq C_\nu \|\varphi\|_\nu \|\psi\|_\nu e^{-c_\nu t}.$$

A flow on a manifold of odd dimension $2k + 1$ is *contact* if it leaves invariant a 1-form α such that $\alpha \wedge (d\alpha)^k$ is everywhere nonzero. The most important examples are the restrictions of geodesic flows to hypersurfaces of constant kinetic energy, α being a primitive of the symplectic form ω (that is, $d\alpha = \omega$) restricted to the hypersurface. So, in particular, Liverani's theorem implies exponential decay of correlations for the geodesic flow of every negatively curved manifold with any dimension.

Little seems to be know about the speed of mixing for flows which are not uniformly hyperbolic. The natural place to start are the singular hyperbolic (Lorenz-like) attractors discussed in Chapter 9:

Problem E.4. Prove exponential decay of correlations for singular hyperbolic attractors and their SRB measures. In the same setting, prove that $f^t_*(\varphi m)$ converges exponentially to a multiple of the SRB measure, for any Lipschitz function φ supported in a neighborhood of the attractor, where m denotes Lebesgue measure in the ambient space.

E.6 Non-uniformly hyperbolic systems

One-dimensional transformations:

Exponential decay of correlations for unimodal maps of the interval, and the corresponding absolutely continuous measures, was first proved by Keller, Nowicki [239] and Young [458], independently. Their results apply under similar assumptions, including exponential growth of the derivative (Collet-Eckmann condition) and exponential control of the recurrence on the orbit of the critical point c:

$$|Df^n(f(c))| \geq \sigma^n \quad \text{and} \quad |f^n(c) - c| \geq e^{-\alpha n} \quad \text{for all } n \geq 1$$

where $\sigma > 1$ and $\alpha > 0$ is small. Both approaches rely on constructing tower extensions to the interval map $f : M \to M$. The one in Keller, Nowicki [239] is a delicate Markov construction that provides information on many other important aspects of the dynamics, such as the distribution of periodic orbits and the dynamical zeta function. Young [458] proposes a lighter non-Markov tower, which was later employed in [43] to prove stochastic stability. Her construction, which is inspired by the idea of binding period in Benedicks, Carleson [52] (see Section 4.1), may be sketched as follows.

One fixes $\delta > 0$, much smaller than α, and $\beta \approx 2\alpha$. The tower space $\hat{M}$ is the (disjoint) union of all $\Delta_k \times \{k\}$, $k \geq 0$, where

$$\Delta_k = \begin{cases} \text{the whole interval } M, \text{ for } k = 0 \\ \text{the neighborhood of radius } e^{-\beta k} \text{ around } f^k(c), \text{ for } k \geq 1. \end{cases}$$

Let $\pi : \hat{M} \to M$ be the natural projection. The tower map $\hat{f} : \hat{M} \to \hat{M}$ is defined by

$$\hat{f}(x, k) = \begin{cases} (f(x), k+1) \text{ if } f(x) \in \Delta_{k+1} \text{ and } k \geq 1 \\ (f(x), k+1) \text{ if } |x - c| < \delta \text{ and } k = 0 \\ (f(x), 0) \quad \text{ in all other cases.} \end{cases}$$

In simple terms: points climb the tower, one level at the time, for as long as that is compatible with the rule $\pi \circ \hat{f} = f \circ \pi$; otherwise, they fall directly to the zeroth level; moreover, points are allowed to leave the zeroth level only if they are very close to the critical point. The crucial property is that return maps to the zeroth level are uniformly expanding:

$$\hat{f}^p(x, 0) = (f^p(x), 0) \quad \Rightarrow \quad |(f^p)'(x)| \geq \sigma^{p/3}.$$

This permits to introduce a Riemannian metric on $\hat{M}$, with a singularity at the critical value $(f(c), 1)$, relative to which the tower map $\hat{f}$ itself is uniformly expanding. This brings the problem to the realm of piecewise expanding maps, with infinitely many intervals of smoothness, so that the theory developed for such maps [252, 398] can be applied.

One defines the transfer operator $\hat{\mathcal{L}}$ associated to the tower map, with Lebesgue measure $\hat{m}$ as the reference measure [5], and proves that it satisfies a variation of the Lasota-Yorke inequality (E.3). From this one deduces that $\hat{\mathcal{L}}$ admits a fixed point $\hat{\phi}$ and has a spectral gap in the space $\hat{\mathcal{H}}$ of functions with bounded variation on $\hat{M}$. Take $\hat{\phi}$ with average 1 relative to $\hat{m}$. Then $\hat{\mu} = \hat{\phi}\hat{m}$ is

[5]This measure is finite, as the length of the k:th tower level is bounded by $2e^{-\beta k}$.

an $\hat{f}$-invariant absolutely continuous probability, and $(\hat{f}, \hat{\mu})$ has exponential decay of correlations in $\hat{\mathcal{H}}$. It follows that $\mu = \pi_*\hat{\mu}$ is the invariant absolutely continuous probability of the original interval map f. Embedding the space $\mathcal{H}$ of functions of bounded variation on M inside $\hat{\mathcal{H}}$ by lifting $\varphi \mapsto \varphi \circ \pi$, one deduces that (f, μ) has exponential decay of correlations in $\mathcal{H}$.

These results have been substantially improved by Bruin, Luzzatto, van Strien [99], who considered much more general multimodal maps, with no explicit condition on the recurrence of the critical orbits and a much weaker expansion condition:

$$\sum_n |(f^n)'(f(c_i))|^{-1/(2\ell-1)} < \infty$$

for every critical point c_i, where ℓ is the order of the critical points ($\ell = 2$ in the generic case). It is known from Nowicki, Sands [329] that for unimodal maps and the corresponding absolutely continuous invariant measures exponential decay of correlations is equivalent to the Collet-Eckmann condition. So, in this generality one must expect also slower types of decay. Based on the framework of Young [461] that we discuss in the next section, Bruin, Luzzatto, van Strien [99] proved that decay of correlations in the space of Hölder functions is, at least, of the same type (polynomial, stretched exponential, exponential) as the growth of the derivative on the critical orbits.

Problem E.5. Absolutely continuous invariant measures exist under the weaker condition $\sum_n |(f^n)'(f(c_i))|^{-1/\ell_{max}} < \infty$, $\ell_{max} = $ maximal order of all critical points, by [100, 330]. What are the optimal rates of decay of correlation in this generality? Are these maps stochastically stable?

Tower extensions are also central to the proof of exponential decay of correlations for Hénon-like maps, by Benedicks, Young [57], as we have seen in Section 4.4. Their construction was then axiomatized by Young [460] and has been applied in several other situations. We mention two main ones:

Dispersing billiards:

One striking application was the proof of exponential decay of correlations for dispersing billiards (Sinai billiards). We start by giving some background information. The monograph of Katok, Strelcyn [235] is the classical general reference on smooth systems with singularities, including billiards.

A planar *billiard* corresponds to the elastic motion of a point inside a bounded connected domain B with piecewise smooth boundary in $\mathbb{R}^2$ or $\mathbb{T}^2$: the motion takes place along straight lines, with elastic reflections (angle of incidence = angle of reflection) at the boundary ∂B. See Figure E.2. This is readily generalized to Riemannian manifolds in any dimension, with geodesics in the place of straight lines. Another interesting generalization are the so-called *dissipative billiards*, obeying different reflection laws or subject to external

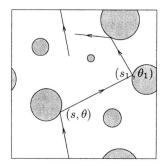

Fig. E.2. Torus billiard with convex scatterers

fields. However, here we shall restrict ourselves to the classical conservative case.

The billiard map $f : (s, \theta) \mapsto (s_1, \theta_1)$ describes the time evolution of the point: it is defined on a subset of $\partial B \times [-\pi/2, \pi/2]$ and assigns to each collision (s, θ) with the boundary the next one (s_1, θ_1), where the coordinate s parametrizes ∂B (by arc-length, say) and the coordinate θ describes the angle of incidence. Note that f may not be defined on the whole $\partial B \times [-\pi/2, \pi/2]$ (e.g. if the boundary is not smooth) and is usually not continuous. On the other hand, it is well-known that f preserves a probability

$$d\mu = a \cos \theta dr d\theta$$

(a is the normalizing constant) absolutely continuous with respect to Lebesgue measure.

The billiard is called *dispersing* if the boundary is strictly convex, seen from the interior of B, at every regular point. See Figure E.2. In this example the boundary is also completely smooth, which can not happen for dispersing billiards in the plane (there must be corner points). The theory of dispersing billiards originated from the fundamental work of Sinai [415], who proved ergodicity and mixing for torus billiards with convex scatterers. A crucial fact behind these results is that such billiards are uniformly hyperbolic, in the sense that there exists an invariant splitting $E^u \oplus E^s$ defined almost everywhere and with *uniform* rates of contraction and expansion along E^s and E^u, respectively.

On the other hand, these systems are only piecewise smooth, and that makes the dynamics much harder to analyze than in the Axiom A case. For one thing, it makes the issue of decay of correlations much more delicate: while hyperbolicity implies some exponential loss of memory at the infinitesimal level, it is not at all clear to what extent that is reflected at the global level, because for these systems stable and unstable manifolds may be very short (they get "chopped" at the discontinuities). Nevertheless, it was shown by Bunimovich, Chernov, Sinai [102, 101] that many planar dispersing billiards with finite horizon [6] have, at least, stretched exponential decay of correlations. For a while it was believed that this might be optimal, the presence of singularities overriding the tendency of uniform hyperbolicity to produce exponential mixing.

However, a few years later Liverani [263] proved exponential decay of correlations for certain two-dimensional uniformly hyperbolic area-preserving maps. A similar result had just been obtained by Chernov [124] for certain piecewise affine models. Neither result applies to billiards, because the derivative is assumed to be bounded, but this showed that in principle singularities need not slow down the rate of mixing. Then Young [460] proved, for the first time, exponential decay of correlations for planar billiards with convex scatterers and finite horizon. This was an application of an abstract framework that also allowed her to extend the result of Liverani [263] to the dissipative case, still in dimension 2. The method and the results were then further extended by Chernov, to very general planar dispersing billiards [126] and to uniformly hyperbolic maps with singularities in any dimension [127]. We refer the reader to Chernov, Young [123] for a most clear and informative exposition, including billiards with convex scatterers in higher dimensions.

Partially hyperbolic maps:

Other far-reaching applications of the tower approach developed in [57, 460] have been to partially hyperbolic diffeomorphisms. The first results were in the setting of maps with mostly contracting central direction treated in Section 11.2.4. As we have seen there, these maps admit a finite number of SRB measures and the union of the corresponding basins contains Lebesgue almost every point. Castro [122, 121] and Dolgopyat [167] considered subclasses for which the SRB measure is unique, and they proved exponential decay of correlations and the central limit theorem in the space of Hölder functions. Roughly speaking, Castro treated partially hyperbolic diffeomorphisms derived from Anosov by deformation, whereas Dolgopyat dealt with perturbations of skew-products over Anosov diffeomorphisms.

More recently, Alves, Luzzatto, Pinheiro [13] considered non-uniformly expanding maps modelled after [12, 440] and related to partially hyperbolic diffeomorphisms with mostly expanding central directions. These maps may have singularities or critical points. For the corresponding SRB (absolutely continuous) measures, Alves, Luzzatto, Pinheiro [13] proved super-polynomial decay of correlations

$$|C_n(\varphi, \psi)| \leq C_\alpha(\varphi, \psi)\, n^{-\alpha} \qquad \forall \alpha > 0$$

---

[6]A technical condition meaning that the number of consecutive *tangential* collisions with the boundary is uniformly bounded.

in the space of Hölder functions. Even more recently, the conclusion was strengthened to stretched exponential decay, $|C_n(\varphi, \psi)| \leq C(\varphi, \psi)\, e^{-c\sqrt{n}}$, for a related class of transformations, by Baladi, Gouezel [42, 196]. This includes the open class of cylinder transformations introduced by Viana [440] (see Section 11.1.3).

E.7 Non-exponential convergence

Invariant cones:

The two methods we mentioned before, via spectral gap and via invariant cones, rely on (projective) hyperbolicity of the transfer operator, and so they are naturally suited to proving *exponential* decay of correlations. Nevertheless, Maume-Deschamps [289] has shown how the cone method can be adapted to deal with other speeds of decay as well. In a few words, her strategy is to construct a sequence of cones C_j, with projective metrics Θ_j, such that $\mathcal{L}(C_{j-1})$ is contained in C_j and has finite Θ_j-diameter, for every j. By Proposition E.2, this implies contraction of the projective metrics

$$\Theta_j(\mathcal{L}(\phi_1), \mathcal{L}(\phi_2)) \leq \lambda_j \Theta_{j-1}(\phi_1, \phi_2), \quad \exists \lambda_j < 1$$

and so

$$\Theta_n(\mathcal{L}^n(\phi_1), \mathcal{L}^n(\phi_2)) \leq \Big(\prod_{j=1}^{n} \lambda_j \Big) \Theta_0(\phi_1, \phi_2).$$

Interpreting the expression of Θ_n, in each specific case, this inequality gives decay of correlations at a speed determined by the product $\prod_{j=1}^{n} \lambda_j$. Moreover, this product can be related to the asymptotics of the return time R^* (see below), to prove that if $m(\{R^* > n\})$ decreases polynomially, stretched exponentially, or exponentially, then so does the correlation sequence $C_n(\varphi, \psi)$, for any Hölder functions. Her estimate for the polynomial case falls short of being optimal; see (E.10) below. On the other hand, this invariant cone approach gives dependence of $C(\varphi, \psi)$ on ψ through the L^1 norm, which is most useful for certain applications.

A non-exponential tower:

A new approach, designed for dealing with all types of decay alike, was proposed by Young [461] and has been used by several authors. It is an extension of [57, 459] in that it is based on a similar tower model, but the analysis of the dynamics on the tower is new and does not involve the transfer operator. Instead Young uses a coupling argument, borrowed from probability theory, to show that *the type of decay of correlations is essentially determined by the tail of the recurrence time*, that is, by the asymptotics of the measure of the sets of points with large recurrence times. Let us outline this argument.

Let $(\Delta_0, \mathcal{B}, m)$ be a probability space and $R : \Delta_0 \to \mathbb{N}$ be a measurable function. Call R the *return time* and m the *reference measure*; in typical situations m could be Lebesgue measure, for instance. The associated *tower* is

$$\Delta = \{(x, \ell) \in \Delta_0 \times \mathbb{N} : R(x) > \ell\} = \bigcup_{\ell=0}^{\infty} (\Delta_\ell \times \{\ell\})$$

where $\Delta_\ell = \{x \in \Delta_0 : R(x) > \ell\}$. Call $\Delta_\ell \times \{\ell\}$ the *ℓ:th level* of the tower. Let $F : \Delta \to \Delta$ be a transformation of the form

$$F(x, \ell) = \begin{cases} (x, \ell + 1) & \text{if } \ell + 1 < R(x) \\ (G(x), 0) & \text{if } \ell + 1 = R(x) \end{cases}$$

where $G : \Delta_0 \to \Delta_0$ is measurable. Noticing that $(G(x), 0) = F^{R(x)}(x, 0)$, let us henceforth write $G = F^R$ and call it the *return map* (to the zeroth level). Assume

(A) (integrability of returns) We have $\int R \, dm = \sum_{\ell=0}^{\infty} m(\Delta_\ell) < \infty$.
(B) (Markov property) There exists a countable partition $\{\Delta_0^i\}$ of Δ_0 up to zero m-measure, such that the function $R(\cdot)$ is constant on each atom Δ_0^i and F^R maps each atom bijectively to Δ_0, with measurable inverse.
(C) (irreducibility) The greatest common divisor of $\{R(\Delta_0^i)\} = 1$.
(D) (weak hyperbolicity) The partition $\mathcal{P} = \{\Delta_0^i \times \{\ell\} : R(\Delta_0^i) > \ell\}$ of Δ is generating, in the sense that $\bigvee_{j=0}^{\infty} F^{-j}(\mathcal{P})$ is the partition into points.
(E) (Hölder Jacobian) For every i, the map $F^R : \Delta_0^i \to \Delta_0$ and its inverse are non-singular with respect to m, so that the Jacobian JF^R exists and is positive almost everywhere. Moreover, there exist C and $\beta < 1$ such that

$$\left| \frac{JF^R(x)}{JF^R(y)} - 1 \right| \leq C\beta^{s(F^R(x), F^R(y))} \quad \text{for all } x, y \in \Delta_0^i \text{ and every } i.$$

Here $s : \Delta_0 \times \Delta_0 \to \mathbb{N} \cup \{0, \infty\}$ is the *separation time*, defined by

$$s(x, y) = \min\{n \geq 0 : (F^R)^n(x) \text{ and } (F^R)^n(y) \text{ are in different } \Delta_0^i\}.$$

Condition (D) ensures that $s(x, y)$ is finite if $x \neq y$. Let $\beta < 1$ be fixed, and call $\beta^{s(x,y)}$ the *separation distance* between the two points.

Recurrence times and rates of decay:

To state the main result in [461], it is useful to extend some of these objects to the whole tower. To begin with, the σ-algebra $\mathcal{B}$ and the measure m extend naturally to Δ, since each level is naturally identified with a subset Δ_ℓ of Δ_0. Note that

$$m(\Delta) = \sum_{\ell=0}^{\infty} m(\Delta_\ell) = \int_{\Delta_0} R \, dm$$

is finite, by condition (A). Finally, let $R^*(x, \ell) = R(x) - \ell$ be the smallest $n \geq 1$ such that $F^n(x, \ell)$ is in the zeroth level. Note that

$$m(\{z \in \Delta : R^*(z) > n\}) = \sum_{\ell > n} m(\Delta_\ell) = \int_{\{R > n\}} R \, dm.$$

Define $\mathcal{H}_\beta$ to be the set of functions $\varphi : \Delta \to \mathbb{R}$ Hölder continuous relative to the separation distance: there exists $C > 0$ satisfying

$$|\varphi(x, \ell) - \varphi(y, k)| \leq \begin{cases} C\beta^{s(x,y)} & \text{if } k = \ell \\ C & \text{in all cases.} \end{cases}$$

The last condition ensures that every $\varphi \in \mathcal{H}_\beta$ is bounded. Then let $\mathcal{D}_\beta$ be the set of functions $\varphi \in \mathcal{H}_\beta$ such that, for each atom $P = \Delta_0^i \times \{\ell\}$ of the partition $\mathcal{P}$, either φ is identically zero or it is everywhere positive on P and, in the latter case,

$$\left| \frac{\varphi(x, \ell)}{\varphi(y, \ell)} - 1 \right| \leq C\beta^{s(x,y)} \quad \text{for all } x, y \in \Delta_0^i. \tag{E.6}$$

We call C a *distortion constant* for φ. By a slight abuse of language we say that a measure $\eta \in \mathcal{D}_\beta$ if it is absolutely continuous with respect to m and the density $d\eta/dm \in \mathcal{D}_\beta$.

Let $\|\lambda\|$ represent the total variation of a signed measure λ (see e.g. [389]).

Theorem E.6 (Young [461]).

1. *The map $F : \Delta \to \Delta$ admits an invariant probability measure $\mu \in \mathcal{D}_\beta$ with density bounded away from zero. Moreover, μ is unique and exact* [7] *(hence, mixing).*

2. *The tail of the recurrence time gives an upper bound for the rate of convergence to equilibrium of any measure $\eta \in \mathcal{D}_\beta$: if $m(\{z \in \Delta : R^*(z) > n\})$ decays polynomially, respectively stretched exponentially, respectively exponentially, then so does $\|F_*^n \eta - \mu\|$.*

3. *The same is true for decay of correlations of pairs of functions $\varphi \in \mathcal{H}_\beta$ and $\psi \in L^\infty(m)$: if $m(\{z \in \Delta : R^*(z) > n\})$ decays polynomially, respectively stretched exponentially, respectively exponentially, then so does $|C_n(\varphi, \psi)|$.*

The irreducibility condition (C) is not strictly necessary for the first sentence in part (1) of the theorem. In part (2) we mean that [8]

$$m(\{z \in \Delta : R^*(z) > n\}) = \mathcal{O}(n^{-\alpha}) \Rightarrow \|F_*^n \eta - \mu\| = \mathcal{O}(n^{-\alpha})$$
$$m(\{z \in \Delta : R^*(z) > n\}) = \mathcal{O}(\lambda^{(n^\theta)}) \Rightarrow \|F_*^n \eta - \mu\| = \mathcal{O}(\lambda^{(n^{\tilde{\theta}})})$$
$$m(\{z \in \Delta : R^*(z) > n\}) = \mathcal{O}(\lambda^n) \Rightarrow \|F_*^n \eta - \mu\| = \mathcal{O}(\tilde{\lambda}^n)$$

---

[7] A measure in a measurable space $(M, \mathcal{B})$ is *exact* for a measurable transformation $f : M \to M$ if the σ-algebra $\cap_n f^{-n}(\mathcal{B})$ has only zero or full measure sets.

[8] Recall that $a_n = \mathcal{O}(b_n)$ means there exists $C > 0$ such that $a_n \leq Cb_n$ for all n.

where $\tilde{\theta}$ and $\tilde{\lambda}$ are independent of η. Analogously in (3), with constants independent of φ and ψ. Observe that the estimates are particularly neat in the polynomial case: one gets the same exponent α.

Remark E.7. The sequence $m(\{z \in \Delta : R^*(z) > n\}$ also provides a *lower* bound for the rate of convergence to equilibrium: there exist probabilities $\eta \in \mathcal{D}_\beta$ and constants $c > 0$ such that

$$\|F_*^n \eta - \mu\| \geq cm(\{z \in \Delta : R^*(z) > n\}) \quad \text{for all } n.$$

Indeed, just take η such that $d\eta/dm \geq d\mu/dm + c$ outside the zeroth level $\Delta_0 \times \{0\}$, where $c > 0$ is some small constant depending only on $1 - \mu(\Delta_0 \times \{0\})$. Using that μ is invariant and $JF \equiv 1$ outside $F^{-1}(\Delta_0 \times \{0\})$,

$$\frac{d(F_*^n \eta)}{dm} \geq \frac{d\mu}{dm} + c \quad \text{on} \quad \bigcup_{\ell > n}(\Delta_\ell \times \{\ell\}),$$

and this implies

$$\|F_*^n \eta - \mu\| = \int \left| \frac{d(F_*^n \eta)}{dm} - \frac{d\mu}{dm} \right| dm \geq c \sum_{\ell > n} m(\Delta_\ell) = cm(\{R^* > n\}).$$

The coupling argument:

We are going to review some main ideas in the proof of Theorem E.6, in the special case when the Jacobian JF^R is constant on each Δ_0^i. See Remark E.8 below for quick comments on the general case.

Together with the Markov property (B), this simplifying assumption implies that the set $\mathcal{M}(\mathcal{P})$ of probabilities absolutely continuous with respect to m and whose densities are constant on each atom of $\mathcal{P}$ and on the entire zeroth level is invariant under forward iteration by F. Now an elementary argument gives that, for any $\eta \in \mathcal{M}(\mathcal{P})$, the sequence $(1/n)\sum_{j=1}^n F_*^j \eta$ has accumulation points μ which are also in $\mathcal{M}(\mathcal{P})$. It is a well-known basic fact that any accumulation point is an F-invariant measure. Using the irreducibility assumption (C) one sees that the density of $F_*^j \eta$ is uniformly bounded away from zero on $\Delta_0 \times \{0\}$ for all large j. This implies [9] that the same is true for the density of μ. Actually, recalling that $JF \equiv 1$ outside the pre-image of the zeroth level, one concludes that $d\mu/dm$ is bounded from zero on the whole tower Δ. This corresponds to part (1) of the theorem.

Now we outline the key coupling argument to prove part (2), for measures in $\mathcal{M}(\mathcal{P})$. Let us consider the transformation $F \times F$ and the measure $m \times m$, defined on the product space $\Delta \times \Delta$. The Jacobian of $F \times F$ relative to $m \times m$ is given by $J(F \times F) = JF \times JF$. We also consider the partition $\mathcal{P} \times \mathcal{P}$ of $\Delta \times \Delta$, and denote

---

[9] Assumption (C) can be avoided at this point. On the other hand, it is crucial for proving that μ is unique and exact.

$$(\mathcal{P} \times \mathcal{P})_n = \bigvee_{j=0}^{n-1} (F \times F)^{-j} (\mathcal{P} \times \mathcal{P}).$$

Let $\pi_i : \Delta \times \Delta \to \Delta$ be the natural projection to the i:th coordinate, $i = 1, 2$. Define the first *simultaneous return time* $T : \Delta \times \Delta \to \mathbb{N}$ by

$$T(z_1, z_2) = \min\{n \geq 1 : F^n(z_1) \text{ and } F^n(z_2) \text{ are both in } \Delta_0 \times \{0\}\}.$$

In view of the Markov property (B), $T(z_1, z_2) = n$ if and only if $T = n$ on the whole element P_n of $(\mathcal{P} \times \mathcal{P})_n$ that contains (z_1, z_2), and $(F \times F)^n$ maps P_n bijectively to $\Delta_0 \times \Delta_0$. Now let η_1 and η_2 be any probabilities in $\mathcal{M}(\mathcal{P})$. It is clear that

$$F_*^n \eta_i = \pi_{i,*}(F \times F)_*^n (\eta_1 \times \eta_2).$$

Since $J(F \times F)^n$ is constant on P_n, the density of $\pi_{i,*}(F \times F)_*^n (\eta_1 \times \eta_2 \mid P_n)$ is constant on the zeroth level. In other words,

$$\pi_{i,*}(F \times F)_*^n (\eta_1 \times \eta_2 \mid P_n) = \frac{m \mid \Delta_0}{m(\Delta_0)} (\eta_1 \times \eta_2)(P_n). \tag{E.7}$$

The point with this equality is that the right hand side does not depend on i. We say that η_1 and η_2 *match at time n*, restricted to P_n: observe that

$$\pi_{1,*}(F \times F)_*^k (\eta_1 \times \eta_2 \mid P_n) = \pi_{2,*}(F \times F)_*^k (\eta_1 \times \eta_2 \mid P_n)$$

for all $k \geq n$. Therefore,

$$\|F_*^n \eta_1 - F_*^n \eta_2\|$$
$$= \|\pi_{1,*}(F \times F)_*^n (\eta_1 \times \eta_2) - \pi_{2,*}(F \times F)_*^n (\eta_1 \times \eta_2)\|$$
$$= \|\pi_{1,*}(F \times F)_*^n (\eta_1 \times \eta_2 \mid \{T > n\}) - \pi_{2,*}(F \times F)_*^n (\eta_1 \times \eta_2 \mid \{T > n\})\|.$$

This proves that

$$\|F_*^n \eta_1 - F_*^n \eta_2\| \leq 2(\eta_1 \times \eta_2)(\{T > n\}) \quad \text{for every } n \geq 1. \tag{E.8}$$

Remark E.8. In the general case, when the Jacobian is non-constant, one can not expect a perfect matching as in (E.7). However, using $\eta_i \in \mathcal{D}_\beta$ together with the bounded distortion assumption (E), one can prove that both measures $\pi_{i,*}(F \times F)_*^n (\eta_1 \times \eta_2 \mid P_n)$, $i = 1, 2$ are in $\mathcal{D}_\beta$, with bounded distortion constants; recall (E.6). In particular, unless the densities vanish identically, one can find $\kappa > 0$ such that

$$\pi_{i,*}(F \times F)_*^n (\eta_1 \times \eta_2 \mid P_n) \geq \kappa(m \mid \Delta_0) \quad \text{for } i = 1, 2.$$

One says that this part $\kappa(m \mid \Delta_0)$ of both measures matches at time n, and keeps iterating the remainder until the next simultaneous return. See Figure E.3. It is tempting to take κ as large as possible (the infimum of the densities), to try to speed up the matching. However, this is not allowed because subtracting the matching part worsens the distortion constant and, in order to be able to repeat the argument at the next step, we need the remaining measures

$$\pi_{i,*}(F \times F)^n_*(\eta_1 \times \eta_2 \mid P_n) - \kappa(m \mid \Delta_0)$$

to be in $\mathcal{D}_\beta$, with not too large a distortion constant. This is achieved by choosing a convenient smaller κ instead.

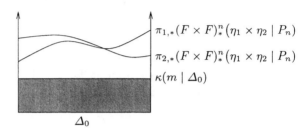

$$\pi_{1,*}(F \times F)^n_*(\eta_1 \times \eta_2 \mid P_n)$$

$$\pi_{2,*}(F \times F)^n_*(\eta_1 \times \eta_2 \mid P_n)$$

$$\kappa(m \mid \Delta_0)$$

Δ_0

Fig. E.3. Matching measures

So far, in (E.8), we have been able to bound the correlation decay by the tail of the simultaneous return time T. In fact, the previous arguments to obtain (E.8) remain valid if we take T to be any other simultaneous return time, not necessarily the first one. Our next goal is to relate $(\eta_1 \times \eta_2)(\{T > n\})$ to the tail $m(\{R^* > n\})$ of the recurrence time. For this, it is convenient to change the definition of T slightly, as follows. First, using the irreducibility condition (C) and the Markov condition (B), we may find $p \geq 1$ and $\gamma > 0$ such that

$$m\big((\Delta_0 \times \{0\}) \cap F^{-n}(\Delta_0 \times \{0\})\big) \geq \gamma \quad \text{for all } n \geq p. \tag{E.9}$$

For any (z_1, z_2) consider the sequence of return times $0 = \tau_0 < \tau_1 < \cdots$ defined by

$$\tau_{j+1} = \begin{cases} \tau_j + p + R^*\big(F^{\tau_j+p}(z_1)\big) & \text{if } j \text{ is even} \\ \tau_j + p + R^*\big(F^{\tau_j+p}(z_2)\big) & \text{if } j \text{ is odd.} \end{cases}$$

Then define T to be the first of these τ_j such that both $F^{\tau_j}(z_1)$ and $F^{\tau_j}(z_2)$ are in the zeroth level.

Observe that if we were to remove the delay term p then τ_{j+1} would be the first return of $F^{\tau_j}(z_i)$ to the zeroth level after time τ_j, with $i = 1, 2$ depending on the parity of j, and T would indeed be the first simultaneous return time. The reason the delay term is added is to allow time for a definite fraction of the mass to return to the zeroth level after time τ_j, so that statistics start afresh for those points. This is important to prove that

(a) the distribution of next return times $\tau_{j+1} - \tau_j$ is bounded by that of first return times:

$$(\eta_1 \times \eta_2)(\tau_{j+1} - \tau_j > p + n \mid \tau_j) \le Cm(\{R^* > n\})$$

(b) there is a definite positive probability the next return will be simultaneous:

$$(\eta_1 \times \eta_2)(T = \tau_{j+1} \mid \tau_j) \ge c > 0.$$

The constants C and c depend on $\eta_1 \times \eta_2$ but this dependence may be removed by considering τ_j for large j only (another adjustment of the definition of T).

Combining (a)-(b) with (E.8) one can prove that the sequence $\|F_*^n \eta_1 - F_*^n \eta_2\|$ has, at least, the same type of decay (polynomial, stretched exponential, exponential) as $m(\{R^* > n\})$. Taking $\eta_1 = \eta$ and $\eta_2 = \mu$, and recalling that μ is F-invariant, this gives the conclusion of part (2) of the theorem.

Part (3) is a consequence. Indeed, let $\varphi \in \mathcal{H}_\beta$ and $\psi \in L^\infty(m)$. It is no restriction to suppose $\int \varphi \, d\mu = 1$. Then

$$|C_n(\varphi, \psi)| = \left| \int (\psi \circ F^n) \varphi \frac{d\mu}{dm} \, dm - \int \psi \, d\mu \right|$$

$$= \left| \int \psi \, d[F_*^n(\varphi \frac{d\mu}{dm} m)] - \int \psi \, d\mu \right| \le \|\psi\|_\infty \|F_*^n \eta - \mu\|,$$

where $\eta = \varphi(d\mu/dm)m$. Suppose $\varphi \in \mathcal{H}_\beta$ is positive and bounded from zero (it is easy to reduce the general case to this one). Then $\varphi \in \mathcal{D}_\beta$ and, since the density $d\mu/dm$ is also in $\mathcal{D}_\beta$, the same is true for $\varphi(d\mu/dm)$. In other words, $\eta \in \mathcal{D}_\beta$. This means that we may apply part (2) to the last term, to conclude the argument.

E.8 Maps with neutral fixed points

Maps of the interval exhibiting indifferent fixed points have been studied from several different points of view: intermittency, bifurcation theory, fractal dimensions, dynamical zeta functions, invariants of smooth conjugacy, renormalization theory, equilibrium states, to mention just a few. They are also the simplest interesting examples of dynamical systems with polynomial mixing, and that is the reason they concern us at this point.

Let $f : [0,1] \to [0,1]$ be a piecewise C^2 map satisfying

1. there exist $a, \gamma > 0$ such that $f(x) = x + ax^{1+\gamma} +$ *higher order terms* near the origin (in particular $f(0) = 0$ and $f'(0) = 1$);
2. there exist $0 = c_0 < c_1 < \cdots < c_{k-1} < c_k = 1$ such that f is C^2 on every (c_{j-1}, c_j) with $|f''|$ bounded outside any neighborhood of 0;
3. $|f'(x)| > 1$ for every $x \ne 0$ and $|f'|$ is bounded from 1 outside any neighborhood of 0.

The following Markov property is not necessary for the conclusions, but it greatly simplifies our discussion:

4. each (c_{j-1}, c_j) is mapped bijectively to $(0, 1)$.

It is well-known that such a map f admits an invariant measure absolutely continuous with respect to Lebesgue measure m, unique up to multiplication by a constant. The order of contact with the identity at the fixed point determines whether such a measure is finite ($\gamma < 1$) or just σ-finite ($\gamma \geq 1$). In the latter case, Dirac measure at 0 is the physical measure of f. In what follows we only consider the former (finite) case.

Let μ be the invariant absolutely continuous probability. A number of authors have undertaken to estimate the rate of decay of correlations of (f, μ), independently and at about the same time. Liverani, Saussol, Vaienti [266] used a random perturbations approach (zero noise limit) inspired from Chernov [125], to prove

$$|C_n(\varphi, \psi)| \leq C(\varphi, \psi) n^{1-1/\gamma} (\log n)^{1/\gamma} \qquad (E.10)$$

for $\varphi \in C^1$ and $\psi \in L^\infty$. Maume-Deschamps [289] obtained a similar estimate, for $\varphi \in$ Lipschitz and $\psi \in L^1$. This is not optimal, however, because of the logarithm factor. The sharp upper bound

$$|C_n(\varphi, \psi)| \leq C(\varphi, \psi) n^{1-1/\gamma} \qquad (E.11)$$

was obtained by Hu [220] for $\varphi \in$ Lipschitz and $\psi \in L^\infty$, and by Young [460] for $\varphi \in$ Hölder and $\psi \in L^\infty$. In the remainder of the section we explain how this can be deduced using the framework in the previous section. A similar bound had been obtained by Isola [226] for affine approximations of the map.

Given $f : [0, 1] \to [0, 1]$ satisfying 1 – 4 above, let f_0 be its restriction to $[0, c_1)$ and f_0^{-1} be the corresponding inverse. Define $c_j^i = f_0^{-i+1}(c_j)$ for $1 \leq j < k$ and $i \geq 1$. Moreover, let $c_1^0 = 1$. Now define $R : (0, 1] \to \mathbb{N}$ by stating that R is constant equal to ℓ on each interval $(c_1^\ell, c_1^{\ell-1}]$. We are going to consider the tower extension $F : \Delta \to \Delta$ associated to this return time R. Let us check the conditions in the previous section.

Notice first that $\Delta_\ell = (0, c_1^\ell]$ for every $\ell \geq 0$. The main technical estimate, which the reader may carry out as an exercise, is

$$m(\Delta_\ell) = c_1^\ell \approx \frac{1}{\ell^{1/\gamma}} \qquad (E.12)$$

where $\approx$ means that the quotient of the two terms is bounded from zero and infinity. Since $\gamma < 1$, this implies that $\sum_\ell m(\Delta_\ell)$ is finite, as required in (A). The Markov condition (B) follows directly from property 4: take $\mathcal{P}$ to be the partition determined by the points c_j^i. The irreducibility condition (C) is obvious in this case, because R is surjective. The map F^R is uniformly expanding: the derivative is bounded below by the infimum of $|f'|$ on (c_1^2, c_1^1).

This implies the partition $\mathcal{P}$ is generating as required in (D). Finally, the distortion condition (E) uses the assumption that $|f''|$ is bounded far from zero (property 2) together with the local form of f near zero (property 1), especially the estimate

$$|c_1^\ell - c_1^{\ell-1}| \approx \frac{1}{\ell^{1+1/\gamma}}\,.$$

Notice that (E.12) gives $m(\{R^* > n\}) \approx \sum_{\ell>n} \ell^{-1/\gamma} \approx n^{-1/\gamma+1}$. Thus, by Theorem E.6, the map F admits an invariant probability μ absolutely continuous with respect to m, and (F,μ) has polynomial decay of correlations:

$$|C_n(\varphi, \psi)| = \mathcal{O}(n^{-1/\gamma+1}) \quad \text{for } \varphi \in \mathcal{H}_\beta \text{ and } \psi \in L^\infty(m). \tag{E.13}$$

Then $\pi_*\mu$ is the absolutely continuous f-invariant probability in $[0,1]$, where $\pi : \Delta \to \Delta_0$ is the natural projection. Note also that every ν-Hölder continuous function $\varphi : \Delta_0 \to \mathbb{R}$ lifts, via $\varphi \mapsto \varphi \circ \pi$, to a function in some $\mathcal{H}_{\beta(\nu)}$. Therefore, (E.13) implies that the correlations of $(f, \pi_*\mu)$ also decay like $n^{-1/\gamma+1}$, as stated in (E.11).

E.9 Central limit theorem

Let X_n, $n \geq 1$ be independent identically distributed random variables in a probability space $(M, \mathcal{F}, \mu)$. The classical central limit theorem states that, assuming the mean $\bar{X} = E(X_n)$ and the variance $\sigma^2 = V(X_n)$ are finite, the deviation of the average $n^{-1} \sum_{j=1}^n X_j$ from the mean $\bar{X}$, scaled by $1/\sqrt{n}$, converges in distribution to a Gaussian law: for every $z \in \mathbb{R}$,

$$\lim_{n\to\infty} \mu\left(\frac{1}{\sqrt{n}} \sum_{j=1}^n (X_j - \bar{X}) \leq z\right) = \frac{1}{\sqrt{2\pi}\sigma} \int_{-\infty}^z e^{\frac{-t^2}{2\sigma^2}}\, dt\,.$$

Given a dynamical system $f : (M, \mu) \to (M, \mu)$, the successive measurements $\varphi \circ f^n$, $n \geq 0$ of a function φ are not independent variables, of course. Nevertheless, we are going to see that if φ satisfies a fast mixing property then the sequence $\varphi \circ f^n$ does share this (and other) properties of independent random variables. We restrict ourselves to a few general facts that are relevant in connection with other results in the text, referring the reader to Denker [149], Viana [441], Baladi [40], for more detailed expositions and other concrete applications.

Let $f : M \to M$ be a measurable map for which μ is invariant and ergodic. Replacing φ by $\phi = \varphi - \int \varphi\, d\mu$ when necessary, we may restrict ourselves to functions with zero mean.

Theorem E.9 (Gordin [195]). *Let $\phi \in L^2(M, \mathcal{F}, \mu)$ be such that $\int \phi\, d\mu = 0$, and let $\mathcal{F}_n$ be the non-increasing sequence of σ-algebras $\mathcal{F}_n = f^{-n}(\mathcal{F})$, $n \geq 0$. Assume that*

$$\sum_{n=0}^{\infty} \|E(\phi \mid \mathcal{F}_n)\|_2 < \infty. \tag{E.14}$$

Then $\sigma^2 = \int \phi^2 \, d\mu + 2 \sum_{n=1}^{\infty} \int \phi(\phi \circ f^n) \, d\mu$ is finite and nonnegative. If $\sigma > 0$ then

$$\lim_{n \to \infty} \mu\left(\frac{1}{\sqrt{n}} \sum_{j=0}^{n-1} \phi \circ f^j \leq z\right) = \frac{1}{\sqrt{2\pi}\sigma} \int_{-\infty}^{z} e^{-\frac{t^2}{2\sigma^2}} \, dt ,$$

for any $z \in \mathbb{R}$. Also, $\sigma = 0$ if and only if $\phi = u \circ f - u$ for some $u \in L^2(\mu)$.

Let us explain the meaning of assumption (E.14) and what it has to do with the speed of mixing.

The *conditional expectation* $E(X \mid \mathcal{G})$ of a random variable X with respect to a σ-algebra $\mathcal{G} \subset \mathcal{F}$ is the Radon-Nikodym derivative of the measure μ_X defined by

$$\mu_X(B) = \int_B X \, d\mu \quad \text{for every } B \in \mathcal{G}$$

relative to the restriction of μ to $\mathcal{G}$. In other words, $E(X \mid \mathcal{G})$ is the essentially unique $\mathcal{G}$-measurable function satisfying $\int_B X \, d\mu = \int_B E(X \mid \mathcal{G}) \, d\mu$ for every $B \in \mathcal{G}$ or, equivalently (take linear combinations and pointwise limits),

$$\int X \psi \, d\mu = \int E(X \mid \mathcal{G}) \psi \, d\mu \tag{E.15}$$

for every $\mathcal{G}$-measurable function ψ. For variables $X \in L^2(M, \mathcal{F}, \mu)$, the conditional expectation coincides with the orthogonal projection to the subspace $L^2(M, \mathcal{G}, \mu)$. Indeed, (E.15) means that $X - E(X \mid \mathcal{G})$ is orthogonal to every $\mathcal{G}$-measurable function and, in particular, to $\psi = E(X \mid \mathcal{G})$. Hence, $X = E(X \mid \mathcal{G}) + (X - E(X \mid \mathcal{G}))$ is a decomposition of X into two orthogonal terms. This implies that $E(X \mid \mathcal{G})$ is in $L^2(M, \mathcal{G}, \mu)$ and is the orthogonal projection of X to $L^2(M, \mathcal{G}, \mu)$.

In particular, the expression $E(\phi \mid \mathcal{F}_n)$ in condition (E.14) corresponds to the orthogonal projection of ϕ to the subspace $L^2(M, \mathcal{F}_n, \mu)$ of $\mathcal{F}_n$-measurable L^2 functions. This means that (recall that $\int \phi \, d\mu = 0$),

$$\|E(\phi \mid \mathcal{F}_n)\|_2 = \sup\left\{\int \xi\phi \, d\mu : \xi \in L^2(M, \mathcal{F}_n, \mu) \text{ with } \|\xi\|_2 = 1\right\}$$

$$= \sup\left\{\int (\psi \circ f^n)\phi \, d\mu : \psi \in L^2(M, \mathcal{F}, \mu) \text{ with } \|\psi\|_2 = 1\right\}$$

$$= \sup\{C_n(\phi, \psi) : \psi \in L^2(M, \mathcal{F}, \mu) \text{ with } \|\psi\|_2 = 1\}.$$

Therefore, assumption (E.14) is fulfilled by any mean zero function ϕ with *summable decay of correlations*, in the following sense: there exist $C_n(\phi) \geq 0$ such that

$$|C_n(\phi, \psi)| \leq C_n(\phi)\|\psi\|_2 \quad \forall \psi \in L^2 \quad \text{with} \quad \sum_n C_n(\phi) < \infty. \qquad \text{(E.16)}$$

This is the case, in particular, if $C_n(\phi)$ decreases exponentially, stretched exponentially, or even polynomially, $C(\varphi)n^{-\alpha}$ *with exponent* $\alpha > 2$.

In other situations where such an L^2 bound is not available, it may still be possible to get the central limit property from the following

Theorem E.10 (Liverani [265]). *Let* $\phi \in L^\infty(M, \mathcal{F}, \mu)$ *have mean zero, and let* $\mathcal{F}_n$ *be the non-increasing sequence of* σ-*algebras* $\mathcal{F}_n = f^{-n}(\mathcal{F})$, $n \geq 0$. *Take measurable functions* ϕ_n *such that* $E(\phi \mid \mathcal{F}_n) = \phi_n \circ f^n$ *and assume that*

(a) $\sum_{n=0}^{\infty} |\phi_n|$ *converges almost everywhere*

(b) and $\sum_{n=0}^{\infty} |\int \phi \, (\phi \circ f^n) \, d\mu|$ *converges.*

Then all the conclusions of Theorem E.9 are valid for ϕ.

The functions ϕ_n exist because $E(\phi \mid \mathcal{F}_n)$ is $\mathcal{F}_n$-measurable; it is clear that they are essentially unique. Note that $\|\phi_n\|_p = \|E(\phi \mid \mathcal{F}_n)\|_p$ for any $p \in [1, \infty]$, because the probability μ is f-invariant. Moreover,

$$\Big| \int \phi \, (\phi \circ f^n) \, d\mu \Big| = \Big| \int E(\phi \mid \mathcal{F}_n) \, (\phi \circ f^n) \, d\mu \Big| \leq \|E(\phi \mid \mathcal{F}_n)\|_p \|\phi\|_\infty \, .$$

Hence, conditions (a) and (b) hold if $\|E(\phi \mid \mathcal{F}_n)\|_p$ is summable for some $p \in [1, \infty]$. In particular, both conditions are weaker than (E.14).

The strategy of the proof is to reduce Theorem E.9 to the central limit theorem for *reversed martingale differences*, that is, sequences of random variables X_n such that each X_n is $\mathcal{F}_n$-measurable and $E(X_n \mid \mathcal{F}_{n+1}) = 0$ for every $n \geq 0$. Using (E.14) one finds a measurable function $\zeta \in L^2(M, \mathcal{F}, \mu)$ such that

$$E(\zeta \mid \mathcal{F}_1) - \zeta \circ f = E(\phi \mid \mathcal{F}_1).$$

Then $\eta = \phi + \zeta \circ f - \zeta \in L^2$ and $E(\eta \mid \mathcal{F}_1) = 0$, and the latter implies $X_n = \eta \circ f^n$ is a reversed martingale difference. Observe that

$$\sum_{j=0}^{n-1} \phi \circ f^j = \zeta - \zeta \circ f^n + \sum_{j=0}^{n-1} \eta \circ f^j$$

so that the central limit theorem for ϕ follows from the corresponding statement for η. See [441] for the detailed arguments, and [175] or [318] for the statement and proof of the central limit theorem for martingale differences. The strategy for Theorem E.10 is similar, but the situation is more delicate because we can not expect ζ and η to be in L^2. This difficulty is bypassed by Liverani using an idea from [244].

Theorem E.9 is of little use when the map f is invertible, because the sequence $(\mathcal{F}_n)_n$ is constant and so (E.14) can never be fulfilled, unless $\phi \equiv 0$. The following version of Gordin's theorem is designed for invertible maps. A proof can be found in [441], inspired by [174, 265]. Liverani [265] also contains a version of Theorem E.10 for invertible maps.

Theorem E.11. *Let* $\phi \in L^2(N, \mathcal{F}, \mu)$ *be such that* $\int \phi \, d\mu = 0$. *Suppose the map* $f \colon M \to M$ *is invertible, and there exists some* σ*-algebra* $\mathcal{G}_0 \subset \mathcal{F}$ *such that* $\mathcal{G}_n = f^{-n}(\mathcal{G}_0)$, $n \in \mathbb{Z}$, *is a non-increasing sequence, and*

$$\sum_{n=0}^{\infty} \|E(\phi \mid \mathcal{G}_n)\|_2 < \infty \quad and \quad \sum_{n=0}^{\infty} \|\phi - E(\phi \mid \mathcal{G}_{-n})\|_2 < \infty.$$

Then all the conclusions of Theorem E.9 are valid for ϕ.

For example, for Anosov diffeomorphisms f one may take $\mathcal{G}_0$ to be the σ-algebra of measurable sets which are unions of entire local stable leaves, meaning, entire connected components of the intersections of stable manifolds with the rectangles of some Markov partition. This σ-algebra contains its preimage $f^{-1}(\mathcal{G}_0)$, strictly, and, in fact, satisfies the hypothesis of Theorem E.11, relative to any equilibrium state of f associated to a Hölder potential. See [263] or [441, Section 4.4].

F

Conclusion

We have reviewed a number of recent developments that outline a global perspective of dynamical systems. Much progress went on as this book was being written and, no doubt, that is going to continue. As we approach the end of our survey, it seems useful to indicate some of what is left to do.

For smooth interval or circle maps, the program is very much advanced, and, while the global behavior of multimodal maps is not yet completely understood, one may expect that the questions we touched upon will be solved in the near future. More general transformations, with discontinuities or singularities, and other 1-dimensional dynamics, such as iterated functions systems, have yet to be analyzed to a similar depth.

There is also very substantial progress in the context of surface diffeomorphisms, as we have seen. In the C^1 topology the main remaining question is whether uniformly hyperbolic maps form a dense subset or, equivalently, whether there exists a non-empty open set where coexistence of infinitely many periodic attractors or repellers is generic. A key problem in this direction is to decide whether homoclinic tangencies associated to the homoclinic class of a periodic point can persist on a whole C^1 open set.

For higher topologies, the dynamical landscape is less understood. It seems that non-hyperbolic behavior always emanates from homoclinic tangencies, but this has yet to be established rigorously. For instance, at this point we can not exclude the possibility of a C^2 open set of diffeomorphisms exhibiting an attractor that contains no periodic attractors. Such pathologies apart, one main problem is whether the phenomenon of coexistence of infinitely many attractors or repellers may occur with positive probability in parameter space, for generic families of diffeomorphisms.

On the other hand, the study of the Hénon model has unveiled the dynamics of a large class of dynamical systems whose non-hyperbolic character is directly linked to homoclinic tangencies, especially strongly dissipative ones, showing that they are amenable to a rather complete dynamical description. It may be that the attractors of most surface diffeomorphisms are either hyperbolic or Hénon-like, but this is presently not yet known. Another main open problem is to extend the theory of Hénon-like systems to the weakly dissipative systems, and even to conservative systems, such as the standard map.

A good picture is also available for flows in 3-dimensions. In the C^1 topology, we now know that singular, or Lorenz-like, attractors and repellers are the only new phenomenon, relative to the case of surface diffeomorphisms. The theory of singular hyperbolicity, which encompasses both uniform hyperbolicity and Lorenz-like behavior, is already well advanced. For higher topologies, several questions remain open, especially concerning the interaction between singular and critical (homoclinic) behaviors: a number of examples have been explored in detail, but a unifying theory is still missing.

The picture becomes mistier as we move up in dimension, for both discrete time and continuous times systems. Yet, for C^1 generic diffeomorphisms, we now have appropriate notions of dynamical ("spectral") decomposition and of elementary dynamical piece. For tame systems (finitely many elementary pieces), they can be analyzed separately, and then the global dynamics can be obtained by assembling these semi-local dynamics together. To analyze the dynamics of isolated elementary pieces, including those of tame diffeomorphisms, is a crucial task, that has yet to be tackled: we know that these sets are volume-hyperbolic and robustly chain recurrent, but these properties have not yet been fully exploited.

Furthermore, the study of wild diffeomorphisms remains wide open. The previous comments apply equally to vector fields where, in addition, we have yet to understand how singular behavior can be inserted into the concepts of partial hyperbolicity and dominated splitting, that is, we must still find appropriate notions of singular volume hyperbolicity and singular dominated splitting.

For smoother situations, most recent progress concerned the statistical properties of certain classes of systems, especially the construction of SRB measures, control of their basins of attractions, and study of their mixing properties. Most available results depend on assumptions of weak hyperbolicity, such as partial hyperbolicity and non-uniform hyperbolicity. A fundamental open problem is whether SRB measures, with basins that cover almost every point, exist for a large (full probability in parameter space, generic, or at least dense) subset of dynamical systems.

The study of Lyapunov exponents has been pursued recently, as a means to try to understand how common such properties of weak hyperbolicity are among dynamical systems, either conservative or dissipative. For conservative C^1 diffeomorphisms, we know that non-uniform hyperbolicity (non-zero Lyapunov exponents) is, actually, a rather rigid property: generically, it implies domination of the Oseledets splitting. A corresponding generic theory of Lyapunov exponents for smoother dynamical systems is wide open, but recent results on linear cocycles suggest one should expect a much different picture.

These developments give us the right to, once more, believe a global theory of dynamical systems may be within reach. On the way, we have discovered that, much beyond the classical set-up, the basic idea of hyperbolicity, in weaker and more flexible forms, remains at the heart of persistent dynamical behavior. Which other fundamental ingredients will be necessary to found such a theory is a basic question that has yet to be clarified. In any event, this is a very exciting project indeed, and we can only hope our text will inspire the reader to join it!

References

1. F. Abdenur. Generic robustness of spectral decompositions. *Annales Scient. E.N.S.*, 36:212–224, 2003.
2. F. Abdenur, A. Avila, and J. Bochi. Robust transitivity and topological mixing for C^1-flows. *Proc. Amer. Math. Soc.*, 132:699–705, 2004.
3. F. Abdenur and L. J. Díaz. Shadowing in the C^1 topology. Preprint 2004.
4. R. Abraham and S. Smale. Nongenericity of Ω-stability. In *Global analysis*, volume XIV of *Proc. Sympos. Pure Math. (Berkeley 1968)*. Amer. Math. Soc., 1970.
5. V. S. Afraimovich, V. V. Bykov, and L. P. Shil'nikov. On the appearence and structure of the Lorenz attractor. *Dokl. Acad. Sci. USSR*, 234:336–339, 1977.
6. V. S. Afraimovich, S.-C. Chow, and W. Liu. Lorenz-type attractors from codimension one bifurcation. *J. Dynam. Differential Equations*, 7:375–407, 1995.
7. J. F. Alves. SRB measures for non-hyperbolic systems with multidimensional expansion. *Ann. Sci. École Norm. Sup.*, 33:1–32, 2000.
8. J. F. Alves. *Statistical analysis of non-uniformly expanding dynamical systems.* Lecture Notes 24th Braz. Math. Colloq. IMPA, Rio de Janeiro, 2003.
9. J. F. Alves and V. Araújo. Random perturbations of nonuniformly expanding maps. *Astérisque*, 286:25–62, 2003.
10. J. F. Alves, V. Araújo, and B. Saussol. On the uniform hyperbolicity of some nonuniformly hyperbolic systems. *Proc. Amer. Math. Soc.*, 131:1303–1309 (electronic), 2003.
11. J. F. Alves, V. Araújo, and C. Vásquez. Random perturbations of diffeomorphisms with dominated splitting. Preprint Porto 2004.
12. J. F. Alves, C. Bonatti, and M. Viana. SRB measures for partially hyperbolic systems whose central direction is mostly expanding. *Invent. Math.*, 140:351–398, 2000.
13. J. F. Alves, S. Luzzatto, and V. Pinheiro. Markov structures and decay of correlations for non-uniformly expanding dynamical systems. Preprint University of Porto, 2002.
14. J. F. Alves and M. Viana. Statistical stability for a robust class of multidimensional non-hyperbolic attractors. *Ergod. Th. & Dynam. Sys.*, 21:1–32, 2001.
15. A. Andronov and L. Pontryagin. Systèmes grossiers. *Dokl. Akad. Nauk. USSR*, 14:247–251, 1937.

16. D. V. Anosov. Geodesic flows on closed Riemannian manifolds of negative curvature. *Proc. Steklov Math. Inst.*, 90:1–235, 1967.
17. N. Aoki. The set of Axiom A diffeomorphisms with no cycles. *Bol. Soc. Bras. Mat., Nova Sér.*, 23 (1–2):21–25, 1992.
18. V. Araújo. Attractors and time averages for random maps. *Annales de l'Inst. Henri Poincaré - Analyse Non-linéaire*, 17:307–369, 2000.
19. V. Araújo. Infinitely many stochastically stable attractors. *Nonlinearity*, 14:583–596, 2001.
20. V. Araújo and A. Tahzibi. Stochastic stability at the boundary of expanding maps. Preprint Porto and arxiv.org/abs/math.DS/0404368, 2004.
21. A. Arbieto and J. Bochi. L^p-generic cocycles have one-point Lyapunov spectrum. *Stoch. Dyn.*, 3:73–81, 2003.
22. A. Arbieto and C. Matheus. A Pasting Lemma I : The case of vector fields. Preprint www.preprint.impa.br 2003.
23. M.-C. Arnaud. Le "closing lemma" en topologie C^1. *Mém. Soc. Math. Fr.*, 74:vi+120, 1998.
24. M.-C. Arnaud. Difféomorphismes symplectiques de classe C^1 en dimension 4. *C. R. Acad. Sci. Paris Sér. I Math.*, 331:1001–1004, 2000.
25. M.-C. Arnaud. Création de connexions en topologie C^1. *Ergod. Th. & Dynam. Systems*, 21:339–381, 2001.
26. M.-C. Arnaud. The generic C^1 symplectic diffeomorphisms of symplectic 4-dimensional manifolds are hyperbolic, partially hyperbolic, or have a completely elliptic point. *Ergod. Th. & Dynam. Sys.*, 22:1621–1639, 2002.
27. M.-C. Arnaud. Approximation des ensembles ω-limites des difféomorphismes par des orbites périodiques. *Annales Sci. E. N. S.*, 36:173–190, 2003.
28. M.-C. Arnaud, C. Bonatti, and S. Crovisier. Dynamique symplectique générique. Preprint Dijon 2004.
29. A. Arnéodo, P. Coullet, and C. Tresser. Possible new strange attractors with spiral structure. *Comm. Math. Phys.*, 79:573–579, 1981.
30. L. Arnold. *Random dynamical systems*. Springer-Verlag, 1998.
31. L. Arnold and N. D. Cong. On the simplicity of the Lyapunov spectrum of products of random matrices. *Ergod. Th. & Dynam. Sys.*, 17(5):1005–1025, 1997.
32. V. I. Arnold. Small denominators I: On the mapping of a circle to itself. *Izv. Akad. Nauk. Math. Series*, 25:21–86, 1961. Transl. A.M.S. 46 (1965).
33. A. Arroyo and F. Rodriguez-Hertz. Homoclinic bifurcations and uniform hyperbolicity for three-dimensional flows. Preprint IMPA, 2002.
34. A. Avila. *Bifurcations of unimodal maps: the topological and metric picture.* PhD thesis, IMPA, 2001.
35. A. Avila and J. Bochi. A formula with some applications to the theory of Lyapunov exponents. *Israel J. Math.*, 131:125–137, 2002.
36. A. Avila, M. Lyubich, and W. de Melo. Regular or stochastic dynamics in real analytic families of unimodal maps. *Inventiones Math.*, 154:451–550, 2003.
37. A. Avila and C. G. Moreira. Statistical properties of unimodal maps: periodic orbits and pathological foliations. Preprint 2003.
38. A. Avila and C. G. Moreira. Statistical properties of unimodal maps: the quadratic family. *Annals of Math.*
39. A. Avila and C. G. Moreira. Statistical properties of unimodal maps: smooth families with negative Schwarzian derivative. *Astérisque*, 286:81–118, 2003.

40. V. Baladi. *Positive transfer operators and decay of correlations*. World Scientific Publishing Co. Inc., 2000.

41. V. Baladi, C. Bonatti, and B. Schmitt. Abnormal escape rates from nonuniformly hyperbolic sets. *Ergod. Th. & Dynam. Sys.*, 19:1111–1125, 1999.

42. V. Baladi and S. Gouezel. A note on stretched exponential decay of correlations for the Viana-Alves map. Preprint Jussieu and arxiv.org, 2003.

43. V. Baladi and M. Viana. Strong stochastic stability and rate of mixing for unimodal maps. *Ann. Sci. École Norm. Sup.*, 29:483–517, 1996.

44. R. Bamón, R. Labarca, R. Mañé, and M.J. Pacifico. The explosion of singular cycles. *Publ. Math. IHES*, 78:207–232, 1993.

45. A. Baraviera and C. Bonatti. Removing zero central Lyapunov exponents. *Ergod. Th. & Dynam. Sys.*, 23:1655–1670, 2003.

46. L. Barreira and Ya. Pesin. Lectures on Lyapunov exponents and smooth ergodic theory. In *Smooth ergodic theory and its applications (Seattle, WA, 1999)*, volume 69 of *Proc. Sympos. Pure Math.*, pages 3–106. Amer. Math. Soc., 2001. Appendix A by M. Brin and Appendix B by D. Dolgopyat, H. Hu and Ya. Pesin.

47. L. Barreira and Ya. Pesin. *Lyapunov exponents and smooth ergodic theory*, volume 23 of *Univ. Lecture Series*. Amer. Math. Soc., 2002.

48. L. Barreira, Ya. Pesin, and J. Schmeling. Dimension and product structure of hyperbolic measures. *Ann. of Math.*, 149:755–783, 1999.

49. S. Bautista, C. Morales, and M. J. Pacifico. The spectral decomposition theorem fails for singular hyperbolic sets. Preprint www.preprint.impa.br, 2004.

50. T. Bedford and M. Urbański. The box and Hausdorff dimension of self-affine sets. *Ergod. Th. & Dynam. Sys.*, 10:627–644, 1990.

51. M. Benedicks and L. Carleson. On iterations of $1 - ax^2$ on $(-1, 1)$. *Annals of Math.*, 122:1–25, 1985.

52. M. Benedicks and L. Carleson. The dynamics of the Hénon map. *Annals of Math.*, 133:73–169, 1991.

53. M. Benedicks and M. Viana. Random perturbations and statistical properties of Hénon-like maps. Preprint www.preprint.impa.br 2002.

54. M. Benedicks and M. Viana. Solution of the basin problem for Hénon-like attractors. *Invent. Math.*, 143:375–434, 2001.

55. M. Benedicks and L.-S. Young. Absolutely continuous invariant measures and random perturbations for certain one-dimensional maps. *Ergod. Th. & Dynam. Sys.*, 12:13–37, 1992.

56. M. Benedicks and L.-S. Young. SBR-measures for certain Hénon maps. *Invent. Math.*, 112:541–576, 1993.

57. M. Benedicks and L.-S. Young. Markov extensions and decay of correlations for certain Hénon maps. *Astérisque*, 261:13–56, 2000. Géométrie complexe et systèmes dynamiques (Orsay, 1995).

58. G. Birkhoff. *Lattice theory*, volume 25. A.M.S. Colloq. Publ., 1967.

59. M. Blank and G. Keller. Stochastic stability versus localization in one-dimensional chaotic dynamical systems. *Nonlinearity*, 10:81–107, 1997.

60. M. Blank, G. Keller, and C. Liverani. Ruelle-Perron-Frobenius spectrum for Anosov maps. *Nonlinearity*, 15:1905–1973, 2002.

61. M. L. Blank. Stochastic properties of deterministic dynamical systems. *Sov. Sci. Rev. C Maths/Phys*, 6:243–271, 1987.

62. A. M. Blokh and M. Yu. Lyubich. Ergodicity of transitive maps of the interval. *Ukrainian Math. J.*, 41:985–988, 1989.

63. J. Bochi. Genericity of zero Lyapunov exponents. *Ergod. Th. & Dynam. Sys.*, 22:1667–1696, 2002.
64. J. Bochi, B. Fayad, and E. Pujals. A remark on conservative dynamics. Preprint Jussieu 2004.
65. J. Bochi and M. Viana. Lyapunov exponents: How often are dynamical systems hyperbolic ? In *Advances in Dynamical Systems*. Cambridge Univ. Press.
66. J. Bochi and M. Viana. The Lyapunov exponents of generic volume preserving and symplectic systems. *Annals of Mathematics*.
67. J. Bochi and M. Viana. Uniform (projective) hyperbolicity or no hyperbolicity: a dichotomy for generic conservative maps. *Annales de l'Inst. Henri Poincaré - Analyse Non-linéaire*, 19:113–123, 2002.
68. C. Bonatti and L. J. Díaz. On maximal transitive sets of generic diffeomorphisms. *Publ. Math. IHES*, 96:171–197, 2002.
69. C. Bonatti and S. Crovisier. Recurrence et généricité. *Inventiones Math.*
70. C. Bonatti and S. Crovisier. Recurrence and genericity. *C. R. Math. Acad. Sci. Paris*, 336:839–844, 2003.
71. C. Bonatti and L. J. Díaz. Nonhyperbolic transitive diffeomorphisms. *Annals of Math.*, 143:357–396, 1996.
72. C. Bonatti and L. J. Díaz. Connexions heterocliniques et genericité d'une infinité de puits ou de sources. *Ann. Sci. École Norm. Sup.*, 32:135–150, 1999.
73. C. Bonatti, L. J. Díaz, and E. Pujals. A C^1-generic dichotomy for diffeomorphisms: weak forms of hyperbolicity or infinitely many sinks or sources. *Annals of Math.*, 158:355–418, 2003.
74. C. Bonatti, L. J. Díaz, E. Pujals, and J. Rocha. Robust transitivity and heterodimensional cycles. *Astérisque,* 286:187–222, 2003.
75. C. Bonatti, L. J. Díaz, and G. Turcat. Pas de shadowing lemma pour les dynamiques partiellement hyperboliques. *C. R. Acad. Sci. Paris Sér. I Math.*, 330:587–592, 2000.
76. C. Bonatti, L. J. Díaz, and R. Ures. Minimality of strong stable and unstable foliations for partially hyperbolic diffeomorphisms. *J. Inst. Math. Jussieu*, 1:513–541, 2002.
77. C. Bonatti, L. J. Díaz, and M. Viana. Discontinuity of the Hausdorff dimension of hyperbolic sets. *C. R. Acad. Sci. Paris Sér. I Math.*, 320:713–718, 1995.
78. C. Bonatti, L. J. Díaz, and F. Vuillemin. Cubic tangencies and hyperbolic diffeomorphisms. *Bol. Soc. Brasil. Mat.*, 29:99–144, 1998.
79. C. Bonatti, X. Gomez-Mont, and M. Viana. Généricité d'exposants de Lyapunov non-nuls pour des produits déterministes de matrices. *Annales de l'Inst. Henri Poincaré - Analyse Non-linéaire*, 20:579–624, 2003.
80. C. Bonatti and R. Langevin. Difféomorphismes de Smale des surfaces. *Astérisque*, 250:viii+235, 1998. With the collaboration of E. Jeandenans.
81. C. Bonatti, C. Matheus, M. Viana, and A. Wilkinson. Abundance of stable ergodicity. *Comentarii Math. Helvetici*.
82. C. Bonatti, A. Pumariño, and M. Viana. Lorenz attractors with arbitrary expanding dimension. *C. R. Acad. Sci. Paris Sér. I Math.*, 325:883–888, 1997.
83. C. Bonatti and M. Viana. Lyapunov exponents with multiplicity 1 for deterministic products of matrices. *Ergod. Th. & Dynam. Sys.*, 24.
84. C. Bonatti and M. Viana. SRB measures for partially hyperbolic systems whose central direction is mostly contracting. *Israel J. Math.*, 115:157–193, 2000.
85. C. Bonatti and A. Wilkinson. Transitive partially hyperbolic diffeomorphisms on 3-manifolds. In preparation.

86. R. Bowen. *Equilibrium states and the ergodic theory of Anosov diffeomorphisms*, volume 470 of *Lect. Notes in Math.* Springer Verlag, 1975.
87. R. Bowen. A horseshoe with positive measure. *Invent. Math.*, 29:203–204, 1975.
88. R. Bowen. Mixing Anosov flows. *Topology*, 15:77–79, 1977.
89. R. Bowen. Hausdorff dimension of quasicircles. *Publ. Math. Inst. Hautes Études Sci.*, 50:11–25, 1979.
90. R. Bowen and D. Ruelle. The ergodic theory of Axiom A flows. *Invent. Math.*, 29:181–202, 1975.
91. A. Boyarsky and P. Góra. *Invariant measures and dynamical systems in one dimension.* Probability and its Applications. Birkhäuser, 1997.
92. X. Bressaud. Subshifts on an infinite alphabet. *Ergod. Th. & Dynam. Sys.*, 19:1175–1200, 1999.
93. M. Brin. Topological transitivity of a certain class of dynamical systems, and flows of frames on manifolds of negative curvature. *Funkcional. Anal. i Priložen.*, 9:9–19, 1975.
94. M. Brin. Bernoulli diffeomorphisms with $n-1$ nonzero exponents. *Ergod. Th. & Dynam. Sys.*, 1:1–7, 1981.
95. M. Brin, D. Burago, and S. Ivanov. On partially hyperbolic diffeomorphisms on 3-manifolds with commutative fundamental group. In *Advances in Dynamical Systems.* Cambridge Univ. Press.
96. M. Brin, J. Feldman, and A. Katok. Bernoulli diffeomorphisms and group extensions of dynamical systems with nonzero characteristic exponents. *Annals of Math.*, 113:159–179, 1981.
97. M. Brin and Ya. Pesin. Partially hyperbolic dynamical systems. *Izv. Acad. Nauk. SSSR*, 1:177–212, 1974.
98. H. Bruin and G. Keller. Equilibrium states for S-unimodal maps. *Ergod. Th. & Dynam. Sys.*, 18:765–789, 1998.
99. H. Bruin, S. Luzzatto, and S. van Strien. Decay of correlations in one-dimensional dynamics. *Ann. Sci. École Norm. Sup.*, 36:621–646, 2003.
100. H. Bruin and S. van Strien. Expansion of derivatives in one-dimensional dynamics. *Israel J. Math.*, 137:223–263, 2003.
101. L. Bunimovich, N. Chernov, and Ya. Sinai. Statistical properties of two-dimensional hyperbolic billiards. *Russian Math. Surveys*, 46:47–106, 1991.
102. L. Bunimovich and Ya. Sinai. Statistical properties of lorentz gas with periodic configuration of scatterers. *Comm. Math. Phys.*, 78:479–497, 1981.
103. L. A. Bunimovich and Ya.G. Sinai. Stochasticity of the attractor in the Lorenz model. In *Nonlinear waves*, pages 212–226. Proc. Winter School, Moscow, Nauka Publ., 1980.
104. K. Burns, D. Dolgopyat, and Ya. Pesin. Partial hyperbolicity, Lyapunov exponents and stable ergodicity. *J. Statist. Phys.*, 108:927–942, 2002. Dedicated to David Ruelle and Yasha Sinai on the occasion of their 65th birthdays.
105. K. Burns, C. Pugh, M. Shub, and A. Wilkinson. Recent results about stable ergodicity. In *Smooth ergodic theory and its applications (Seattle WA, 1999)*, volume 69 of *Procs. Symp. Pure Math.*, pages 327–366. Amer. Math. Soc., 2001.
106. K. Burns, C. Pugh, and A. Wilkinson. Stable ergodicity and Anosov flows. *Topology*, 39:149–159, 2000.
107. K. Burns and A. Wilkinson. Stable ergodicity of skew products. *Ann. Sci. École Norm. Sup. (4)*, 32:859–889, 1999.

108. G. Buzzard. Infinitely many periodic attractors for holomorphic maps. *Annals of Math.*, 145:389–417, 1997.

109. J. Buzzi. Markov extensions for multi-dimensional dynamical systems. *Israel J. Math.*, 112:357–380, 1999.

110. J. Buzzi. A.c.i.m's for arbitrary expanding piecewise ℝ-analytic mappings of the plane. *Ergod. Th. & Dynam. Sys.*, 20:697–708, 2000.

111. J. Buzzi. Thermodynamical formalism for piecewise invertible maps: absolutely continuous invariant measures as equilibrium states. In *Smooth ergodic theory and its applications (Seattle, WA, 1999)*, volume 69 of *Proc. Sympos. Pure Math.*, pages 749–783. Amer. Math. Soc., 2001.

112. J. Buzzi and V. Maume-Deschamps. Decay of correlations for piecewise invertible maps in higher dimensions. *Israel J. Math.*, 131:203–220, 2002.

113. J. Buzzi and O. Sarig. Uniqueness of equilibrium measures for countable Markov shifts and multidimensional piecewise expanding maps. *Ergod. Th. & Dynam. Sys.*, 23:1383–1400, 2003.

114. J. Buzzi, O. Sester, and M. Tsujii. Weakly expanding skew-products of quadratic maps. *Ergod. Th. & Dynam. Sys.*, 23:1401–1414, 2003.

115. Y. Cao. Lyapunov exponents and uniform hyperbolicity. Preprint 2002.

116. Y. Cao and J. M. Mao. The non-wandering set of some Hénon maps. *Chaos, Solitons & Fractals*, 11:2045–2053, 2000.

117. C. Carballo and C. Morales. Homoclinic classes and finitude of attractors for vector fields on *n*-manifolds. *Bull. London Math. Soc.*, 35:85–91, 2003.

118. C. Carballo, C. Morales, and M. J. Pacifico. Maximal transitive sets with singularities for generic C^1 vector fields. *Bull. Braz. Math. Soc.*, 31:287–303, 2000.

119. C. Carballo, C. Morales, and M. J. Pacifico. Homoclinic classes for generic C^1 vector fields. *Ergod. Th. & Dynam. Syst.*, 23:403–415, 2003.

120. M. Carvalho. First homoclinic tangencies in the boundary of Anosov diffeomorphisms. *Disc. & Contin. Dynam. Sys.*, 4:765–782, 1998.

121. A. A. Castro. Fast mixing for attractors with mostly contracting central direction. *Ergod. Th. & Dynam. Sys.*

122. A. A. Castro. Backward inducing and exponential decay of correlations for partially hyperbolic attractors with mostly contracting central direction. *Israel J. Math*, 130:29–75, 2002.

123. N. Chernov and L.-S. Young. Decay of correlations for Lorentz gases and hard balls. In *Hard ball systems and the Lorentz gas*, volume 101 of *Encyclopaedia Math. Sci.*, pages 89–120. Springer, 2000.

124. N. I. Chernov. Ergodic and statistical properties of two-dimensional hyperbolic billiards. *J. Stat. Phys.*, 69:111–134, 1992.

125. N. I. Chernov. Markov approximations and decay of correlations for anosov flows. *Annals of Math.*, 147:269–324, 1998.

126. N. I. Chernov. Decay of correlations and dispersing billiards. *J. Statist. Phys.*, 94:513–556, 1999.

127. N. I. Chernov. Statistical properties of piecewise smooth hyperbolic systems in high dimensions. *Discrete Cont. Dynam. Sys.*, 5:425–448, 1999.

128. T. Cherry. Analytic quasi-periodic curves of discontinuous type in a torus. *Proc. London Math. Soc.*, 44, 1938.

129. P. Collet. A remark about uniform de-correlation prefactors. Preprint École Polytechnique, Paris.

130. P. Collet and J. P. Eckmann. On the abundance of aperiodic behaviour for maps of the interval. *Comm. Math. Phys.*, 73:115–160, 1980.

131. P. Collet and J. P. Eckmann. Positive Lyapunov exponents and absolute continuity for maps of the interval. *Ergod. Th. & Dynam. Sys.*, 3:13–46, 1983.

132. P. Collet, H. Epstein, and G. Gallavotti. Perturbations of geodesic flows on surfaces of constant negative curvature and their mixing properties. *Comm. Math. Phys.*, 95:61–112, 1984.

133. E. Colli. Infinitely many coexisting strange attractors. *Ann. Inst. H. Poincaré Anal. Non Linéaire*, 15:539–579, 1998.

134. W. Colmenárez. SRB measures for singular hyperbolic attractors. Preprint 2002.

135. C. Conley. *Isolated invariant sets and the Morse index*, volume 38 of *CBMS Regional Conference Series in Mathematics*. American Mathematical Society, 1978.

136. M. J. Costa. Saddle-node horseshoes giving rise to global Hénon-like attractors. *Anais Acad. Bras. Ciências*, 70:393–400, 1998.

137. M. J. Costa. Chaotic behaviour of one-dimensional saddle-node horseshoes. *Discrete Contin. Dyn. Syst.*, 9:505–548, 2003.

138. P. Coullet and C. Tresser. Itérations d'endomorphims et groupe de renormalization. *C. R. Acad. Sci. Paris*, 287, Série I:577–580, 1978.

139. E. Coven, I. Kan, and J. Yorke. Pseudo-orbit shadowing in the family of tent maps. *Trans. Amer. Math. Soc.*, 308:227–241, 1988.

140. W. Cowieson. Absolutely continuous invariant measures for most piecewise smooth expanding maps. *Ergod. Th. & Dynam. Sys.*, 22:1061–1078, 2002.

141. W. Cowieson and L.-S. Young. SRB measures as zero-noise limits. Preprint Courant Institute, 2004.

142. S. Crovisier. Saddle-node bifurcations for hyperbolic sets. Preprint Orsay 2000.

143. S. Crovisier. Une remarque sur les ensembles hyperboliques localement maximaux. *C. R. Acad. Sci. Paris*, 334:401–404, 2002.

144. P. Cvitanović, G. Gunaratne, and I. Procaccia. Topological and metric properties of Hénon-type strange attractors. *Phys. Rev. A*, 38:1503–1520, 1988.

145. A. de Carvalho and T. Hall. How to prune a horseshoe. *Nonlinearity*, 15:R19–R68, 2002.

146. W. de Melo. Structural stability of diffeomorphisms on two-manifolds. *Invent. Math.*, 21:233–246, 1973.

147. W. de Melo and S. van Strien. *One-dimensional dynamics*. Springer Verlag, 1993.

148. J.-P. Dedieu and M. Shub. On random and mean exponents for unitary invariant probability measures on $\mathbb{GL}(n, \mathbb{C})$. *Astérisque*, 287:1–18, 2003.

149. M. Denker. The central limit theorem for dynamical systems. In *Dynamical Systems and Ergodic Theory*, volume 23. Banach Center Publ., Warsaw, 1989.

150. M. Denker and M. Urbański. The dichotomy of Hausdorff measures and equilibrium states for parabolic rational maps. In *Ergodic theory and related topics, III (Güstrow, 1990)*, volume 1514 of *Lecture Notes in Math.*, pages 90–113. Springer Verlag, 1992.

151. L. J. Díaz. Persistence of cycles and nonhyperbolic dynamics at heteroclinic bifurcations. *Nonlinearity*, 8:693–713, 1995.

152. L. J. Díaz. Robust nonhyperbolic dynamics and heterodimensional cycles. *Ergod. Th. & Dynam. Sys.*, 15:291–315, 1995.

153. L. J. Díaz, E. Pujals, and R. Ures. Partial hyperbolicity and robust transitivity. *Acta Math.*, 183:1–43, 1999.

154. L. J. Díaz, I. L. Rios, and M. Viana. The intermittency route to chaotic dynamics. In *Global analysis of dynamical systems*, pages 309–327. IOP Publ., 2001.

155. L. J. Díaz and J. Rocha. Heterodimensional cycles, partial hyperbolicity, and limit dynamics. Preprint 2000.

156. L. J. Díaz and J. Rocha. Nonconnected heteroclinic cycles: bifurcation and stability. *Nonlinearity*, 5:1315–1341, 1992.

157. L. J. Díaz and J. Rocha. Large measure of hyperbolic dynamics when unfolding heterodimensional cycles. *Nonlinearity*, 10:857–884, 1997.

158. L. J. Díaz and J. Rocha. Partial hyperbolicity and transitive dynamics generated by heteroclinic cycles. *Ergod. Th. & Dynam. Sys.*, 21:25–76, 2001.

159. L. J. Díaz, J. Rocha, and M. Viana. Strange attractors in saddle-node cycles: prevalence and globality. *Invent. Math.*, 125:37–74, 1996.

160. L. J. Díaz and B. Santoro. Collision, explosion, and collapse of homoclinic classes. *Nonlinearity*.

161. L. J. Díaz and R. Ures. Persistent homoclinic tangencies at the unfolding of cycles. *Ann. Inst. H. Poincaré, Anal. Non-linéaire*, 11:643–659, 1996.

162. C. I. Doering. Persistently transitive vector fields on three-dimensional manifolds. In *Procs. on Dynamical Systems and Bifurcation Theory*, volume 160, pages 59–89. Pitman, 1987.

163. D. Dolgopyat. Limit theorems for partially hyperbolic diffeomorphisms. Preprint 2000.

164. D. Dolgopyat. On differentiability of SRB states. Preprint 2000.

165. D. Dolgopyat. On decay of correlations in Anosov flows. *Ann. of Math.*, 147:357–390, 1998.

166. D. Dolgopyat. Prevalence of rapid mixing in hyperbolic flows. *Ergod. Th. & Dynam. Sys.*, 18:1097–1114, 1998.

167. D. Dolgopyat. On dynamics of mostly contracting diffeomorphisms. *Comm. Math. Phys*, 213:181–201, 2000.

168. D. Dolgopyat, H.Hu, and Ya. Pesin. An example of a smooth hyperbolic measure with countably many ergodic components. In *Smooth ergodic theory and its applications (Seattle, WA, 1999)*, volume 69 of *Proc. Sympos. Pure Math.* Amer. Math. Soc., 2001.

169. D. Dolgopyat and Ya. Pesin. Every manifold supports a completely hyperbolic diffeomorphism. *Ergod. Th. & Dynam. Sys.*, 22:409–437, 2002.

170. D. Dolgopyat and A. Wilkinson. Stable accessibility is C^1 dense. *Astérisque*, 287:33–60, 2003.

171. P. Duarte. Plenty of elliptic islands for the standard family of area preserving maps. *Ann. Inst. H. Poincaré Anal. Non. Linéaire*, 11:359–409, 1994.

172. P. Duarte. Abundance of elliptic isles at conservative bifurcations. *Dynam. Stability Systems*, 14:339–356, 1999.

173. F. Dumortier, H. Kokubu, and H. Oka. A degenerate singularity generating geometric Lorenz attractors. *Ergodic Theory Dyn. Syst*, 15(5):833–856, 1995.

174. D. Dürr and S. Goldstein. Remarks on the central limit theorem for weakly dependent random variables. In *Stochastic Processes – Mathematics and Physics*, volume 1158 of *Lect. Notes in Math.*, pages 104–118. Springer Verlag, 1986.

175. R. Durret. *Probability: theory and examples*. Wadsworth & Brooks Publ., 1996. 2nd edition.

176. J.-P. Eckmann and D. Ruelle. Ergodic theory of chaos and strange attractors. *Rev. Mod. Phys.*, 57:617–656, 1985.
177. H. Enrich. A heteroclinic bifurcation of Anosov diffeomorphisms. *Ergod. Th. & Dynam. Sys.*, 18:567–608, 1998.
178. K. Falconer. *Fractal geometry.* John Wiley & Sons Ltd., 1990. Mathematical foundations and applications.
179. A. Fathi, M. Herman, and J.-C. Yoccoz. A proof of Pesin's stable manifold theorem. In *Geometric dynamics (Rio de Janeiro 1981)*, volume 1007 of *Lect. Notes in Math.*, pages 177–215. Springer Verlag, 1983.
180. B. Fayad. Non uniform hyperbolicity and elliptic dynamics. Preprint Université Paris 13, 2003.
181. M. Feigenbaum. Qualitative universality for a class of nonlinear transformations. *J. Stat. Phys.*, 19:25–52, 1978.
182. P. Ferrero and B. Schmitt. Ruelle's Perron-Frobenius theorem and projective metrics. In *Coll. Math. Soc. János Bolyai*, volume 27, pages 333–336. Banach Center, 1979.
183. T. Fisher. Hyperbolic sets that are not isolated. Preprint Northwestern University 2003.
184. J. E. Fornaess and E. Gavosto. Existence of generic homoclinic tangencies for Hénon mappings. *J. Geom. Anal.*, 2:429–444, 1992.
185. J. Franks. Necessary conditions for the stability of diffeomorphisms. *Trans. A.M.S.*, 158:301–308, 1971.
186. H. Furstenberg. Non-commuting random products. *Trans. Amer. Math. Soc.*, 108:377–428, 1963.
187. H. Furstenberg. Boundary theory and stochastic processes on homogeneous spaces. In *Harmonic analysis in homogeneous spaces*, volume XXVI of *Proc. Sympos. Pure Math. (Williamstown MA, 1972)*, pages 193–229. Amer. Math. Soc., 1973.
188. H. Furstenberg and H. Kesten. Products of random matrices. *Ann. Math. Statist.*, 31:457–469, 1960.
189. J.M. Gambaudo and J. Rocha. Maps of the sphere at the boundary of chaos. *Nonlinearity*, 7:1251–1259, 1999.
190. S. Gan and L. Wen. Heteroclinic cycles and homoclinic closures for generic diffeomorphisms. Preprint Peking University 2001.
191. P. Glendinning and C. Sparrow. T-points: A codimension two heteroclinic bifurcation. *Jour. Stat. Phys.*, 43:479–488, 1986.
192. I. Ya. Gold'sheid and G. A. Margulis. Lyapunov indices of a product of random matrices. *Uspekhi Mat. Nauk.*, 44:13–60, 1989.
193. S. V. Gonchenko, L. P. Shil'nikov, and D. V. Turaev. Dynamical phenomena in systems with structurally unstable Poincaré homoclinic orbits. *Chaos*, 6:15–31, 1996.
194. P. Góra and A. Boyarsky. Absolutely continuous invariant measures for piecewise expanding C^2 transformations in $\mathbb{R}^n$. *Israel Jour. Math.*, 67:272–286, 1989.
195. M. I. Gordin. The central limit theorem for stationary processes. *Dokl. Akad. Nauk. SSSR*, 188:1174–1176, 1969.
196. S. Gouezel. Décorrélations des applications non-uniformément dilatantes. Preprint 2003.
197. N. Gourmelon. Difféomorphismes robustement transitifs. Master's thesis, Univ. Grenoble, 2003.

198. J. Graczyk and G. Swiatek. Generic hyperbolicity in the logistic family. *Annals of Math.*, 146:1–52, 1997.
199. J. Graczyk and G. Swiatek. Smooth unimodal maps in the 1990s. *Ergod. Th. & Dynam. Sys.*, 19:263–287, 1999.
200. M. Grayson, C. Pugh, and M. Shub. Stably ergodic diffeomorphisms. *Annals of Math.*, 140:295–329, 1994.
201. J. Guckenheimer and R. F. Williams. Structural stability of Lorenz attractors. *Publ. Math. IHES*, 50:59–72, 1979.
202. Y. Guivarc'h and A. Raugi. Products of random matrices : convergence theorems. *Contemp. Math.*, 50:31–54, 1986.
203. C. Gutierrez. A counter-example to a C^2 closing lemma. *Ergod. Th. & Dynam. Sys.*, 7:509–530, 1987.
204. M. Hall. On the sums and products of continued fractions. *Annals of Math.*, 48:966–993, 1947.
205. P. Halmos. On automorphisms of compact groups. *Bulletin Amer. Math. Soc.*, 49:619–624, 1943.
206. S. Hayashi. Hyperbolicity and homoclinic bifurcations generating nonhyperbolic dynamics. Preprint 2002.
207. S. Hayashi. Diffeomorphisms in $\mathcal{F}^1(m)$ satisfy Axiom A. *Ergod. Th. & Dynam. Sys.*, 12 (2):233–253, 1992.
208. S. Hayashi. Connecting invariant manifolds and the solution of the C^1 stability and Ω-stability conjectures for flows. *Annals of Math.*, 145:81–137, 1997.
209. S. Hayashi. A C^1 make or break lemma. *Bol. Soc. Brasil. Mat.*, 31:337–350, 2000.
210. M. Hénon. A two dimensional mapping with a strange attractor. *Comm. Math. Phys.*, 50:69–77, 1976.
211. M. Hénon and Y. Pomeau. Two strange attractors with a simple structure. In *Turbulence and Navier-Stokes equations*, volume 565, pages 29–68. Springer Verlag, 1976.
212. M. Herman. Sur la conjugaison différentiable des difféomorphismes du cercle a des rotations. *Publ. Math. IHES*, 103–104, 1979.
213. M. Herman. Une méthode nouvelle pour minorer les exposants de Lyapunov et quelques exemples montrant le caracère local d'un théorème d'Arnold et de Moser sur le tore de dimension 2. *Comment. Math. Helvetici*, 58:453–502, 1983.
214. M. Herman. Différentiabilité optimale et contre-exemples à la fermeture en topologie C^∞ des orbites récurrentes de flots hamiltoniens. *C. R. Acad. Sci. Paris Sér. I Math.*, 313:49–51, 1991.
215. M. Herman. Exemples de flots hamiltoniens dont aucune perturbation en topologie C^∞ n'a d'orbites périodiques sur un ouvert de surfaces d'énergies. *C. R. Acad. Sci. Paris Sér. I Math.*, 312:989–994, 1991.
216. M. Hirsch, C. Pugh, and M. Shub. *Invariant manifolds*, volume 583 of *Lect. Notes in Math.* Springer Verlag, 1977.
217. F. Hofbauer and G. Keller. Quadratic maps without asymptotic measure. *Comm. Math. Phys.*, 127:319–337, 1990.
218. E. Hopf. Statistik der geodätischen Linien in Mannigfaltigkeiten negativer Krümmung. *Ber. Verh. Sächs. Akad. Wiss. Leipzig*, 91:261–304, 1939.
219. V. Horita, N. Muniz, and P. Sabini. Nonperiodic bifurcations of one-dimensional maps. Preprint 2004.

220. H. Hu. Statistical properties of some almost hyperbolic systems. In *Smooth ergodic theory and its applications (Seattle, WA, 1999)*, volume 69 of *Proc. Sympos. Pure Math.*, pages 367–384. Amer. Math. Soc., 2001.

221. B. Hunt and V. Kaloshin. A stretched exponential bound on the rate of growth of the number of periodic points for prevalent diffeomorphisms I-II. *Electron. Res. Announc. A. M. S.*, 7:17–27 and 28–36, 2001.

222. M. Hurley. Attractors: persistence and density of their basins. *Trans. AMS*, 269:247–271, 1982.

223. M. Hurley. Generic homeomorphisms have no smallest attractors. *Proc. Amer. Math. Soc.*, 123:1277–1280, 1995.

224. M. Hurley. Properties of attractors of generic homeomorphisms. *Ergod. Th. & Dynam. Systems*, 16:1297–1310, 1996.

225. M. C. Irwin. *Smooth dynamical systems*, volume 94 of *Pure and Applied Mathematics*. Academic Press Inc., 1980.

226. S. Isola. Renewal sequences and intermittency. *J. Stat. Phys.*, 97:263–280, 1999.

227. M. Jakobson. Absolutely continuous invariant measures for one-parameter families of one-dimensional maps. *Comm. Math. Phys.*, 81:39–88, 1981.

228. V. Kaloshin. Generic diffeomorphisms with superexponential growth of the number of periodic orbits. *Comm. Math. Phys.*, 211:253–271, 2000.

229. I. Kan. Open sets of diffeomorphisms having two attractors, each with an everywhere dense basin. *Bull. Amer. Math. Soc.*, 31:68–74, 1994.

230. A. Katok. Bernoulli diffeomorphisms. *Annals of Math.*, 110:529–547, 1979.

231. A. Katok. Lyapunov exponents, entropy and periodic points of diffeomorphisms. *Publ. Math. IHES*, 51:137–173, 1980.

232. A. Katok and B. Hasselblatt. *Introduction to the modern theory of dynamical systems*. Cambridge University Press, 1995.

233. A. Katok and Yu. Kifer. Random perturbations of transformations of an interval. *J. Analyse Math.*, 47:193–237, 1986.

234. A. Katok and A. Kononenko. Cocycle's stability for partially hyperbolic systems. *Math. Res. Lett.*, 3:191–210, 1996.

235. A. Katok and J. M. Strelcyn. *Invariant manifolds, entropy and billiards. Smooth maps with singularities*, volume 1222 of *Lect. Notes in Math.* Springer Verlag, 1986.

236. G. Keller. Ergodicité et mesures invariantes pour les transformations dilatantes par morceaux d'une région bornée du plan. *C.R. Acad. Sci. Paris*, A 289:625–627, 1979.

237. G. Keller. Stochastic stability in some chaotic dynamical systems. *Monatsh. Math.*, 94:313–333, 1982.

238. G. Keller. Markov extensions, zeta functions, and Fredholm determinants for piecewise invertible dynamical systems. *Trans. Amer. Math. Soc.*, 314:433–497, 1989.

239. G. Keller and T. Nowicki. Spectral theory, zeta functions and the distribution of periodic points for Collet-Eckmann maps. *Comm. Math. Phys.*, 149:31–69, 1992.

240. Yu. Kifer. On small random perturbations of some smooth dynamical systems. *Math. USSR Izv.*, 8:1083–1107, 1974.

241. Yu. Kifer. *Ergodic theory of random perturbations*. Birkhäuser, 1986.

242. Yu. Kifer. *Random perturbations of dynamical systems*. Birkhäuser, 1988.

243. J. Kingman. The ergodic theorem of subadditive stochastic processes. *J. Royal Statist. Soc.*, 30:499–510, 1968.

244. C. Kipnis and S. R. S. Varadhan. Central limit theorem for additive functions of reversible Markov processes and applications to simple exclusions. *Comm. Math. Phys*, 104:1–19, 1986.

245. O. Knill. Positive Lyapunov exponents for a dense set of bounded measurable $SL(2, \mathbf{R})$-cocycles. *Ergod. Th. & Dynam. Sys.*, 12:319–331, 1992.

246. O. Kozlovski. Getting rid of the negative Schwarzian derivative condition. *Ann. of Math.*, 152:743–762, 2000.

247. O. Kozlovski. Axiom A maps are dense in the space of unimodal maps in the C^k topology. *Ann. of Math.*, 157:1–43, 2003.

248. O. Kozlovski, S. van Strien, and W. Shen. Rigidity for real polynomials. Preprint Warwick 2003.

249. H. Kunita. *Stochastic flows and stochastic differential equations.* Cambridge University Press, 1990.

250. R. Labarca. Bifurcation of contracting singular cycles. *Ann. Sci. Éc. Norm. Supér.*, 28:705–745, 1995.

251. R. Labarca and M.J. Pacifico. Stability of singular horseshoes. *Topology*, 25:337–352, 1986.

252. A. Lasota and J.A. Yorke. On the existence of invariant measures for piecewise monotonic transformations. *Trans. Amer. Math. Soc.*, 186:481–488, 1973.

253. F. Ledrappier. Propriétés ergodiques des mesures de Sinaï. *Publ. Math. I.H.E.S.*, 59:163–188, 1984.

254. F. Ledrappier. Quelques propriétés des exposants caractéristiques. *Lect. Notes in Math.*, 1097:305–396, 1984.

255. F. Ledrappier, M. Shub, C. Simó, and A. Wilkinson. Random versus deterministic exponents in a rich family of diffeomorphisms. *J. Stat. Phys*, 113:85–149, 2003.

256. F. Ledrappier and J.-M. Strelcyn. A proof of the estimation from below in pesin's entropy formula. *Ergod. Th & Dynam. Sys*, 2:203–219, 1982.

257. F. Ledrappier and L.-S. Young. The metric entropy of diffeomorphisms. I. Characterization of measures satisfying Pesin's entropy formula. *Ann. of Math.*, 122:509–539, 1985.

258. F. Ledrappier and L.-S. Young. The metric entropy of diffeomorphisms. II. Relations between entropy, exponents and dimension. *Ann. of Math.*, 122:540–574, 1985.

259. S.-T. Liao. Obstruction sets. II. *Beijing Daxue Xuebao*, 2:1–36, 1981.

260. S.-T. Liao. Hyperbolicity properties of the non-wandering sets of certain 3-dimensional systems. *Acta Math. Scientia*, 3:361–368, 1983.

261. S.T. Liao. On the stability conjecture. *Chinese Annals of Math*, 1:9–30, 1980.

262. C. Liverani. On contact Anosov flows. *Annals of Math.*

263. C. Liverani. Decay of correlations. *Annals of Math.*, 142:239–301, 1995.

264. C. Liverani. Decay of correlations for piecewise expanding maps. *J. Stat. Phys*, 78:1111–1129, 1995.

265. C. Liverani. Central limit theorem for deterministic systems. In *Procs. Intern. Conf. on Dynamical Systems (Montevideo 1995) – A tribute to Ricardo Mañé*, pages 56–75. Pitman, 1996.

266. C. Liverani. A probabilistic approach to intermittency. *Ergod. Th. & Dynam. Sys.*, 19:1399–1420, 1999.

267. E. N. Lorenz. Deterministic nonperiodic flow. *J. Atmosph. Sci.*, 20:130–141, 1963.

268. S. Luzzatto and W. Tucker. Non-uniformly expanding dynamics in maps with singularities and criticalities. *Inst. Hautes Études Sci. Publ. Math.*, 89:179–226 (2000), 1999.

269. S. Luzzatto and M. Viana. Lorenz-like attractors without invariant foliations. In preparation.

270. S. Luzzatto and M. Viana. Positive Lyapunov exponents for Lorenz-like families with criticalities. *Astérisque*, 261:201–237, 2000. Géométrie complexe et systèmes dynamiques (Orsay, 1995).

271. S. Luzzatto and M. Viana. Parameter exclusions in Hénon-like systems. *Russian Math. Surveys*, 58:1053–1092, 2003.

272. M. Lyubich. Almost every real quadratic map is either regular or stochastic. To appear Annals of Math.

273. M. Lyubich. Dynamics of quadratic maps I-II. *Acta Math.*, 178:185–247, 248–297, 1997.

274. M. Lyubich. Feigenbaum-Coullet-Tresser, universality, and Milnor's hairiness conjecture. *Annals of Math.*, 149:319–420, 1999.

275. M. Lyubich. The quadratic family as a qualitatively solvable model of chaos. *Notices Amer. Math. Soc.*, 47:1042–1052, 2000.

276. M. Lyubich. Dynamics of quadratic maps III: Parapuzzle and SRB measure. *Astérisque*, 261:173–200, 2001. Géométrie complexe et systèmes dynamiques (Orsay, 1995).

277. R. Mañé. Contributions to the stability conjecture. *Topology*, 17:383–396, 1978.

278. R. Mañé. An ergodic closing lemma. *Annals of Math.*, 116:503–540, 1982.

279. R. Mañé. Oseledec's theorem from the generic viewpoint. In *Procs. International Congress of Mathematicians, Vol. 1, 2 (Warsaw, 1983)*, pages 1269–1276, Warsaw, 1984. PWN Publ.

280. R. Mañé. Hyperbolicity, sinks and measure in one-dimensional dynamics. *Comm. Math. Phys.*, 100:495–524, 1985.

281. R. Mañé. *Ergodic theory and differentiable dynamics.* Springer Verlag, 1987.

282. R. Mañé. On the creation of homoclinic points. *Publ. Math. I.H.E.S.*, 66:139–159, 1988.

283. R. Mañé. A proof of the C^1 stability conjecture. *Publ. Math. I.H.E.S.*, 66:161–210, 1988.

284. R. Mañé. The Lyapunov exponents of generic area preserving diffeomorphisms. In *International Conference on Dynamical Systems (Montevideo, 1995)*, pages 110–119. Longman, 1996.

285. A. Manning and H. McCluskey. Hausdorff dimension of horseshoes. *Ergod. Th. & Dynam. Sys.*, 3:251–260, 1983.

286. J. M. Marstrand. Some fundamental geometrical properties of plane sets of fractional dimensions. *Proc. London Math. Soc.*, 4:257–302, 1954.

287. M. Martens and T. Nowicki. Invariant measures for lebesgue typical quadratic maps. *Astérisque*, 261:239–252, 2001. Géométrie complexe et systèmes dynamiques (Orsay, 1995).

288. J. Mather. Commutators of diffeomorphisms. *Comm. Math. Helv.*, 48:195–233, 1973.

289. V. Maume-Deschamps. Projective metrics and mixing properties on towers. *Trans. Amer. Math. Soc.*, 353:3371–3389 (electronic), 2001.

290. C. McMullen. *Renormalization and 3-manifolds which fiber over the circle,* volume 142 of *Annals of Math. Studies.* Princeton Univ. Press, 1996.
291. R. Metzger. Sinai-Ruelle-Bowen measures for contracting Lorenz maps and flows. *Ann. Inst. H. Poincaré Anal. Non Linéaire,* 17:247–276, 2000.
292. R. Metzger. Stochastic stability for contracting Lorenz maps and flows. *Comm. Math. Phys.,* 212:277–296, 2000.
293. J. Milnor and W. Thurston. On iterated maps of the interval: I, II. In *Dynamical Systems – Maryland 1986-87,* volume 1342, pages 465–563. Springer Verlag, 1988.
294. M. Misiurewicz. Absolutely continuous invariant measures for certain maps of the interval. *Publ. Math. IHES,* 53:17–51, 1981.
295. Y. Mitsumatsu. Foliations and contact structures on 3-manifolds. In *Geometry and dynamics (Warsaw 2000),* pages 75–125. World Scientific Publ., 2002.
296. L. Mora and C. Martin-Ribas. A complement to Hayashi's connecting lemma. Preprint 2001.
297. L. Mora and M. Viana. Abundance of strange attractors. *Acta Math.,* 171:1–71, 1993.
298. C. Morales. On the explosion of singular-hyperbolic attractors. *Ergod. Th. & Dynam. Sys.*
299. C. Morales. Singular-hyperbolic attractors with handlebody basins. Preprint www.preprint.impa.br 2003.
300. C. Morales. Lorenz attractor through saddle-node bifurcations. *Ann. Inst. H. Poincaré Anal. Non Linéaire,* 13:589–617, 1996.
301. C. Morales and M. J. Pacifico. New singular strange attractors arising from hyperbolic flows. Preprint.
302. C. Morales and M. J. Pacifico. On the dynamics of singular hyperbolic systems. Preprint www.preprint.impa.br 2003.
303. C. Morales and M. J. Pacifico. Sufficient conditions for robustness of attractors. *Pacific J. Math.*
304. C. Morales and M. J. Pacifico. Mixing attractors for 3-flows. *Nonlinearity,* 14:359–378, 2001.
305. C. Morales and M. J. Pacifico. A dichotomy for three-dimensional vector fields. *Ergod. Th. & Dynam. Sys.,* 23:1575–1600, 2003.
306. C. Morales, M. J. Pacifico, and E. Pujals. Robust transitive singular sets for 3-flows are partially hyperbolic attractors and repellers. *Annals of Math.*
307. C. Morales, M. J. Pacifico, and E. Pujals. Global attractors from the explosion of singular cycles. *C. R. Acad. Sci. Paris Sér. I Math.,* 325:1317–1322, 1997.
308. C. Morales, M. J. Pacifico, and E. Pujals. On C^1 robust singular transitive sets for three-dimensional flows. *C. R. Acad. Sci. Paris Sér. I Math.,* 326:81–86, 1998.
309. C. Morales, M. J. Pacifico, and E. Pujals. Singular hyperbolic systems. *Proc. Amer. Math. Soc.,* 127:3393–3401, 1999.
310. C. Morales, M. J. Pacifico, and E. Pujals. Strange attractors across the boundary of hyperbolic systems. *Comm. Math. Phys.,* 211:527–558, 2000.
311. C. Morales and E. Pujals. Singular strange attractors on the boundary of Morse-Smale systems. *Ann. Sci. École Norm. Sup.,* 30:693–717, 1997.
312. C. G. Moreira. Geometric properties of the Markov and Lagrange spectra. Preprint www.preprint.impa.br, 2002.
313. C. G. Moreira. Stable intersections of Cantor sets and homoclinic bifurcations. *Ann. Inst. H. Poincaré Anal. Non Linéaire,* 13:741–781, 1996.

314. C. G. Moreira, J. Palis, and M. Viana. Homoclinic tangencies and fractal invariants in arbitrary dimension. *C. R. Acad. Sci. Paris*, 333:475–480, 2001.

315. C. G. Moreira and J.-C. Yoccoz. Tangences homocliniques stables pour les ensembles hyperboliques de grande dimension fractale. In preparation.

316. C. G. Moreira and J.-C. Yoccoz. Stable intersections of regular Cantor sets with large Hausdorff dimensions. *Ann. of Math.*, 154:45–96, 2001. Corrigendum in Annals of Math. 154:527, 2001.

317. N. Muniz. *Hénon-like families in higher dimensions: SRB measures and basin problem*. PhD thesis, IMPA, 2000.

318. J. Neveu. *Mathematical foundations of the calculus of probability*. Holden-Day Publ., 1965.

319. S. Newhouse. Nondensity of Axiom A(a) on S^2. In *Global analysis*, volume XIV of *Proc. Sympos. Pure Math. (Berkeley 1968)*, pages 191–202. Amer. Math. Soc., 1970.

320. S. Newhouse. Diffeomorphisms with infinitely many sinks. *Topology*, 13:9–18, 1974.

321. S. Newhouse. The abundance of wild hyperbolic sets and nonsmooth stable sets for diffeomorphisms. *Publ. Math. I.H.E.S.*, 50:101–151, 1979.

322. S. Newhouse. Continuity properties of entropy. *Annals of Math.*, 129:215–235, 1990. Errata in Annals of Math. 131:409–410, 1990.

323. S. Newhouse and J. Palis. Bifurcations of Morse-Smale dynamical systems. In *Dynamical systems (Proc. Sympos. Univ. Bahia, Salvador, 1971)*, pages 303–366. Academic Press, 1973.

324. S. Newhouse and J. Palis. Cycles and bifurcation theory. *Astérisque*, 31:44–140, 1976.

325. S. Newhouse, J. Palis, and F. Takens. Bifurcations and stability of families of diffeomorphisms. *Publ. Math. I.H.E.S.*, 57:5–71, 1983.

326. V. Niţică and A. Török. An open dense set of stably ergodic diffeomorphisms in a neighborhood of a non-ergodic one. *Topology*, 40:259–278, 2001.

327. S. Novikov. The topology of foliations. *Trudy Moskov. Mat. Obsc*, 14:248–278, 1965.

328. T. Nowicki. A positive Lyapunov exponent for the critical value of an S-unimodal mapping implies uniform hyperbolicity. *Ergod. Th. & Dynam. Sys.*, 8:425–435, 1988.

329. T. Nowicki and D. Sands. Non-uniform hyperbolicity and universal bounds for S-unimodal maps. *Inventiones Math.*, 132:633–680, 1998.

330. T. Nowicki and S. van Strien. Invariant measures exist under a summability condition for unimodal maps. *Invent. Math.*, 105:123–136, 1991.

331. B. Oksendal. *Stochastic differential equations. An introduction with applications*. Universitext. Springer Verlag, 1995.

332. K. Oliveira. Equilibrium states for certain non-uniformly hyperbolic systems. *Ergod. Th. & Dynam. Sys.*, 23:1891–1906, 2003.

333. K. Oliveira and M. Viana. Thermodynamical formalism for a robust class of non-uniformly hyperbolic maps. Preprint www.preprint.impa.br 2004.

334. V. I. Oseledets. A multiplicative ergodic theorem: Lyapunov characteristic numbers for dynamical systems. *Trans. Moscow Math. Soc.*, 19:197–231, 1968.

335. J. C. Oxtoby and S. M. Ulam. Measure-preserving homeomorphisms and metrical transitivity. *Ann. of Math.*, 42:874–920, 1941.

336. M. J. Pacifico and A. Rovella. Unfolding contracting singular cycles. *Annales Sci. École. Norm. Sup.*, 26:691–700, 1993.

337. M. J. Pacifico, A. Rovella, and M. Viana. Persistence of global spiraling attractors. In preparation.

338. M. J. Pacifico, A. Rovella, and M. Viana. Infinite-modal maps with global chaotic behavior. *Ann. of Math.*, 148:441–484, 1998. Corrigendum in Annals of Math. 149, page 705, 1999.

339. J. Palis. A note on Ω-stability. In *Global analysis*, volume XIV of *Proc. Sympos. Pure Math. (Berkeley 1968)*, pages 221–222. Amer. Math. Soc., 1970.

340. J. Palis. On the C^1 Ω-stability conjecture. *Publ. Math. I.H.E.S.*, 66:211–215, 1988.

341. J. Palis. A global view of Dynamics and a conjecture on the denseness of finitude of attractors. *Astérisque*, 261:335–347, 2000. Géométrie complexe et systèmes dynamiques (Orsay, 1995).

342. J. Palis and W. de Melo. *Geometric Theory of Dynamical Systems. An introduction*. Springer Verlag, 1982.

343. J. Palis and S. Smale. Structural stability theorems. In *Global analysis*, volume XIV of *Proc. Sympos. Pure Math. (Berkeley 1968)*, pages 223–232. Amer. Math. Soc., 1970.

344. J. Palis and F. Takens. Hyperbolicity and the creation of homoclinic orbits. *Annals of Math.*, 125:337–374, 1987.

345. J. Palis and F. Takens. *Hyperbolicity and sensitive-chaotic dynamics at homoclinic bifurcations*. Cambridge University Press, 1993.

346. J. Palis and M. Viana. On the continuity of Hausdorff dimension and limit capacity for horseshoes. In *Dynamical systems, Valparaiso 1986*, pages 150–160. Springer Verlag, 1988.

347. J. Palis and M. Viana. High dimension diffeomorphisms displaying infinitely many periodic attractors. *Ann. of Math.*, 140:207–250, 1994.

348. J. Palis and J.-C. Yoccoz. Homoclinic tangencies for hyperbolic sets of large Hausdorff dimension. *Acta Math.*, 172:91–136, 1994.

349. J. Palis and J.-C. Yoccoz. Fers à cheval non uniformément hyperboliques engendrés par une bifurcation homocline et densité nulle des attracteurs. *C. R. Acad. Sci. Paris Sér. I Math.*, 333:867–871, 2001.

350. J. Palis and J.-C. Yoccoz. Implicit formalism for affine-like maps and parabolic composition. In *Global analysis of dynamical systems*, pages 67–87. IOP Publ., 2001.

351. M. Peixoto. Structural stability on two-dimensional manifolds. *Topology*, 1:101–120, 1962.

352. Ya. Pesin. Dynamical systems with generalized hyperbolic attractors: hyperbolic, ergodic and topological properties. *Ergod. Th. & Dynam. Sys.*, 12:123–151, 1992.

353. Ya. Pesin. *Dimension theory in dynamical systems*. University of Chicago Press, 1997. Contemporary views and applications.

354. Ya. Pesin and Ya. Sinai. Gibbs measures for partially hyperbolic attractors. *Ergod. Th. & Dynam. Sys.*, 2:417–438, 1982.

355. Ya. B. Pesin. Families of invariant manifolds corresponding to non-zero characteristic exponents. *Math. USSR. Izv.*, 10:1261–1302, 1976.

356. Ya. B. Pesin. Characteristic Lyapunov exponents and smooth ergodic theory. *Russian Math. Surveys*, 324:55–114, 1977.

357. S. Yu. Pilyugin and O. B. Plamenevskaya. Shadowing is generic. *Topology Appl.*, 97:253–266, 1999.

358. V. Pliss. The location of separatrices of periodic saddle-point motion of systems of second-order differential equations. *Diff. Uravnenija*, 7:906–927, 1971.

359. V. Pliss. On a conjecture due to Smale. *Diff. Uravnenija*, 8:262–268, 1972.

360. R.V. Plykin. Sources and currents of A-diffeomorphisms of surfaces. *Math USSR Sbornik*, 94:243–264, 1974.

361. M. Pollicott. On the rate of mixing of Axiom A flows. *Invent. Math.*, 81:413–426, 1985.

362. M. Pollicott. *Lectures on ergodic theory and Pesin theory on compact manifolds.* Cambridge University Press, 1993.

363. C. Pugh. The closing lemma. *Amer. J. of Math.*, 89:956–1009, 1967.

364. C. Pugh. An improved closing lemma and a general density theorem. *Amer. J. of Math.*, 89:1010–1021, 1967.

365. C. Pugh and C. Robinson. The C^1 closing lemma, including Hamiltonians. *Ergod. Th. & Dynam. Sys.*, 3:261–313, 1983.

366. C. Pugh and M. Shub. Ergodic attractors. *Trans. Amer. Math. Soc.*, 312:1–54, 1989.

367. C. Pugh and M. Shub. Stably ergodic dynamical systems and partial hyperbolicity. *J. Complexity*, 13:125–179, 1997.

368. C. Pugh and M. Shub. Stably ergodicity and julienne quasi-conformality. *J. Europ. Math. Soc.*, 2:1–52, 2000.

369. E. Pujals. *On saddle-node focus singular cycles.* PhD thesis, IMPA, 1995.

370. E. Pujals and M. Sambarino. The dynamics of dominated splitting. Preprint 2001.

371. E. Pujals and M. Sambarino. Homoclinic tangencies and hyperbolicity for surface diffeomorphisms. *Annals of Math.*, 151:961–1023, 2000.

372. E. Pujals and M. Sambarino. On homoclinic tangencies, hyperbolicity, creation of homoclinic orbits, and variation of entropy. *Nonlinearity*, 13:921–926, 2000.

373. A. Pumariño and A. Rodriguez. *Persistence and coexistence of strange attractors in homoclinic saddle-focus connections*, volume 1658 of *Lect. Notes in Math.* Springer Verlag, 1997.

374. A. Pumariño and J. A. Rodriguez. Coexistence and persistence of infinitely many strange attractors. *Ergod. Th. & Dynam. Sys.*, 21:1511–1523, 2001.

375. M. Ratner. The rate of mixing for geodesic and horocycle flows. *Ergod. Th. & Dynam. Sys.*, 7:267–288, 1987.

376. I. L. Rios. Unfolding homoclinic tangencies inside horseshoes: hyperbolicity, fractal dimensions and persistent tangencies. *Nonlinearity*, 14:431–462, 2001.

377. J. Robbin. A structural stability theorem. *Annals of Math.*, 94:447–493, 1971.

378. C. Robinson. C^r structural stability implies Kupka-Smale. In *Dynamical Systems (Proc. Sympos. Univ. Bahia, Salvador, 1971)*, pages 443–449. Academic Press, 1971.

379. C. Robinson. Structural stability of vector fields. *Annals of Math.*, 99:154–175, 1974. Errata in Annals of Math. 101:368, 1975.

380. C. Robinson. Structural stability of C^1 diffeomorphisms. *J. Diff. Equ.*, 22:28–73, 1976.

381. C. Robinson. Homoclinic bifurcation to a transitive attractor of Lorenz type. *Nonlinearity*, 2:495–518, 1989.

382. C. Robinson. *Dynamical systems.* Studies in Advanced Mathematics. CRC Press, 1999. Stability, symbolic dynamics, and chaos.

383. C. Robinson and L. S. Young. Nonabsolutely continuous foliations for an Anosov diffeomorphism. *Invent. Math.*, 61:159–176, 1980.

384. F. Rodriguez-Hertz. *Stable ergodicity of certain linear automorphisms of the torus*. PhD thesis, IMPA, 2001. Preprint premat.fing.edu.uy 2001.

385. V.A. Rokhlin. Selected topics from the metric theory of dynamical systems. *A. M. S. Transl.*, 49:171–240, 1966. Transl. from Uspekhi Mat. Nauk. 4 - 2 (1949), 57–128.

386. V.A. Rokhlin. Lectures on the entropy theory of measure-preserving transformations. *Russ. Math. Surveys*, 22 -5:1–52, 1967. Transl. from Uspekhi Mat. Nauk. 22 - 5 (1967), 3–56.

387. N. Romero. Persistence of homoclinic tangencies in higher dimensions. *Ergod. Th. & Dynam. Sys.*, 15:735–757, 1995.

388. A. Rovella. The dynamics of perturbations of the contracting Lorenz attractor. *Bull. Braz. Math. Soc.*, 24:233–259, 1993.

389. W. Rudin. *Real and complex analysis*. McGraw-Hill, 3 edition, 1987.

390. D. Ruelle. A measure associated with Axiom A attractors. *Amer. J. Math.*, 98:619–654, 1976.

391. D. Ruelle. An inequality for the entropy of differentiable maps. *Bull. Braz. Math. Soc.*, 9:83–87, 1978.

392. D. Ruelle. Ergodic theory of diferentiable dynamical systems. *Publ. Math. IHES*, 50:27–58, 1981.

393. D. Ruelle. Flots qui ne mélangent pas exponentiellement. *C. R. Acad. Sci. Paris*, 296:191–193, 1983.

394. D. Ruelle. *Elements of differentiable dynamics and bifurcation theory*. Academic Press, 1989.

395. D. Ruelle. The thermodynamical formalism for expanding maps. *Comm. Math. Phys.*, 125:239–262, 1989.

396. D. Ruelle and A. Wilkinson. Absolutely singular dynamical foliations. *Comm. Math. Phys.*, 219:481–487, 2001.

397. H. H. Rugh. Generalized Fredhomm determinants and Selberg zeta functions for Axiom A dynamical systems. *Ergod. Th. & Dynam. Sys.*, 16:805–819, 1996.

398. M. Rychlik. Bounded variation and invariant measures. *Studia Math.*, 76:69–80, 1983.

399. M. Rychlik. Lorenz attractors through Shil'nikov-type bifurcation. Part 1. *Erg. Th. & Dynam. Syst.*, 10:793–821, 1990.

400. P. Sabini. *Nonperiodic bifurcations on the boundary of hyperbolic systems*. PhD thesis, IMPA, 2001.

401. K. Sakai. Pseudo-orbit tracing property and strong transversality of diffeomorphisms on closed manifolds. *Osaka J. Math.*, 31:373–386, 1994.

402. B. SanMartín. Contracting singular cycles. *Ann. Inst. Henri Poincaré, Anal. Non Linéaire*, 15:651–659, 1998.

403. A. Sannami. An example of a regular Cantor whose arithmetic difference is a Cantor set with positive measure. *Hokkaido Math. Journal*, 21:7–23, 1992.

404. O. Sarig. Existence of Gibbs measures for countable Markov shifts. *Proc. Amer. Math. Soc.*, 131:1751–1758 (electronic), 2003.

405. E. A. Sataev. Invariant measures for hyperbolic maps with singularities. *Russ. Math. Surveys*, 471:191–251, 1992.

406. B. Saussol. Absolutely continuous invariant measures for multi-dimensional expanding maps. *Israel J. Math*, 116:223–248, 2000.

407. W. Shen. The metric properties of multimodal interval maps and c^2 density of Axiom A. Preprint Warwick, 2001.

408. L. P. Shil'nikov. A case of the existence of a denumerable set of periodic motions. *Sov. Math. Dokl.*, 6:163–166, 1965.

409. M. Shub. Endomorphisms of compact differentiable manifolds. *Amer. Journal of Math.*, 91:129–155, 1969.

410. M. Shub. Topologically transitive diffeomorphisms on T^4. In *Dynamical Systems*, volume 206 of *Lect. Notes in Math.*, page 39. Springer Verlag, 1971.

411. M. Shub. *Global stability of dynamical systems*. Springer Verlag, 1987.

412. M. Shub and A. Wilkinson. Pathological foliations and removable zero exponents. *Invent. Math.*, 139:495–508, 2000.

413. M. Shub and A. Wilkinson. Stably ergodic approximation: two examples. *Ergod. Th. & Dynam. Sys.*, 20:875–893, 2000.

414. R. Simon. A 3-dimensional Abraham-Smale example. *Procs. Amer. Math. Soc.*, 34:629–630, 1972.

415. Ya. Sinai. Dynamical systems with elastic reflections: ergodic properties of scattering billiards. *Russian Math. Surveys*, 25:137–189, 1970.

416. Ya. Sinai. Gibbs measures in ergodic theory. *Russian Math. Surveys*, 27:21–69, 1972.

417. D. Singer. Stable orbits and bifurcations of maps of the interval. *SIAM J. Appl. Math.*, 35:260–267, 1978.

418. S. Smale. Diffeomorphisms with many periodic points. *Bull. Am. Math. Soc.*, 73:747–817, 1967.

419. J. Smillie. Dynamics in two complex dimensions. In *Proceedings of the International Congress of Mathematicians, Vol. III (Beijing, 2002)*, pages 373–382. Higher Ed. Press, 2002.

420. V. V. Solodov. Components of topological foliations. *Math. Sb.*, 119 (161):340–354, 1982.

421. J. Sotomayor. Generic bifurcations of dynamical systems. In *Dynamical Systems (Procs. Sympos., Univ. Bahia, Salvador, 1971)*, pages 561–582. Academic Press, 1973.

422. C. Sparrow. *The Lorenz equations: bifurcations, chaos and strange attractors*, volume 41 of *Applied Mathematical Sciences*. Springer Verlag, 1982.

423. D. Sullivan. *Bounds, quadratic differentials, and renormalization conjectures*, volume II of *A.M.S. Centennial Publications*. Amer. Math. Soc., 1992.

424. A. Tahzibi. Stably ergodic diffeomorphisms which are not partially hyperbolic. *Israel J. Math.*

425. F. Takens. Intermittancy: Global aspects. In *Dynamical systems, Valparaiso 1986*, volume 1331 of *Lecture Notes in Mathematics*, pages 213–239. Springer Verlag, 1988.

426. L. Tedeschini-Lalli and J. Yorke. How often do simple dynamical systems have infinitely many coexisting sinks ? *Comm. Math. Phys.*, 106:635–657, 1986.

427. C. Tresser. About some theorems by L. P. Shil'nikov. *Ann. Inst. Henri Poincaré*, 40:441–461, 1984.

428. M. Tsujii. Physical measures for partially hyperbolic endomorphisms. Preprint 2002.

429. M. Tsujii. Absolutely continuous invariant measures for piecewise real-analytic expanding maps on the plane. *Comm. Math. Phys.*, 208:605–622, 2000.

430. M. Tsujii. Fat solenoidal attractors. *Nonlinearity*, 14:1011–1027, 2001.

431. W. Tucker. The Lorenz attractor exists. *C. R. Acad. Sci. Paris Sér. I Math.*, 328:1197–1202, 1999.

432. W. Tucker. A rigorous ODE solver and Smale's 14th problem. *Found. Comput. Math.*, 2:53–117, 2002.

433. M. Urbański. Hausdorff measures versus equilibrium states of conformal infinite iterated function systems. *Period. Math. Hungar.*, 37:153–205, 1998. International Conference on Dimension and Dynamics (Miskolc, 1998).

434. R. Ures. On the approximation of Hénon-like attractors by homoclinic tangencies. *Ergod. Th. & Dynam. Sys.*, 15, 1995.

435. S. Ushiki, H. Oka, and H. Kokubu. Existence of strange attractors in the unfolding of a degenerate singularity of a vector field in invariant by translation. *C.R. Acad. Sci. Paris., Sér I*, 298:39–42, 1984.

436. C. Vásquez. *Statistical stability for diffeomorphisms with a dominated splitting.* PhD thesis, IMPA, 2003. Preprint www.preprint.impa.br.

437. M. Viana. Almost all cocycles over any hyperbolic system have non-vanishing Lyapunov exponents. Preprint www.preprint.impa.br 2004.

438. M. Viana. Strange attractors in higher dimensions. *Bull. Braz. Math. Soc.*, 24:13–62, 1993.

439. M. Viana. Global attractors and bifurcations. In *Nonlinear dynamical systems and chaos (Groningen, 1995)*, pages 299–324. Birkhäuser, 1996.

440. M. Viana. Multidimensional nonhyperbolic attractors. *Publ. Math. I.H.E.S.*, 85:63–96, 1997.

441. M. Viana. *Stochastic dynamics of deterministic systems.* Lecture Notes 21st Braz. Math. Colloq. IMPA, Rio de Janeiro, 1997.

442. M. Viana. Dynamics: a probabilistic and geometric perspective. In *Proceedings of the International Congress of Mathematicians, Vol. I (Berlin, 1998)*, pages 557–578, 1998.

443. M. Viana. What's new on Lorenz strange attractors ? *Math. Intelligencer*, 22(3):6–19, 2000.

444. T. Vivier. Flots robustement transitifs sur les variétés compactes. *Comptes Rendus Acad. Sci. Paris*, 337:791–796, 2003.

445. P. Walters. *An introduction to ergodic theory.* Springer Verlag, 1982.

446. Q. Wang and L.-S. Young. Strange attractors with one direction of instability in higher dimensions. Preprint 2002.

447. Q. Wang and L.-S. Young. Strange attractors with one direction of instability. *Comm. Math. Phys.*, 218:1–97, 2001.

448. Q. Wang and L.-S Young. From invariant curves to strange attractors. *Comm. Math. Phys.*, 225:275–304, 2002.

449. L. Wen. Homoclinic tangencies and dominated splittings. *Nonlinearity*, 15:1445–1469, 2002.

450. L. Wen and Z. Xia. C^1 connecting lemmas. *Trans. Amer. Math. Soc.*, 352:5213–5230, 2000.

451. A. Wilkinson. Stable ergodicity of the time-one map of a geodesic flow. *Ergod. Th. & Dynam. Sys.*, 18:1545–1587, 1998.

452. R.F. Williams. The structure of the Lorenz attractor. *Publ. Math. IHES*, 50:73–99, 1979.

453. Z. Xia. Homoclinic points in symplectic and volume-preserving diffeomorphisms. *Comm. Math. Phys.*, 177(2):435–449, 1996.

454. J.-C. Yoccoz. Travaux de Herman sur les tores invariants. *Astérisque*, 206:Exp. No. 754, 4, 311–344, 1992. Séminaire Bourbaki, Vol. 1991/92.

455. Y. Yomdin. Volume growth and entropy. *Israel J. Math.*, 57:285–300, 1987.

456. L.-S. Young. Dimension, entropy and Lyapunov exponents. *Ergod. Th. & Dynam. Sys.*, 2:109–124, 1982.
457. L.-S. Young. Stochastic stability of hyperbolic attractors. *Ergod. Th. & Dynam. Sys.*, 6:311–319, 1986.
458. L.-S. Young. Decay of correlations for certain quadratic maps. *Comm. Math. Phys.*, 146:123–138, 1992.
459. L.-S. Young. Lyapunov exponents for some quasi-periodic cocycles. *Ergod. Th. & Dynam. Sys.*, 17(2):483–504, 1997.
460. L.-S. Young. Statistical properties of dynamical systems with some hyperbolicity. *Annals of Math.*, 147:585–650, 1998.
461. L.-S. Young. Recurrence times and rates of mixing. *Israel J. of Math.*, 110:153–188, 1999.
462. G.-C. Yuan and J. Yorke. An open set of maps for which every point is absolutely nonshadowable. *Procs. Amer. Math. Soc.*, 128:909–918, 2000.
463. M. Yuri. Thermodynamic formalism for certain nonhyperbolic maps. *Ergod. Th. & Dynam. Sys.*, 19:1365–1378, 1999.
464. M. Yuri. Weak Gibbs measures for certain non-hyperbolic systems. *Ergod. Th. & Dynam. Sys.*, 20:1495–1518, 2000.
465. M. Yuri. Thermodynamical formalism for countable to one Markov systems. *Trans. Amer. Math. Soc.*, 335:2949–2971, 2003.
466. E. Zehnder. Note on smoothing symplectic and volume preserving diffeomorphisms. *Lect. Notes in Math.*, 597:828–854, 1977.

Index

absolutely continuous
  foliation **138**, 140, 150, 302
  holonomy 295
  measure 19, 69, 214, 217, 224, 243,
    245, 270, **304**, 333
  random noise 315
  SRB measure 240
  strong unstable foliation 230
accessibility **140**, 150, 153
  class 153
  essential **140**, 150, 151, 153, 155
  property 153, 230
  stable **140**, 150, 154
action of a group
  independent 264, 265
  transitive 262
adding machine 33, 211
almost homoclinic sequence 104
Anosov
  derived from **124**, 145
  diffeomorphism 124, 142, 145, 147,
    216, 220, 253, 256, 258, 287
    volume preserving 138, 154, 329
  flow 124, 142, 145, 147, 170, 172,
    184, 287, 329–331
    contact 331
    pseudo- 152
aperiodic
  chain recurrence class 192
  elementary dynamical piece 208,
    212
arithmetic difference 35, 38, 49
Arnold family 14

attached singularity 178
attainability relation 196
attractor **7**, 192
  basin of **7**, 192, 206, 212, 231, 241,
    314
  espiral 188
  finiteness conjecture *9–11*
  Hénon-like 33, 58, **66**, 91, 187
  hyperbolic **4**, 321
  intermingled basins 214
  Lorenz **157**, 159, 179, 321
    multidimensional 163, 179
  Lorenz-like 169
  multidimensional
    robust 217
  partially hyperbolic 211
  Plykin 169
  quasi- 33, **207**, 209
  robust 158
  singular 105, **159**, 168, 174, 186
    hyperbolic 158, 175, 179, 181
    persistent 186
    robust 163, 168
  solenoid-type 329
  strange 33, 61, 158, 159
  topological **206**, 223
  transitive 159
  uniformly hyperbolic 211
automorphisms of the torus 152
Axiom A viii, **2**, 100
  diffeomorphism 193, 198, 204, 205
  singular **183**, 184, 185

Printing: Mercedes-Druck, Berlin
Binding: Stein + Lehmann, Berlin

Lightning Source UK Ltd.
Milton Keynes UK
UKOW04n2133020514

231023UK00001B/30/P